DEEP GREEN RESISTANCE
Strategy to Save the Planet

DEEP GREEN RESISTANCE
Strategy to Save the Planet

Aric McBay, Lierre Keith, and Derrick Jensen

Seven Stories Press
NEW YORK

A Seven Stories Press First Edition

Seven Stories Press
140 Watts Street
New York, NY 10013
www.sevenstories.com

College professors may order examination copies of Seven Stories Press titles for a free six-month trial period. To order, visit http://www.sevenstories.com/textbook or send a fax on school letterhead to (212) 226-1411.

Book design by Jon Gilbert

Library of Congress Cataloging-in-Publication Data

McBay, Aric.
 Deep green resistance : strategy to save the planet / Aric McBay, Lierre Keith, and Derrick Jensen.
     p. cm.
 Includes bibliographical references.
 ISBN 978-1-58322-929-3 (pbk.)
  1. Environmentalism. 2. Sustainable living. 3. Global environmental change. I. Keith, Lierre. II. Jensen, Derrick, 1960- III. Title.
 GE195.M385 2011
 333.72--dc22

                                    2011007287
Printed in the United States

9 8 7 6 5 4

Contents

Figures

I found it was better to fight, always,
no matter what.
—Andrea Dworkin

AUTHORS' NOTE

Before we started writing this book, the three authors—Aric, Lierre, and Derrick—decided to divide the material we wanted to cover among ourselves, so that every chapter would have one main author. The "I" in each chapter refers to the person responsible for writing it. The chapters conclude with Derrick's answers to questions he is frequently asked on the subject of resistance.

Preface

by Derrick Jensen

Somebody, after all, had to make a start. What we wrote and said is also believed by many others. They just do not dare express themselves as we did.
—Sophie Scholl, The White Rose Society

This book is about fighting back. The dominant culture—civilization—is killing the planet, and it is long past time for those of us who care about life on earth to begin taking the actions necessary to stop this culture from destroying every living being.

By now we all know the statistics and trends: 90 percent of the large fish in the oceans are gone, there is ten times as much plastic as phytoplankton in the oceans, 97 percent of native forests are destroyed, 98 percent of native grasslands are destroyed, amphibian populations are collapsing, migratory songbird populations are collapsing, mollusk populations are collapsing, fish populations are collapsing, and so on. Two hundred species are driven extinct each and every day. If we don't know those statistics and trends, we should.

This culture destroys landbases. That's what it *does*. When you think of Iraq, is the first thing that comes to mind cedar forests so thick that sunlight never touched the ground? One of the first written myths of this culture is about Gilgamesh deforesting the hills and valleys of Iraq to build a great city. The Arabian Peninsula used to be oak savannah. The Near East was heavily forested (we've all heard of the cedars of Lebanon). Greece was heavily forested. North Africa was heavily forested.

We'll say it again: this culture destroys landbases.

And it won't stop doing so because we ask nicely.

We don't live in a democracy. And before you gasp at this blasphemy, ask yourself: Do governments better serve corporations or living beings? Does the judicial system hold CEOs accountable for their destructive, often murderous acts?

Here are a couple of riddles that aren't very funny—Q: What do you

11

get when you cross a long drug habit, a quick temper, and a gun? A: Two life terms for murder, earliest release date 2026. Q: What do you get when you cross two nation-states, a large corporation, forty tons of poison, and at least 8,000 dead human beings? A: Retirement, with full pay and benefits (Warren Anderson, CEO of Union Carbide, which caused the mass murder at Bhopal).

Do the rich face the same judicial system as you or I? Does life on earth have as much standing in a court as does a corporation?

We all know the answers to these questions.

And we know in our bones, if not our heads, that this culture will not undergo any sort of voluntary transformation to a sane and sustainable way of living. We—Aric, Lierre, and Derrick—have asked thousands upon thousands of people from all walks of life, from activists to students to people we meet on buses and planes, whether they believe this culture will undergo that voluntary transformation. Almost no one ever says *yes*.

If you care about life on this planet, and if you believe this culture won't voluntarily cease to destroy it, how does that belief affect your methods of resistance?

Most people don't know, because most people don't talk about it.

This book talks about it: this book is about that shift in strategy, and tactics.

This book is about fighting back.

We must put our bodies and our lives between the industrial system and life on this planet. We must start to fight back. Those who come after, who inherit whatever's left of the world once this culture has been stopped—whether through peak oil, economic collapse, ecological collapse, or the efforts of brave women and men resisting in alliance with the natural world—are going to judge us by the health of the landbase, by what we leave behind. They're not going to care how you or I lived our lives. They're not going to care how hard we tried. They're not going to care whether we were nice people. They're not going to care whether we were nonviolent or violent. They're not going to care whether we grieved the murder of the planet. They're not going to care whether we were enlightened or not. They're not going to care what sort of excuses we had to not act (e.g., "I'm too stressed to think about it," or "It's too big and scary," or "I'm too busy," or "But those in power will kill us if we effectively act against them," or "If we fight back, we run the risk of becoming like they are," or "But I recycled," or any of a thousand other excuses we've all heard too many times). They're not going to care how simply we lived. They're not going to care how pure we were in thought or action. They're not going to care if we became the change we wished to see. They're not going to care whether we voted Democrat, Republican, Green, Libertarian, or not at all. They're not going to care if we wrote really big books about it. They're not going to care whether we had "compassion" for the CEOs and politicians running this deathly economy.

They're going to care whether they can breathe the air and drink the water. We can fantasize all we want about some great turning, but if the people (including the nonhuman people) can't breathe, it doesn't matter.

§ § §

Every new study reveals that global warming is happening far more quickly than was previously anticipated. Staid scientists are now suggesting the real possibility of billions of human beings being killed off by what some are calling a Climate Holocaust. A recently released study suggests an increase in temperatures of 16°C (30°F) by the year 2100.

We are not talking about this culture killing humans, and indeed the planet, sometime in the far-distant future. This is the future that children born today will see, and suffer, in their lifetimes.

Honestly, is this culture worth more than the lives of your own children?

§ § §

In *The Nazi Doctors*, Robert Jay Lifton explored how it was that men who had taken the Hippocratic Oath could lend their skills to concentration camps where inmates were worked to death or killed in assembly lines. He found that many of the doctors honestly cared for their charges, and did everything within their power—which means pathetically little—to make life better for the inmates. If an inmate got sick, they might give the inmate an aspirin to lick. They might put the inmate to bed for a day or two (but not for too long or the inmate might be "selected" for murder). If the patient had a contagious disease, they might kill the patient to keep the disease from spreading. All of this made sense within the confines of Auschwitz. The doctors, once again, did everything they could to help the inmates, except for the most important thing of all: They never questioned the existence of Auschwitz itself. They never questioned working the inmates to death. They never questioned starving them to death. They never questioned imprisoning them. They never questioned torturing them. They never questioned the existence of a culture that would lead to these atrocities. They never questioned the logic that leads inevitably to the electrified fences, the gas chambers, the bullets in the brain.

We as environmentalists do the same. We fight as hard as we can to protect the places we love, using the tools of the system the best that we can. Yet we do not do the most important thing of all: We do not question the existence of this deathly culture. We do not question the existence of an economic and social system that is working the world to death, that is starving it to death, that is imprisoning it, that is torturing it. We never question the logic that leads inevitably to clear-cuts, murdered oceans, loss of topsoil, dammed rivers, poisoned aquifers.

And we certainly don't act to stop these horrors.

How do you stop global warming that is caused in great measure by the burning of oil and gas? If you ask any reasonably intelligent seven-year-old, that child should be able to give you the obvious answer. But if you ask any reasonably intelligent thirty-five-year-old who works for a green high-tech consulting corporation, you'll probably receive an answer that helps the corporation more than the real, physical world.

When most people in this culture ask, "How can we stop global warming?" they aren't really asking what they pretend they're asking. They are instead asking, "How can we stop global warming without stopping the burning of oil and gas, without stopping the industrial infrastructure, without stopping this omnicidal system?" The answer: you can't.

Here's yet another way to look at it: What would you do if space aliens had invaded this planet, and they were vacuuming the oceans, and scalping native forests, and putting dams on every river, and changing the climate, and putting dioxin and dozens of other carcinogens into every mother's breast milk, and into the flesh of your children, lover, mother, father, brother, sister, friends, into your own flesh? Would you resist? If there existed a resistance movement, would you join it? If not, why not? How much worse would the damage have to get before you would stop those who were killing the planet, killing those you love, killing you?

Ninety percent of the large fish in the oceans are already gone. Where is your threshold for resistance? Is it 91 percent? 92? 93? 94? Would you wait till they had killed off 95 percent? 96? 97? 98? 99? How about 100 percent? Would you fight back then?

By asking these questions we are in no way implying that people should not try to work within the system to slow this culture's destructiveness. Right now a large energy corporation, state and federal governments, local Indian nations, and various interest groups (from environmental organizations to fishermen to farmers) are negotiating to remove five dams on the Klamath River within the next fifteen years (whether salmon will survive that long is dubious). That's something. That's important.

But there are 2 million dams in the United States alone; 60,000 of those dams are taller than thirteen feet, and 70,000 are taller than six feet. If we only took out one of those 70,000 dams per day, it would take us 200 years. Salmon don't have that time. Sturgeon don't have that time.

If salmon could take on human manifestation, what would they do?

This book is about fighting back.

And what do we mean by fighting back? As we'll explore in this book, it means first and foremost thinking and feeling for ourselves, finding who and what we love, and figuring out how best to defend our beloved, using the means that are appropriate and necessary. The strategy of Deep Green Resistance (DGR) starts by acknowledging the dire circumstances that industrial civilization has created for life on this planet. The goal of DGR is to deprive the rich of their ability to steal from the poor and the powerful of their ability to destroy the planet. It also means defending and rebuilding just and sustainable human communities nestled inside repaired and restored landbases. This is a vast undertaking, but it can be done. Industrial civilization can be stopped.

◙ ◙ ◙

People routinely approach each of this book's authors—Aric, Lierre, and Derrick—and tell us how their hope and despair have merged into one. They no longer want to do everything they can to protect the places they love, everything, that is, except the most important thing of all: to bring down the culture itself. They want to go on the offensive. They want to stop this culture in its tracks. But they don't know how.

This book is about creating a culture of resistance. And it's about creating an actual resistance. It's about creating the conditions for salmon to be able to return, for songbirds to be able to return, for amphibians to be able to return.

This book is about fighting back.

And this book is about winning.

◙ ◙ ◙

Direct actions against strategic infrastructure is a basic tactic of both militaries and insurgents the world over for the simple reason that it works. But such actions alone are never a sufficient strategy for achieving a just outcome. This means that any strategy aiming for a just future must include a call to build direct democracies based on human rights and sustainable material cultures. The different branches of these resistance movements must work in tandem: the aboveground and belowground, the militants and the nonviolent, the frontline activists and the cultural workers. We need it all.

And we need courage. The word "courage" comes from the same root as *coeur*, the French word for heart. We need all the courage of which the human heart is capable, forged into both weapon and shield to defend what is left of this planet. And the lifeblood of courage is, of course, love.

So while this is a book about fighting back, in the end this is a book about love. The songbirds and the salmon need your heart, no matter how weary, because even a broken heart is still made of love. They need your heart because they are disappearing, slipping into that longest night of extinction, and the resistance is nowhere in sight. We will have to build that resistance from whatever comes to hand: whispers and prayers, history and dreams, from our bravest words and braver actions. It will be hard, there will be a cost, and in too many implacable dawns it will seem impossible. But we will have to do it anyway. So gather your heart and join with every living being. With love as our First Cause, how can we fail?

PART I: RESISTANCE

Chapter 1

The Problem
by Lierre Keith

You cannot live a political life, you cannot live a moral life if you're not willing to open your eyes and see the world more clearly. See some of the injustice that's going on. Try to make yourself aware of what's happening in the world. And when you are aware, you have a responsibility to act.

—Bill Ayers, cofounder of the Weather Underground

A black tern weighs barely two ounces. On energy reserves less than a small bag of M&M's and wings that stretch to cover twelve inches, she flies thousands of miles, searching for the wetlands that will harbor her young. Every year the journey gets longer as the wetlands are desiccated for human demands. Every year the tern, desperate and hungry, loses, while civilization, endless and sanguineous, wins.

A polar bear should weigh 650 pounds. Her energy reserves are meant to see her through nine long months of dark, denned gestation, and then lactation, when she will give up her dwindling stores to the needy mouths of her species' future. But in some areas, the female's weight before hibernation has already dropped from 650 to 507 pounds.[1] Meanwhile, the ice has evaporated like the wetlands. When she wakes, the waters will stretch impassably open, and there is no Abrahamic god of bears to part them for her.

The Aldabra snail should weigh something, but all that's left to weigh are skeletons, bits of orange and indigo shells. The snail has been declared not just extinct, but the first casualty of global warming. In dry periods, the snail hibernated. The young of any species are always more vulnerable, as they have no reserves from which to draw. In this case, the adults' "reproductive success" was a "complete failure."[2] In plain terms, the babies died and kept dying, and a species millions of years old is now a pile of shell fragments.

What is your personal carrying capacity for grief, rage, despair? We are living in a period of mass extinction. The numbers stand at 200 species a day.[3] That's 73,000 a year. This culture is oblivious to their passing, feels entitled to their every last niche, and there is no roll call on the nightly news.

There is a name for the tsunami wave of extermination: the Holocene extinction event. There's no asteroid this time, only human behavior, behavior that we could choose to stop. Adolph Eichman's excuse was that no one told him that the concentration camps were wrong. We've all seen the pictures of the drowning polar bears. Are we so ethically numb that we need to be told this is wrong?

There are voices raised in concern, even anguish, at the plight of the earth, the rending of its species. "Only zero emissions can prevent a warmer planet," one pair of climatologists declare.[4] James Lovelock, originator of the Gaia hypothesis, states bluntly that global warming has passed the tipping point, carbon offsetting is a joke, and "individual lifestyle adjustments" are "a deluded fantasy."[5] It's all true, and self-evident. "Simple living" should start with simple observation: if burning fossil fuels will kill the planet, then stop burning them.

But that conclusion, in all its stark clarity, is not the popular one to draw. The moment policy makers and environmental groups start offering solutions is the exact moment when they stop telling the truth, inconvenient or otherwise. Google "global warming solutions." The first paid sponsor, Campaign Earth, urges "No doom and gloom!! When was the last time depression got you really motivated? We're here to inspire realistic action steps and stories of success." By "realistic" they don't mean solutions that actually match the scale of the problem. They mean the usual consumer choices—cloth shopping bags, travel mugs, and misguided dietary advice—which will do exactly nothing to disrupt the troika of industrialization, capitalism, and patriarchy that is skinning the planet alive. As Derrick has pointed out elsewhere, even if every American took *every single action* suggested by Al Gore it would only reduce greenhouse gas emissions by 21 percent.[6] Aric tells a stark truth: even if through simple living and rigorous recycling you stopped your own average American's annual one ton of garbage production, "your per capita share of the industrial waste produced in the US is still almost twenty-six tons. That's thirty-seven times as much waste as you were able to save by eliminating a full 100 percent of your personal waste."[7] Industrialism itself is what has to stop. There is no kinder, greener version that will do the trick of leaving us a living planet. In blunt terms, industrialization is a process of taking entire communities of living beings and turning them into commodities and dead zones. Could it be done more "efficiently"? Sure, we could use a little less fossil fuels, but it still ends in the same wastelands of land, water, and sky. We could stretch this endgame out another twenty years, but the planet still dies. Trace every industrial artifact back to its source— which isn't hard, as they all leave trails of blood—and you find the same devastation: mining, clear-cuts, dams, agriculture. And now tar sands, mountaintop removal, wind farms (which might better be called dead bird and bat farms). No amount of renewables is going to make up for the fossil fuels or change the nature of the extraction, both of which are prerequisites for this way of life. Neither fossil fuels nor extracted substances will ever be sustainable; by definition, they will run out. Bringing a cloth shopping bag to the store, even if you walk there in your Global Warming Flip-Flops, will not stop the tar sands. But since these actions also won't disrupt anyone's life, they're declared both realistic and successful.

The next site's Take Action page includes the usual: buying light bulbs, inflating tires, filling dishwashers, shortening showers, and rearranging the deck chairs. It also offers the ever-crucial Global Warming Bracelets and, more importantly, Flip-Flops. Polar bears everywhere are weeping with relief.

The first noncommercial site is the Union of Concerned Scientists. As one might expect, there are no exclamation points, but instead a statement that "[t]he burning of fossil fuel (oil, coal, and natural gas) alone counts for about 75 percent of annual CO_2 emissions." This is followed by a list of Five Sensible Steps. Step One? No, not stop burning fossil fuels—"Make Better Cars and SUVs." Never mind that the automobile itself is the pollution, with its demands—for space, for speed, for fuel—in complete opposition to the needs of both a viable human community and a living planet. Like all the others, the scientists refuse to call industrial civilization into question. We can have a living planet *and* the consumption that's killing the planet, can't we?

The principle here is very simple. As Derrick has written, "[A]ny social system based on the use of nonrenewable resources is by definition unsustainable."[8] Just to be clear, nonrenewable means *it will eventually run out*. Once you've grasped that intellectual complexity, you can move on to the next level. "Any culture based on the nonrenewable use of renewable resources is just as unsustainable." Trees are renewable. But if we use them faster than they can grow, the forest will turn to desert. Which is precisely what civilization has been doing for its 10,000 year campaign, running through soil, rivers, and forests as well as metal, coal, and oil. Now the oceans are almost dead and their plankton populations are collapsing, populations that both feed the life of the oceans and create oxygen for the planet. What will we fill our lungs with when they are gone? The plastics with which industrial civilization is replacing them? In parts of the Pacific, plastic outweighs plankton 48 to 1.[9] Imagine if it were your blood, your heart, crammed with toxic materials—not just chemicals, but physical gunk—until there was ten times more of it than you. What metaphor is adequate for the dying plankton? Cancer? Suffocation? Crucifixion?

But the oceans don't need our metaphors. They need action. They need industrial civilization to stop destroying and devouring. In other words, they need us to *make* it stop.

Which is why we are writing this book.

◙ ◙ ◙

Most people, or at least most people with a beating heart, have already done the math, added up the arrogance, sadism, stupidity, and denial, and reached the bottom line: a dead planet. Some of us carry that final sum like the weight of a corpse. For others, that conclusion turns the heart to a smoldering coal. But despair and rage have been declared unevolved and unclean, beneath the "spiritual warriors" who insist they will save the planet by "healing" themselves. How this activity will stop the release of carbon and the felling of forests is never actually explained. The answer lies vaguely between being the change we wish to see and a 100th monkey of hope, a monkey that is frankly more Christmas pony than actual possibility.

Given that the culture of America is founded on individualism and awash in privilege, it's no surprise that narcissism is the end result.

The social upheavals of the '60s split along fault lines of responsibility and hedonism, of justice and selfishness, of sacrifice and entitlement. What we are left with is an alternative culture, a small, separate world of the converted, content to coexist alongside a virulent mainstream. Here, one can find workshops on "scarcity consciousness," as if poverty were a state of mind and not a structural support of capitalism. This culture leaves us ill-prepared to face the crisis of planetary biocide that greets us daily with its own grim dawn. The facts are not conducive to an open-hearted state of wonder. To confront the truth as adults, not as faux children, requires an adult fortitude and courage, grounded in our adult responsibilities to the world. It requires those things because the situation is horrific and living with that knowledge will hurt. Meanwhile, I have been to workshops where global warming was treated as an opportunity for personal growth, and no one there but me saw a problem with that.

The word *sustainable*—the "Praise, Jesus!" of the eco-earnest—serves as an example of the worst tendencies of the alternative culture. It's a word that perfectly meshes corporate marketers' carefully calculated upswell of green sentiment with the relentless denial of the privileged. It's a word I can barely stand to use because it has been so exsanguinated by cheerleaders for a technotopic, consumer kingdom come. To doubt the vague promise now firmly embedded in the word—that we can have our cars, our corporations, our consumption, and our planet, too—is both treason and heresy to the emotional well-being of most progressives. But here's the question: Do we want to feel better or do we want to be effective? Are we sentimentalists or are we warriors?

For "sustainable" to mean anything, we must embrace and then defend the bare truth: the planet is primary. The life-producing work of a million species is literally the earth, air, and water that we depend on. No human activity—not the vacuous, not the sublime—is worth more than that matrix. Neither, in the end, is any human life. If we use the word "sustainable" and don't mean that, then we are liars of the worst sort: the kind who let atrocities happen while we stand by and do nothing.

Even if it were possible to reach narcissists, we are out of time. Admitting we have to move forward without them, we step away from the cloying childishness and optimistic white-lite denial of so much of the left and embrace our adult knowledge. With all apologies to Yeats, in knowledge begins responsibilities. It's to you grown-ups, the grieving and the raging, that we address this book.

◻ ◻ ◻

The vast majority of the population will do nothing unless they are led, cajoled, or forced. If the structural determinants are in place for people to live their lives without doing damage—for example, if they're hunter-gatherers with respected elders—then that's what happens. If, on the other hand, the environment has been arranged for cars, industrial schooling is mandatory, resisting war taxes will land you in jail, food is only available through giant corporate enterprises selling giant corporate degradation, and misogynist pornography is only a click away 24/7—well, welcome to the nightmare. This culture is basically conducting a massive Milgram experiment on us, only the electric shocks aren't fake—they're killing off the planet, species by species.

But wherever there is oppression there is resistance. That is true everywhere, and has been forever. The resistance is built body by body from a tiny few, from the stalwart, the brave, the determined, who are willing to stand against both power and social censure. It is our prediction that there will be no mass movement, not in time to save this planet, our home. That tiny percent—Margaret Mead's small group of thoughtful, committed citizens—has been able to shift both the cultural consciousness and the power structures toward justice in times past. It is valid to long for a mass movement, however, no matter how much we rationally know that we're wishing on a star. Theoretically, the human race as a whole could face our situation and make some decisions—tough decisions, but fair ones, that include an equitable distribution of both resources and justice, that respect and embrace the limits of our planet. But none of the institutions that govern our lives, from the economic to the religious, are on the side of justice or sustainability. Theoretically, these institutions could be forced to change. The history of every human rights struggle bears witness to how courage and sacrifice can dismantle power and injustice. But again, it takes time. If we had a thousand years, even a hundred years, building a movement to transform the dominant institutions around the globe would be the task before us. But the Western black rhinoceros is out of time. So is the golden toad, the pygmy rabbit. No one is going to save this planet except us.

So what are our options? The usual approach of long, slow institutional change has been foreclosed, and many of us know that. The default setting for environmentalists has become personal lifestyle "choices." This should have been predictable as it merges perfectly into the demands of capitalism, especially the condensed corporate version mediating our every impulse into their profit. But we can't consume our way out of environmental collapse; consumption is the problem. We might be forgiven for initially accepting an exhortation to "simple living" as a solution to that consumption, especially as the major environmental organizations and the media have declared lifestyle change our First Commandment. Have you accepted compact fluorescents as your personal savior? But lifestyle change is not a solution as it doesn't address the root of the problem.

We have believed such ridiculous solutions because our perception has been blunted by some portion of denial and despair. And those are legitimate reactions. I'm not persuading anyone out of them. But do we want to develop a strategy to manage our emotional state or to save the planet?

And we've believed in these lifestyle solutions because everyone around us insists they're workable, a collective repeating mantra of "renewables, recycling" that has dulled us into belief. Like Eichmann, no one has told us that it's wrong.

Until now. So this is the moment when you will have to decide. Do you want to be part of a serious effort to save this planet? Not a serious effort at collective delusion, not a serious effort to feel better, not a serious effort to save you and yours, but an actual strategy to stop the destruction of everything worth loving. If your answer feels as imperative as instinct, read on.

Q: Won't we just reach a tipping point in public opinion?

Derrick Jensen: In 2004, George W. Bush received more than 62 million votes in the United States. Admittedly, the Democrats are just the good cop in a good cop/bad cop scenario, but that doesn't alter the fact that 62 million people voted for George W. Bush. Now people are camping out overnight to get Sarah Palin's signature. In the small county where I live there are a few issues that will get enough people excited to storm the board of supervisor's office. One is that they want to maintain their ability to grow small amounts of marijuana. Another is that they want the right to drive ORVs anywhere they goddamn please.

People are not rioting over the unwillingness of this government to provide health care. People aren't rioting over the toxification of the total environment and their loved ones dying of cancer. They're not rioting over the United States spending billions of dollars—billions and billions of dollars—to kill people all over the world. And, in fact, one of the smartest political moves that any politician can make is to increase the military budget. That is tremendously popular.

This culture must be undone completely. That's an absolute necessity. Humanity lived without industrialism for most of its existence. And industrialism is killing the planet. Humans cannot exist without the planet. The planet (and sustainable human existence) is more important than industrialism.

Of course, we would all rather have a voluntary transformation, a tipping point. But if this tipping point does not occur, we need a backup plan.

◊ ◊ ◊

Q: I'm a fan of Daniel Quinn. He says we should just walk away. I know there is something wrong here. What do you think?

Derrick Jensen: There are two problems with this. With civilization having metastasized across the globe and bombing the moon, where are you supposed to walk to? Are you supposed to walk to the melting Arctic? Are you supposed to walk to the middle of the ocean, where there's forty-eight times as much plastic as there is phytoplankton? Where are you supposed to go? There is dioxin in every mother's breast milk, so you can't even drink breast milk without getting dioxin. There are carcinogens in every stream in the United States and, presumably, in the world. Where are you supposed to go?

Some respond to this by saying, "Oh, no, it's supposed to be a mental state. We're supposed to walk away emotionally and withdraw." But the real physical world is the basis for all life and you cannot withdraw from that.

Withdrawal in the face of moral complexity is no answer. Withdrawal in the face of atrocity is no answer. Two hundred species went extinct today. When faced with those committing atrocities, it is incumbent upon you to stop those atrocities using any means necessary. If you were being tortured to death in some basement, and I knew this, would you want me to walk away? Would you accept it if I said, "Oh, here's an answer, I will walk away." What would you call me if I did that? What would you call anyone else who did that?

Chapter 2

Civilization and Other Hazards
by Aric McBay

The only defense of this monstrous absurdity [cap and trade schemes] that I have heard is, "Well, you are right, it's no good, but the train has left the station." If the train has left, it had better be derailed soon or the planet, and all of us, will be in deep doo-doo.

—James Hansen, climate scientist

> try telling yourself
> you are not accountable
> to the life of your tribe
> the breath of your planet

—Adrienne Rich, feminist poet and essayist

So what are we up against?

Think for a moment about the ecological legacy of the dominant culture, its wholesale destruction of entire landbases ("impact on the environment," in the mealy-mouthed words of industrial apologists).

The Aral Sea, between what are now Kazakhstan and Uzbekistan, is a perfect example. Its name means "sea of islands," after the thousands of islands scattered across the once-fertile waters. In the 1950s, the USSR instituted an intensive industrial irrigation program meant to turn the Aral Sea's basin into a vast cotton plantation. At the time the sea was still huge—by area it could easily have swallowed Denmark, Sri Lanka, or the Dominican Republic. But the sea shrank rapidly from the 1960s onward, starved of water, and the growing salinity wiped out fish and other creatures. Now less than 10 percent of the sea remains. The moderating effect of the sea is gone; once-temperate summers are hot and dry, the winters long and cold. Where there was once a sea filled with life, there is now a dead and dusty plain, made toxic by decades of accumulated fertilizer and industrial waste. Vozrozhdeniya Island (well, formerly an island) holds the ruins of a Soviet bioweapons facility. Abandoned ships scatter the poisonous plain, their rusting hulks monuments to a time when the sea had fish—and water.

It's hard to think of a better term than *postapocalyptic*. But the apocalypse is not yet *post*; the remnants of the sea continue to shrink. There were three separate salty "lakes" left from the Aral Sea, but as I write one lake has finally succumbed and evaporated. Now only two briny, toxic remnants remain of the vast sea of islands.

What happened in the Aral Sea is happening everywhere, and fast. It took fifty years to turn the Aral Sea into a desert, but that same area of land is lost to desert every single year in the rest of the world. It's not hard to find entire biomes that have been destroyed by this culture. The prairies of the American West. The ancient forests of the Middle East. At this point it's much harder to find a biome that *hasn't* been destroyed.

And in some places those in power are just getting started, like in the case of the Athabasca Tar Sands under much of northern Alberta. The tar sands are subterranean deposits of bitumen mixed with sand, with many of the deposits underlying boreal forest. (If you were looking to find the "least destroyed biome," the world's boreal forest would be a good candidate; pre–global warming, anyway.) To get at the tar sands, oil companies literally scrape away the living forest and soils on the surface. Then they dig out the sands, taking about two tons of sand per barrel of oil they produce. Then, water drained from nearby rivers is used to wash the bitumen out of the sand—several volumes of water are used for every volume of oil—leaving a toxic water-oil by-product that kills fish, birds, and indigenous people living in the area. If you simply hated the land and wanted to destroy it, you would be hard-pressed to find a more vicious way of doing it.

Huge quantities of natural gas are used to cook the bitumen into a synthetic oil. The energy required means that oil produced from tar sands produces at least five times as much greenhouse gases as conventional oil. If you wanted to come up with even *nastier* way to consume fossil fuels, congratulations.

All of this is a clear pattern. The dominant culture eats entire biomes. No, that is too generous, because eating implies a natural biological relationship. This culture doesn't just consume ecosystems, it obliterates them, it murders them, one after another. This culture is an ecological serial killer, and it's long past time for us to recognize the pattern.

◻ ◻ ◻

The crises facing the planet do not stem from human nature,[1] but from, as we previously discussed, the mode of social and political organization we call civilization. What do we need to know about civilization to defeat it?

It is *globalized*. Civilization spans the globe and, despite superficial political boundaries, is integrated infrastructurally and economically. Any local resistance effort faces an opponent with global resources, so effective strategies must be enacted around the world. However, civilization approaches finite limits—83 percent of the biosphere is already under direct human influence.[2]

It is *mechanized*. An industrial civilization requires machines for production. Mechanization has centralized political and economic power by moving the means of production beyond the scale at which human communities function equitably and democratically. It has created a dramatic population spike (through industrial agriculture) and global ecological devastation (through industrial fishing, logging, and so on).[3] Most humans are now dependent on industrial "production," while the system itself is utterly dependent on finite minerals and energy-dense fossil fuels.[4]

It is *very young* on cultural, ecological, and geological timescales, but seems old on a personal timescale. Civilized history spans a few thousand years, human history several millions, and ecological history several billions.[5] But since much traditional knowledge has been lost or destroyed by those in power in order to glorify civilization, normalize their oppression, and render alternative ways of living unthinkable, we have the impression that civilization is as old as time.

It is primarily an *urban phenomenon*. Civilizations emerge from and promote the growth of cities.[6] Cities offer a pool of workers who, crowded together and severed from land, must labor to survive.[7] Urban areas are densely surveilled and policed. Urban areas are epicentres of strife when civilizations fall; as Lewis Mumford wrote, "Each historic civilization . . . begins with a living urban core, the polis, and ends in a common graveyard of dust and bones, a Necropolis, or city of the dead: fire-scorched ruins, shattered buildings, empty workshops, heaps of meaningless refuse, the population massacred or driven into slavery."[8]

It employs an extensive *division of labor* and high degree of *social stratification*. Specialization increases production, but a narrow focus prevents most people from making systemic criticisms of civilization; they are too worried about their immediate lives and problems to look at the big picture. Similarly, social stratification keeps power centralized and maintains an underclass to perform undesirable labor. Modern civilization, with its vast manufacturing capacity, has so far produced a large middle class in the rich nations, a historically unique circumstance. Though such people are unwilling to risk this privilege by challenging industrial society, prolonging collapse will ensure that they lose that privilege—and much more.

It is *militarized*. Civilizations, intrinsically expansionist and voracious, are intensely competitive. The military is prioritized in politics, industry, and science, and this sometimes rears its head as overt fascism. Control of citizens is implemented through police. As anthropologist Stanley Diamond wrote, "Civilization originates in conquest abroad and repression at home."[9] Glorification of the military causes people to identify with the state and its spectacular violence, and advertises the consequences of fighting back.

Closely related, and in spite of feminist advances, civilization is *patriarchal* and exalts masculinity. Civilization systematically oppresses women and celebrates the masculine expression of power and violence.

It is based on *large-scale agriculture*. Hunting, gathering, and horticulture cannot support civilizations. Only intensive, large-scale agriculture can provide the "surplus" to support cities and specialized elites. Historical agriculture was heavily dependent on slavery, serfdom, and cruelties. Industrial agriculture depends upon petroleum, an arrangement that will not last.

From the beginning it has been *predicated on perpetual growth*. This growth is inseparable from agriculture and settlement; settlement requires agriculture, which results in population growth and militarized elites who control the resources, and begins to overburden and destroy the local landbase.

Societies, cultures, and businesses that expand in the short term do so at the expense of entities that grow more slowly (or not at all), regardless of long-term consequences. In other words, civilization is characterized by *short-term thinking*; the structure of civilization rewards those who think in the short term and those who take more than they give back. Because those in power take more than they give back, they often win in the short term. But because ultimately you cannot win by taking more from the land than it gives willingly, they must lose in the long term.

Because of its drive toward war, ecological destructiveness, and perpetual expansion in a finite world, the *history of civilizations is defined by collapse*. Throughout history, civilizations have either collapsed or been conquered, the conquerors going on to meet one or both of those fates. Collapse is the typical, not exceptional, outcome for a civilization. As Gibbon wrote of Rome: "The story of the ruin is simple and obvious; and instead of inquiring why the Roman Empire was destroyed, we should rather be surprised that it subsisted for so long."[10]

Civilization is *hierarchical and centralized* both politically and infrastructurally. This is self-perpetuating; those in power want more power, and they have the means to get it. Superficially, global power is held by a number of different national governments; in the modern day those governments are mostly in the thrall of a corporate capitalist elite. In social terms, civilization's hierarchy is pervasive and standardized; most political and corporate leaders are interchangeable, replaceable components. The corollary of the centralization of power is the externalization of consequences (such as destroying the planet). Wherever possible, the poor and nonhumans are made to experience those consequences so the wealthy can remain comfortable.

Hierarchy and centralization result in *increasing regulation of behavior and increasing regimentation*. With the destruction of traditional kinship systems and methods of conflict resolution caused by the expansion of civilization and the rise of heavily populated urban centers, those in power have imposed their own laws and systems to enforce hierarchy and regulation.

As a means of enforcing hierarchy and regulation, civilization also makes major investments in *monumental architecture and propaganda*. Past civilizations had pyramids, coliseums, and vast military marches to impress or cow their populations. Although modern civilizations still have monumental architecture (especially in the form of superstores and megamalls), the wealthier human population is immersed in virtual architecture—a twenty-four-hour digital spectacle of noise and propaganda.

Civilization also requires *large amounts of human labor*, and is based on either compelling that labor directly or systematically removing feasible livelihood alternatives. We're often told that civilization was a step forward which freed people from the "grind" of subsistence. If that were true, then the history of civilization would not be rife with slavery, conquest, and the spread of religious and political systems by the sword. Spending your life as a laborer for sociopaths is only appealing if equitable land-based communities—and the landbase itself—are destroyed. In other words, civilization perpetuates itself by producing deliberate conditions of scarcity and deprivation.

Civilization is *capable of making Earth uninhabitable* for humans and the majority of living species. Historical civilizations self-destructed before causing global damage, but global industrial civilization has been far more damaging than its predecessors. We no longer have the option of waiting it out. There is nowhere left to go. Civilization will collapse one way or another, and it's our job to insure that something is left afterward.

ⓢ ⓢ ⓢ

The dominant culture isn't only a serial killer—it's also an amnesiac. Entire species and biomes are not just wiped out, but forgotten. And worse, they are deliberately erased, scratched out of history. People don't recognize this culture's pattern of ecocide because they don't mourn for all that has already been lost, been killed.

Everyone knows what a penguin is, right? Well, the name didn't always refer to the cute Antarctic birds. The name, which means *fat one*, formerly referred to the great auk, the seabird that populated Atlantic islands in vast numbers. Only when the great auk was hunted to extinction (and then forgotten by most) did the moniker move to the South Pole.

Cod are another example. Abundant cod swam off the coast of Newfoundland and the Maritimes. They were so numerous that it took a long time to fish them to the brink of extinction.[11] And yet, you can still buy cod at the grocery store. Why? Because the name has been taken for marketing reasons. If you buy something labeled cod, you no longer get true Atlantic cod (*Gadus morhua*). Instead you get something that has been deliberately mislabeled: rockfish (*Sebastes* spp.) or Alaska pollack (*Theragra chalcogramma*) or the poisonous oilfish (*Ruvettus pretiosus*). This constantly happens in the seafood industry—a species is wiped out, and replaced by a renamed or deliberately mislabeled fish. And then *that* one is wiped out and the cycle continues.

All of this gives grocery shoppers and eaters a sense that things are fine. They hear about bad things happening to fish on the news, maybe, but there's still plenty to eat at the store, so what's the problem? But if you take a moment to think about it, this renaming is deeply disturbing. It's like going home to find that a serial killer has murdered your family and replaced them with bystanders plucked off the street, renamed after your dead kin. The killer sits there in your house, grinning, insisting that everything is fine.

We don't need to know every single casualty of this culture to fight back (although every one I learn about fills me with more ardor to do so). But we cannot understand the severity and urgency of our situation, nor can we formulate an appropriate response, without first understanding at least some of these crises.

INDUSTRIAL PRACTICES THAT ARE TOXIC OR INCOMPATIBLE WITH LIFE

Global warming is caused by the emission of greenhouse gases from burning fossil fuels, as well as other industrial activities and land destruction. Concentrations of atmospheric methane have increased by about 250 percent from preindustrial levels. The preindustrial concentration of CO_2 was about 280 ppm (parts per million). In 2005 it passed 379 ppm. In 2010 it stands at 392 ppm. The Intergovernmental Panel on Climate Change (IPCC) estimates that it could reach 541 to 970 ppm by the year 2100. However, many climate scientists believe that levels must be kept beneath 350 ppm to avoid "irreversible catastrophic effects."[12]

Models predict a temperature increase of 2.4 to 6.4°C (4.3 to 11.5°F) during the twenty-first century.[13] An average increase of that amount would be bad enough, but the increase won't be distributed evenly. Instead, some areas will be subjected to smaller increases, while many regions will be subjected to severe temperature increases upward of 8°C (14.4°F). There will also be with year-to-year variation, some years a few degrees cooler, and others a few degrees warmer. These stacked effects will further add to the potential extremes. Rare (every ten years) extreme weather events, such as major storms, could happen every year. Catastrophic events that should happen once in a hundred years could happen every decade.

The effects of greenhouse gas emissions are delayed because it takes time for the extra heat captured by the atmosphere to accumulate. We are only now feeling the effects of decades-old emissions, and current emissions will take decades to have their full effect. Even if emissions stopped immediately, existing gases would contribute to global warming and rising sea levels for at least one thousand years.[14] Furthermore, global warming becomes self-sustaining beyond a certain point. As tundra melts, frozen organic matter will thaw and release great gouts of greenhouse gases. Drastic climate changes will damage many such biomes, causing them to release more carbon.

Projections are one thing, but paleontologists have implicated global warming in all but one of Earth's prehistoric mass extinctions.[15] The most severe mass die-off, dubbed the "Great Dying," happened a quarter of a billion years ago and wiped out 96 percent of all marine species and 70 percent of all land-based vertebrates.[16] A massive release of methane from the ocean floor has been blamed. Currently, in the Arctic Ocean warming has forced methane to bubble up in great, churning plumes.[17] NASA says a tipping point that would lead to "disastrous effects" will be reached by 2017.[18] Others argue that such a tipping point—perhaps one of several—has already been reached.[19] Of course, for many species and cultures on or past the brink of extinction, it has certainly already been reached.

Global warming is most urgent, but more insidious forms of pollution causing the poisoning of the planet lurk. Researchers at Cornell University blamed 40 percent of *all human deaths* on water, air, and soil pollution.[20] Speaking from my experience as a paramedic—and my personal experience seeing friends or loved ones facing cancer and similar diseases—I can tell you that death by pollution is usually a ghastly way to go. It is not quick or painless, but a drawn-out descent into slow asphyxiation (in the case of diseases caused by air pollution), and sores, rashes, and tumors (in the others). This is worse even than the myth of nature red in tooth and claw; being eaten by a bear or a tiger is fast and merciful compared to a gasping, hacking death by coal lung. And think of the sheer numbers of deaths. Every year some 57 million humans die from all causes, which means that 23 million of them are killed by pollution. That's 63,000 per day or the equivalent of twenty-one September 11 attacks every day.

The burden of ecocide is felt most by the poor. In China's burgeoning cities, smoke from coal-burning stoves and cooking oil kills 300,000 people per year.[21] And it has long been known that pollution-spewing industrial facilities and hazardous waste sites are much more likely to be placed where people of color live, rather than in predominantly white areas.[22]

Though agricultural or sanitation problems do cause runoff and water contamination, industry is the main pollution culprit. When industry stops or declines, pollution levels drop immediately. The Northeast Blackout of 2003 caused such a decline in air pollution. Twenty-four hours after the blackout began, sulphur dioxide levels dropped 90 percent, stratospheric ozone levels 50 percent, and light-scattering particulates 70 percent.[23]

More insidious types of pollution aren't so responsive. Persistent organic pollutants, the poisons that accumulate and biomagnify in body fat, have become globally ubiquitous. These pollutants endure for centuries, and on breaking down may release more toxic by-products. This crisis requires immediate action to prevent further accumulation.[24]

An essential dynamic of civilization is the *centralization of power and the externalization of consequences*. The last fifty years have clearly seen a fusion of runaway corporatism, militarism, and the systematic exploitation of the poor, both domestically and internationally. To continue the centralization of power, the expansion of capitalism, and resource extraction, those in power must destroy traditional, land-based cultures and increase social control.

The destruction of indigenous and sustainable cultures is unrelenting. Language is a good indicator. There are some 6,800 human languages, of which 750 are extinct or nearly extinct. Of 300 indigenous North American languages, only 30 are expected to remain by the year 2050. About half of all languages are endangered.[25]

The gap between the rich and the poor has continued to grow rapidly. The income of the richest 1 percent of people equals that of the poorest 57 percent.[26] The three richest people own more than the poorest 10 percent of people combined. This inequality occurs both between and within countries. In 1992 the pay ratio between the CEO and the average American worker was about 42 to 1. By the year 2000 it had grown to 525 to 1.

Civilization is not one hierarchy, but multiple interlocking hierarchies and systems of oppression based on gender, race, and class. For example, women do two-thirds of global work, earn less than 10 percent of wages, and own less than 1 percent of wealth.[27] We can make similar observations about race and class.

Some say that even the poor are wealthier now than ever before in history, which depends on how you measure "wealth." (But that's not very meaningful when the global economy is based on dwindling supplies of finite resources, meaning such "wealth" is short-lived and based on future impoverishment.) The next fifty years aside, the past fifty are telling. In 2007 some 57 percent of 6.5 billion people were malnourished, up from 20 percent of a 2.5 billion population in 1950.[28]

This wealth and well-being gap is partly a by-product of the mantra of profit-at-any-cost, but also from deliberate attempts to harm or impoverish, so that marginalized people are less able to mount resistance against occupation and resource extraction. As Nobel Peace Prize laureate and war criminal Dr. Henry Kissinger infamously advised, "Depopulation should be the highest priority of foreign policy towards the third world, because the US economy will require large and increasing amounts of minerals from abroad, especially from less developed countries."

International policies like structural adjustment programs (SAPs) are just the latest form of colonialism. SAPs force poor countries to increase tax collection and cut government spending, sell off public lands and enterprises to private corporations, and remove restrictions (like those pesky labor and environmental policies) on trade and the generation of profit. SAPs have been criticized from the beginning for dramatically increasing poverty and inequality, reversing land reforms, and forcing people off the land and into urban slums.[29]

These policies often go hand in hand with inducements to borrow money from the industrialized nations to buy infrastructure or commodities from those very countries, one of many practices which has resulted in crushing debt in the third world. In some countries, such as Kenya and Burundi, debt repayment vastly outstrips spending on social services like health care. The cancellation of debt has been shown to result in a prompt and significant increase in social spending.[30] The poor countries of the world pay about $4 million in debt per hour.

Enormous as this may seem when we compare it to our own household budgets, it's small compared to the $58 million the US spends on the military each hour.[31] According to the Stockholm International Peace Research Institute, global military spending now exceeds 1.3 trillion dollars. Although spending dipped after the end of the Cold War, it began to climb more steeply with the so-called War on Terror and has now approached its previous peak.[32] The United States, which uses the majority of its discretionary budget on the military, spends almost as much as all other countries combined, and, after accounting for inflation, recently surpassed its own Cold War record for annual spending.[33]

There have been social advances over the last century, especially in civil rights for people of color and women. But human societies ultimately rest on the foundation of the landbase, and global ecocide threatens to reverse the progress that has been made. Economic crises will occur and worsen, but they are difficult to predict because finance is imaginary. The state of the real world, on the other hand, requires no speculation.

In *Overshoot: The Ecological Basis of Revolutionary Change*, William R. Catton Jr. identifies "drawdown" as "an inherently temporary expedient by which life opportunities [i.e., carrying capacity] for a species are temporarily increased by extracting from the environment for use by that species some significant fraction of an accumulated resource that is not being replaced as fast as it is drawn down." Drawdown means using reserves, rather than income, to meet yearly demand. *Industrial drawdown* increases both the human population and the "overhead" costs of operating industrial society.

The dominant culture is utterly reliant on drawdown, such that it is hard to identify something that's *not* being drawn down at a staggering rate. The most crucial substances to industrial society and human life—soil, water, cheap energy, food stocks—are exactly those being drawn down most rapidly. And as Catton writes, the use of drawdown is an "inescapably dead-end" approach.

Cheap oil undergirds every aspect of industrial society. Without oil, industrial farms couldn't grow food, consumer goods couldn't be transported globally, and superpowers couldn't wage war on distant countries. *Peak oil* is already causing disruption in societies around the world, with cascading effects on everything from food production to the global economy.

Peak oil extraction has passed and extraction will decline from this point onward. No industrial renewables are adequate substitutes. Richard C. Duncan sums it up in his "Olduvai Theory" of industrial civilization. Duncan predicted a gradual per capita energy decline between 1979 and 1999 (the "slope") followed by a "slide" of energy production that "begins in 2000 with the escalating warfare in the Middle East" and that "marks the all-time peak of world oil production." After that is the "cliff," which "begins in 2012 when an epidemic of permanent blackouts spreads worldwide, i.e., first there are waves of brownouts and temporary blackouts, then finally the electric power networks themselves expire."[34] According to Duncan, 2030 marks the end of industrial civilization and a return to "global equilibrium"—namely, the Stone Age.

Natural gas is also near peak production. Other fossil fuels, such as tar sands and coal, are harder to access and offer a poor energy return. The ecological effects of extracting and processing those fuels (let alone the effects of burning them) would be disastrous even compared to petroleum's abysmal record.

Will peak oil avert global warming? Probably not. It's true that cheap oil has no adequate industrial substitute. However, the large use of coal predates petroleum. Even postcollapse, it's possible that large amounts of coal, tar sands, and other dirty fossil fuels could be used.

Although peak oil is a crisis, its effects are mostly beneficial: reduced burning of fossil fuels, reduced production of garbage, and decreased consumption of disposable goods, reduced capacity for superpowers to project their power globally, a shift toward organic food growing methods, a necessity for stronger communities, and so on. The worst effects of peak oil will be secondary—caused not by peak oil, but by the response of those in power.

Suffering a shortage of fossil fuels? Start turning food into fuel or cutting down forests to digest them into synthetic petroleum. Economic collapse causing people to default on their mortgages? Fuel too expensive to run some machines? The capitalists will find a way to kill two birds with one stone and institute a system of debtors prisons that will double as forced labor camps. A large number of prisons in the US and around the world already make extensive use of barely paid prison laborers, after all. Mass slavery, gulags, and the like are common in preindustrial civilizations. You get the idea.

Industrial civilization is heavily dependent on many different *finite resources* and materials, a fact which makes its goal of perpetual growth impossible. In particular, certain metals are in short supply.[35] Running out of cheap platinum wouldn't have much ecological impact. But shortages of more crucial minerals, like copper, will hamper industrial society's ability to cope with its own collapse. Severe shortages and high prices will worsen the social and ecological practices of mining companies (bad as they are now). These shortages would also represent a failure of industrial civilization's fundamental and false promise to expand and bring its benefits to all people in the world. According to one study, upgrading the infrastructure in the "developing world" to the status of the "developed world" would require essentially all of the copper and zinc (and possibly all the platinum) in the earth's crust, as well as near-perfect metal recycling.[36]

The growing global *food crisis* is a severe confluence of economic, political, and ecological factors. Right now plenty of food is being produced, but for economic reasons it isn't being distributed fairly. If, at its apex of production, industrial agriculture can't feed everyone, imagine what will happen when it collapses. Prices for corn and rice are already dramatically increasing, in part because of biofuels, even though the biofuel industry is still small.

The food crisis is going to get worse, but it's not going to be a "Malthusian crisis," in which a crisis exponential population growth outpaces increasing agricultural production. Our crisis is likely to culminate in a *decrease* in agricultural production caused by energy decline and increasing use of biofuels, and worsened by climate change and ecological damage. Sustainable ways of growing food are labor-intensive because they are horticultural and polycultural, rather than agricultural and monocultural. (That is, sustainable methods are small-scale and ecologically diverse, rather than the opposite.) As soil microbiologist Peter Salonius states flatly, "Intensive crop culture for high population[s] is unsustainable."[37] The longer humanity waits before switching to sustainable food sources and reversing population growth, the greater the disparity will be between carrying capacity and population.

The food crisis is deeply tied to two other ecological crises: *water drawdown* and *soil loss*. Industrial water consumption is drying up rivers and swallowing entire aquifers around the world. Although shallow groundwater can gradually be replenished by rainfall, when those supplies become depleted many farms and industries use deep wells with powerful pumps to extract water from fossil aquifers, which aren't replenished by rainfall. This shift to industrial drilling for water—essentially water mining—has caused major drops in water tables. In India, for example, deep electrically pumped wells used by large cash-crop monoculture farms have caused a major drop in water tables. This means small and subsistence farmers who use hand wells are losing

their water supplies, a disaster which has caused a dramatic rise in sui-
cides.[38] Approximately half of hand-dug wells in India—up to 95
percent of all wells in some regions—are now dry, driving an abandonment of rural villages.

In the grain-growing regions of central China, the water table is
dropping about 3 meters (10 feet) per year, and up to twice as fast in
other areas.[39] Chinese wheat production fell by 34 million tons between
1998 and 2005, a gap larger than the annual wheat production of
Canada.[40] In Saudi Arabia (as well as other countries), the technology
being used for well drilling is now a modified version of oil drilling
technology, because many wells must exceed one kilometer in depth to
reach fresh water.

Access to groundwater has always allowed agriculturalists to *occasionally* consume more water than rained down each year, but now
farming around the world has become dependent on its overconsumption. And make no mistake, drawdown of aquifers through deep
drilling and pumping is utterly driven by and dependent on a highly
industrialized culture. Without industrial machinery, even the most
unsustainable society would be limited to drawing the amount of water
that the water table could sustainably recharge each year. Furthermore,
water used by industry and agriculture far outweighs residential water
use, and typically less than 1 percent of residential water is actually used
for drinking.

Among the most threatening crisis is *soil drawdown and desertification*.
It takes a thousand years for the earth to create a few inches of topsoil.
Currently, topsoil is being lost at ten to twenty times the rate at which
it can be replenished. In his book *Dirt: The Erosion of Civilizations*, geol-
ogist David Montgomery traces the collapse of previous civilizations
that destroyed the topsoil upon which they depended. He estimates that
about 1 percent of the world's topsoil is lost each year.[41] According to
United Nations University, by 2025 Africa may only have enough intact
land to feed 25 percent of its human population.[42]

Desertification is primarily caused by overcultivation, deforestation,
overgrazing, and climate change. About 30 percent of Earth's land surface is at risk of desertification, including 70 percent of all drylands.

Fifty-two thousand square kilometers are turned to desert each year; about the area of Hong Kong is turned to desert each week. The UN reports that desertification threatens the livelihood of one billion people in 110 countries.[43]

More land was converted into cropland in the three decades following 1950 than in the fifteen decades following 1700.[44] Cultivated lands now cover about one quarter of the earth's land surface, but about 40 percent of agricultural land in the world has become degraded in the last fifty years.[45] Further expansion of agriculture to move beyond damaged lands is no longer an option—humans already occupy 98 percent of the areas where rice, wheat, or corn can be grown.[46] Canadian research scientist Peter Salonius estimates that once petroleum has been exhausted, the soils of the earth will be so degraded that the planet will only be able to support 100 million to 300 million people.[47]

Per capita seafood consumption has tripled since 1950.[48] Thanks to *overfishing*, between 1950 and 2003, 90 percent of the large fish in the ocean have been wiped out, and those who remain are smaller.[49] Since then, industrial fishing has continued to take more fish each year. By the midpoint of the twenty-first century, scientists estimate, *all* oceanic fish stocks worldwide will have collapsed.[50] Bottom trawling, a form of industrial fishing that involves dragging heavy nets across the sea bottom, obliterates seafloor habitat and seafloor creatures in the "most destructive of any actions that humans conduct in the ocean."[51] Every six months, bottom trawlers drag an area the size of the continental United States.

The orange roughy is just one of the creatures who have been decimated by this practice. These fish may grow to nearly three feet in length, and live up to one and a half centuries. Because they are so long lived and slow to mature, and because they produce few eggs compared to most fish, their populations are slow to rebound from any trouble. The assault of bottom trawling is ceaseless. Schools of orange roughy recently discovered near Australia have declined by 90 percent in a decade.[52]

Orange roughies spend much of their time congregating in large schools. As scientific research has recently confirmed, fish are highly intelligent and social animals. Dr. Culum Brown of the University of Edinburgh writes, "In many areas, such as memory, their cognitive powers match or exceed those of 'higher' vertebrates, including non-human primates."[53] Doctor Brown, along with Doctors Kevin Laland and Jens Krause, go on to say that "fish are steeped in social intelligence, pursuing Machiavellian strategies of manipulation, punishment and reconciliation, exhibiting stable cultural traditions and co-operating to inspect predators and catch food."[54] Furthermore, they recognize their "shoal mates" (that is, their friends) and have long-term relationships, follow the social prestige and relationships of others, and build complex nests. Of course, the rich social lives of fish—the researchers above use the word "culture"—are ignored by those who facilitate their industrial decimation.

As with many resource extraction industries, large-scale commercial fishing would not be economically feasible without heavy government subsidies. Economists have calculated that the expense of catching and marketing fish is almost twice as much as the value of the global catch.[55] None of these figures, of course, include the true ecological costs of destroying biomes that cover the majority of the earth's surface.

And then there's *deforestation*. Global warming–induced mild winters have increased the spread of temperate forest pests like the mountain pine beetle. Massive tree kills caused by the beetle (and industrial logging) have turned many Canadian forests from carbon sinks into carbon emitters.[56] They are now contributing to accelerating warming, worsening the spread of pests like the pine beetle.

Fully half of the mature tropical forests have been wiped out globally, and some areas have been hit especially hard. The Philippines have lost 90 percent of their forests, Haiti has lost 99 percent, and between 1990 and 2005 Nigeria lost 80 percent of its old-growth forest.[57] Without major global action, by 2030 only 10 percent of the tropical forest will remain intact, with another 10 percent in a fragmented and degraded condition.[58] If we don't prevent it, hundreds of thousands of species will go extinct; global warming, drought, soil erosion, and landslides will all worsen severely.

Tropical forests are being wiped out at a rate of 160,000 square kilometers per year, with demand for biofuels driving that number upward.[59] To put this into perspective, imagine lining all of that destruction up into one long swath that stretched from horizon to horizon in width and more than 16,000 kilometers in length.[60] To walk this distance on the globe you would have to start in Cape Town, South Africa, walk the entire length of Africa to Cairo, hike across the Middle East to the tip of the Caspian Sea, and then traverse the entire width of Asia, finally stopping at the Bering Sea near Kamchatka. Or you could string it from the southern-most tip of Argentina all the way to Alaska, the length of South and North America combined. To walk that scar from end to end would take you eighteen months, during which you would see nothing but stumps and ash and dust and ruin. And because it would take you eighteen months to see only twelve months of destruction, you would never be able to see it all.

The year 2005 broke all previous records for woodcutting.[61] The harvesting of wood for fuel and lumber is only one factor. In the Amazon the main factor is clearing land for cattle-grazing pasture. Other causes include government subsidies for settlements, road building, and infrastructure development, and commercial agriculture, mostly of soybeans for export. According to one researcher, "Soybean farms cause some forest clearing directly. But they have a much greater impact on deforestation by consuming cleared land, savanna, and transitional forests, thereby pushing ranchers and slash-and-burn farmers ever deeper into the forest frontier. Soybean farming also provides a key economic and political impetus for new highways and infrastructure projects, which accelerate deforestation by other actors."[62]

As is the case with many forms of fiscally and industrially driven ecocide, analysts have noted that deforestation in Brazil is "strongly correlated" with the "health" of the economy. Periods of economic slowdown match periods of lesser deforestation, while a rapidly growing economy causes much greater deforestation. Writes Rhett Butler: "During lean times, ranchers and developers do not have the cash to rapidly expand their pasturelands and operations, while the government lacks funds to sponsor highways and colonization programs and grant tax breaks and subsidies to forest exploiters."[63] In other words, economic growth is bad for the health of the planet, and economic contraction is good for the health of the planet.

Much of the world's remaining tropical forest is in the Amazon. This enormous rainforest creates the moist climate it needs by transpiring huge amounts of water and affecting air currents over the entire continent. Deforestation stops that transpiration and encourages desertification. This may create a self-perpetuating cycle of drought that kills even the largest trees and further reduces transpiration. Many ecologists believe that there is a tipping point beyond which this cycle would become irreversible and the Amazon would turn into a desert.[64] Some estimates put this tipping point as early as 2007, which would mean that action was required yesterday (or, second best, immediately). There is ample evidence that worsening drought is already well underway.[65] This cascading drought would not be limited to Latin America: "Scientists say that this would spread drought into the Northern Hemisphere, including Britain, and could massively accelerate global warming with incalculable consequences, spinning out of control, a process that might end in the world becoming uninhabitable."[66]

◙ ◙ ◙

The media report on these crises as though they are all separate issues. They are not. They are inextricably entangled with each other and with the culture that causes them. As such, all of these problems have important commonalities, with major implications for our strategy to resist them.

These problems are urgent, severe, and worsening, and the most worrisome hazards share certain characteristics:

They are progressive, not probabilistic. These problems are getting worse. These problems are not hypothetical, projected, or "merely possible" like Y2K, asteroid impacts, nuclear war, or supervolcanoes. These crises are not "possible" or "impending"—they are well underway and will continue to worsen. The only uncertainty is how fast, and thus how long our window of action is.

They are rapid, but not instant. These crises arose rapidly, but often not so rapidly as to trigger a prompt response; people get used to them, a phenomenon called the "shifting baselines syndrome." For example, wildlife populations are often compared to measures from fifty years ago, instead of measures from before civilization, which makes the damage seem much less severe than it actually is.[67] Even trends which appear slow at first glance (like global warming) are extremely rapid when considered over longer timescales, such as the duration of the human race or even the duration of civilization.

They are nonlinear, and sometimes runaway or self-sustaining. The hazards get worse over time, but often in unpredictable ways with sudden spikes or discontinuities. A 10 percent increase of greenhouse gases might produce 10 percent warming or it might cause far more. Also, the various crises interact to create cascading disasters far worse than any one alone. Hurricanes (such as Katrina) may be worsened by global warming and by habitat destruction in their paths (Katrina's impact was worsened by wetlands destruction). The human impact may then be worsened further by poverty and the use of the police, military, and hired mercenaries (like Blackwater)[68] to impede the ability of those poor people to move freely or access basic and necessary supplies.

These crises have long lead or lag times. The problems are often created long before they become a visible issue. They also grow or accelerate exponentially, such that action must be taken well in advance of the crisis to be effective. Although an alert minority is usually aware of the issue, the problem may have become very serious and entrenched before gaining the attention, let alone the action, of the majority. Peak oil was predicted with a high degree of accuracy in 1956.[69] The greenhouse effect was discovered in 1824, and industrially caused global warming was predicted by Swedish scientist Svante Arrhenius in 1896.[70]

Hazards have deeply rooted momentum. These crises are rooted in the most fundamental practices and infrastructure of civilization. Social convention, the concentration of power, and dominant economic systems all prevent the necessary changes. If I ran a corporation and tried to be genuinely sustainable, the company would soon be outcompeted and go bankrupt.[71] If I were a politician and I banned the majority of unsustainable practices, I would promptly be ejected from office (or more likely, assassinated).

They are industrially driven. In virtually all cases, industry is the primary culprit, either because it consumes resources itself (e.g., oil and coal) or permits resource extraction and global trade that would otherwise be extremely difficult (e.g., bottom trawling). Furthermore, industrial capitalism and industrial governments offer artificial subsidies for ecocidal practices that would not otherwise be economically tenable. Factors like overpopulation (as discussed shortly) are secondary or tertiary at best.

They provide benefits to the powerful and costs to the powerless. The acts that cause these crises—all long-standing economic activities—offer short-term benefits to those who are already powerful. But these hazards are most dangerous and damaging to the people who are poorest and most powerless.

They facilitate temporary victories and permanent losses. No successes we might have are guaranteed to last as long as industrial civilization stands. Conversely, most of our losses are effectively permanent. Extinct species cannot be resurrected. Overdrawn aquifers or clear-cut forests will not return to their original states on timescales meaningful to humans. The destruction of land-based cultures, and the deliberate impoverishment of much of humanity, results in major loss and long-term social trauma. With sufficient action, it's possible to solve many of the problems we face, but if that action doesn't materialize in time, the effects are irreversible.

Proposed "solutions" often make things worse. Because of all the qualities noted above, analysis of the hazards tends to be superficial and based on short-term thinking. Even though analysts who look at the big picture globally may use large amounts of *data*, they often refuse to ask deeper or more uncomfortable questions. The hasty enthusiasm for industrial biofuels is one manifestation of this. Biofuels have been embraced by some as a perfect ecological replacement for petroleum. The problems with this are many, but chief among them is the simple fact that growing plants for vehicle fuel takes land the planet simply can't spare. Soy, palm, and sugar cane plantations for oil and ethanol are now driving the destruction of tropical rainforest in the Amazon and Southeast Asia. Critics like Jane Goodall and the Rainforest Action Network argue that the plantations on rainforest land destroy habitat and water cycles, worsen global warming, destroy and pollute the soil, and displace land-based peoples.[72] This so-called solution to the catastrophe of petroleum ends up being just as bad—if not worse—than petroleum.

The hazards do not result from any single program. They tend to result from the underlying structure and essential nature of civilization, not from any particular industry, technology, government, or social attitude. Even global warming, which is caused primarily by burning fossil fuels, is the result of many kinds of industries using many kinds of fossil fuels as well as deforestation and agriculture.

☒ ☒ ☒

So how can we use what we know about the structure of industrial civilization, and about the most urgent problems it has caused, to inform our strategy and tactics? It's clear that some "solutions" can be immediately discounted or deprioritized because they won't work in a reasonable time frame, and there's no time to waste. Unfortunately, most of the solutions offered by apologists for those in power fall into this category.

Ineffective or less effective solutions are likely to have one or more of the following characteristics:

They may *reinforce existing power disparities*. Virtually any solution based on corporate capitalism is likely to meet this criterion. When Monsanto genetically engineers a plant to require less pesticides, they're not doing it to help the planet—they're doing it to make money, and so to increase their power. Carbon trading schemes are a clear example of this problem; they are capitalist shell games that allow corporations to rake in more profits while avoiding any real accountability and passing the costs on to regular people. (If it's not clear to you how this would play out, consider how much money the average person paid in income taxes last year, and ask yourself why General Electric paid zero dollars.)[73]

Ineffective solutions also *suppress autonomy or sustainability* that impedes profit. This is true both now and historically. Another way of phrasing this would be to say any solutions that require those in power to act against their own self-interest or otherwise behave in a way that fundamentally contradicts their known patterns of action will almost undoubtedly be ineffective, because these solutions will not be voluntarily implemented by those in power.

Solutions that *rely primarily on technofixes or technological and political elites* acting through large-scale industrial infrastructure will be ineffective. Adequate technologies already exist (for example, the hand wells in India) to meet human needs, but are either not implemented or are ignored in favor of more damaging technologies. Furthermore, suggested solutions are often stacked on top of (and so, increase dependence on) the existing and destructive infrastructure, rather than routing around it. Photovoltaic solar panels are suggested as a solution to problems caused by industrial civilization, but making those panels requires more industry and doesn't address root causes.

Solutions that *encourage increasing consumption and population growth* as a "solution" to existing problems also won't work. If you've gotten this far, we probably agree that any solution that encourages people to consume more—even if it's a nifty new hybrid SUV—is probably not going to be a suitable answer to our problem. And increasing population as a solution to human problems, is, of course, silly. This course of action is sometimes argued for by suggesting that more humans bring more creativity. But doubling the number of people on the planet will not double the quality or quantity of solutions produced. Twice the number of people will, however, eat twice as much, drink twice as much, use twice as much energy, and so on.

Attempts to solve a single problem without regard to other problems will also be ineffective. This sort of issue crops up often with "solutions" intended to solve energy problems. For example, ethanol from corn has been pitched repeatedly as a replacement for oil. But the widespread use of corn to make ethanol would worsen habitat destruction (by requiring more agricultural land) as well as worsening soil and water drawdown. Furthermore, ethanol from corn produces only a small amount of energy beyond that required to grow and process the corn.

Ditto for solutions *that involve great delays and postpone action until the distant future*—for example, voluntary emissions reductions with a target date of 2050. It's almost impossible to catalogue the consequences of further delay. Each day means more sustainable cultures destroyed, more species rendered extinct, more tipping points passed, more permanent losses. Each day also means an increasing gap between human population and carrying capacity, a gap with which we will have to reckon in the not-too-distant future.

It's true that there is growing interest in ecology and living sustainably in much of the world. But regardless of how you measure it, you cannot reasonably argue that this psychological shift toward sustainability is happening faster than the damage done by industrial civilization. It's great that there is a growing interest in organic gardening in the first world, but, meanwhile, millions of land-based peoples living in the third world are being forced from their land which means they can no longer grow their own food. The first-world organic gardeners are just a trickle compared to that flood. And prior to World War II and the invention of chemical pesticides, all gardening was organic. We aren't exactly gaining ground.

A similar problem applies technologically. Some people argue that we simply have to wait until advanced green technology surpasses unsustainable modern technology, but this doesn't make sense; unsustainable technologies have an economic edge *because* they take more than they give back.[74] Take the problem of overdrawn aquifers in China, where water tables are dropping several meters per year. It may still be possible to use hand-operated pumps in these areas. Let's say we wait a couple of decades for really cheap solar panels and pumps to become accessible to rural Chinese people. The water table will have fallen so far that they will need those solar-powered pumps just to survive because their hand wells will be dry. The purpose of those pumps will be to compensate for the ecological damage caused during the time it took to develop them—in other words, it won't be any easier to get water, and it will require more expense and equipment that they will have to pay for. One step forward and two steps back. Since damage is happening so much faster than recovery can, and is often more severe than even the most optimistic technologies could compensate for, significant delays are not acceptable.

Solutions that *focus on changing individual lifestyles* will also not be effective. As we've already discussed in this book and elsewhere, our problems are primarily of a systemic, not an individual, nature. Furthermore, lifestyle solutions encourage people to think of themselves as consumers and act in the capacity of consumers. This is an extremely limiting approach that distracts us from our identities as human beings, as members of human and living communities, and as living creatures in general. The idea that vast numbers of people would simply withdraw from the capitalist economy is a fantasy. If we had a large enough number of committed people to make a dent in global consumption, we would have a large enough number of committed people to exert serious political force against destructive institutions.

On a closely related note, many ineffective suggested solutions *are primarily based on token, symbolic, or trivial actions, and a superficial approach*. These kinds of solutions are what William R. Catton Jr. calls "cosmeticism"—"faith that relatively superficial adjustments in our activities" will keep the industrial age going—and they result from an acknowledgment of the fact that industrial civilization is destroying the world, but a refusal to accept the full implications of this problem. Though changing to compact fluorescents may offer some relief from guilt, to consider that as any kind of a meaningful solution is to ignore the nature of our predicament.

Others *focus on superficial or secondary causes, rather than the primary causal factor*. An example of this is the central focus that some people and organizations have on overpopulation. Damage caused by humans is primarily the result of overconsumption, not overpopulation. Though they may consume thirty times the resources of a third worlder, by focusing on overpopulation first worlders can displace responsibility for various problems to "those people." This ignores the fact that even very large families of third worlders likely consume less than a single first worlder. Furthermore, the overpopulation that does exist is largely caused by unsustainable industrial technology and the use of resource drawdown and conquest to create phantom carrying capacity.[75]

Arguments around overpopulation are often framed in a racist fashion that places blame on people of color in third world countries. Furthermore, problems like malnutrition or hunger in the third world are often blamed on "backwardness" and a lack of industrial infrastructure or technical knowledge. Of course, the key to reducing damage is, and has long been, reducing consumption and the capacity of industrial civilizations to draw down resources and expand into lands and habitat belonging to others.

That said, the fact that overpopulation isn't the main problem now does not make us immune from the consequences of adding more people. There are more humans on the planet than the planet can support (industrial or otherwise). When drawdown mechanisms cease, we—especially our hypothetical children—will all have to deal with the consequences, and the fewer humans there are at the time the less hardship there will be.

In general, though, the worst shortcoming of most suggested solutions is that they *are not consonant with the severity of the problem, the window of time available for effective action, or the number of people expected to act*. The solution should not be dependent on the assumption that very large numbers of people will act against their initial inclinations if we can't reasonably expect that to happen. If we wanted to back the idea that the solution to a problem like global warming is for everyone to voluntarily stop using fossil fuels, then we would have to reasonably believe that this is a plausible scenario. Unfortunately, it is not.

◙   ◙   ◙

In contrast, effective solutions (or at least, more effective) are likely to share a different set of characteristics:

They *address root problems* and are based on a "big picture" understanding of the situation. They include a long-term view of our situation, a critique of civilization, and a long-term plan.

A corollary of that is that the solutions should involve a higher level of *strategic rigor*. They should not be based on beautiful yet abstract ideas about what might make a better world, but derive from a tangible strategy that proposes a plan of action from point A to point B.

They *enable many different people to work toward addressing the problem*. Rather than being dependent on elites, solutions should enable as many people as possible to participate. This is not the same as requiring everyone to act to take down civilization or requiring the majority of people to act in a way we don't reasonably expect them to act. It does mean, however, that our strategy should include a way for all—from the most restrained to the most militant—to have a role if they desire.

Effective solutions *are suitable to the scale of the problem, and take into account the reasonable lead time required for action and the number of people expected to act.* If we can only expect a small number of people to take serious action, then our plans must only require a small number of people.

They *involve immediate action* AND *planning for further long-term action.* Crises like global warming cannot be addressed too soon. The most immediate action should target the worst contributors to each hazard, and happen as soon as possible. Subsequent actions should work their way down the severity scale.

They *make maximum use of available levers and fulcrums.* Which is to say, they play to our strengths and take advantage of the weaknesses of those who are trying to destroy the world. Each act should make as much impact as possible on as many different problems as possible.

And ultimately, of course, effective solutions must directly or indirectly work toward *taking down civilization.*

◙ ◙ ◙

Q: How do I know that civilization is not redeemable?

Derrick Jensen: Look around. Ninety percent of the large fish in the oceans are gone. Salmon are collapsing. Passenger pigeons are gone. Eskimo curlews are gone. Ninety-eight percent of native forests are gone, 99 percent of wetlands, 99 percent of native grasslands. What standards do you need? What is the threshold at which you will finally acknowledge that it's not redeemable?

In *A Language Older Than Words* I explained how we all are suffering from what Judith Herman would call "Complex Posttraumatic Stress Disorder." Judith Herman asks, "What happens if you are raised in captivity? What happens if you're long-term held in captivity, as in a political prisoner, as in a survivor of domestic violence?" You come to believe that all relationships are based on power, that might makes right, that there is no such thing as fully mutual relationships. That, of course, describes this culture's entire epistemology and this culture's entire way of relating. Indigenous peoples have said that the fundamental difference between Western and indigenous ways of being is that even the most open-minded Westerners view listening to the natural world as a metaphor as opposed to the way the world really works. So the world consists of resources to be exploited, as opposed to other beings to enter into relationship with. We have been so traumatized that we are incapable of perceiving that real relationships are possible. That is one reason that this culture is not redeemable.

Here is another answer. In *The Culture of Make Believe*, I wrote about how this culture is irredeemable because the social reward systems of this culture lead inevitably to atrocity. This culture is based on competition as opposed to cooperation and, as such, will inevitably lead to wars over resources.

Ruth Benedict, the anthropologist, tried to figure out why some cultures are good (to use her word) and some cultures are not good. In a good culture, men treat women well, adults treat children well, people are generally happy, and there's not a lot of competition. She found that the good cultures all have one thing in common. They figured out something very simple: they recognize that humans are both social creatures and selfish, and they merge selfishness and altruism by praising behaviors that benefit the group as a whole and disallowing behaviors that benefit the individual at the expense of the group. The bad cultures socially reward behavior that benefits the individual at the expense of the group. If you reward behavior that benefits the group, that's the sort of behavior you will get. If you reward behavior that is selfish, acquisitive, that's the behavior you will get. This is Behavior Modification 101.

This culture rewards highly acquisitive, psychopathological behavior, and that is the behavior we see. It's inevitable.

Need another answer? In *Endgame* I explained that a culture that requires the importation of resources cannot be sustainable. In order to be sustainable a culture must help the landbase, but if your culture requires the importation of resources, it means you've denuded the landbase of that particular resource. In other words, you have harmed your landbase. This is by definition unsustainable. As cities—which require the importation of resources—grow, they will denude and destroy ever larger areas. Because it is based on the importation of resources, this culture is functionally and inherently unsustainable.

Further, any way of life based on the importation of resources is also functionally based on violence, because if your way of life requires the importation of resources, trade will never be sufficiently reliable: if people in the next watershed over won't trade you for some necessary resource, you will take it, because you need it. So, to bring this to the present, we could all become enlightened, and the US military would still have to be huge: how else will they get access to the oil they need to run the economy, oil that just happens to lie under someone else's land? The point is that no matter what we think of the irredeemability of this culture's mass psychology or system of rewards, this culture—civilization—is also irredeemable on a purely functional level.

Another reason this culture is irredeemably unsustainable is that we can talk all we want about new technologies, but so long as they require copper wiring, they are going to require an industrial infrastructure, and they are going to require a mining infrastructure, and that is inherently unsustainable.

More signs of irredeemability: right now the United States is spending $100 billion a year to invade and occupy Afghanistan. That is $3,500 for every Afghan man, woman, and child, per year. At the same time, everybody from right-wing pundits to the zombies on NPR ask the question, "Is it too expensive to stop global warming?" There is always money to kill people. There is never enough money for life-affirming ends.

I look around in every direction and I see no sign of redeemability in this culture. The real physical world is being murdered. The pattern is there. We need to recognize that pattern, and then we need to stop those who are killing the planet.

Chapter 3

Liberals and Radicals
by Lierre Keith

Pacifism is objectively pro-Fascist. This is elementary common sense. If you hamper the war effort of one side you automatically help that of the other. Nor is there any real way of remaining outside such a war as the present one. . . . others imagine that one can somehow "overcome" the German army by lying on one's back, let them go on imagining it, but let them also wonder occasionally whether this is not an illusion due to security, too much money and a simple ignorance of the way in which things actually happen. . . . Despotic governments can stand "moral force" till the cows come home; what they fear is physical force.

—George Orwell, author and journalist

Can it be done? Can industrial civilization be stopped? Theoretically, any institution built by humans can be taken apart by humans. That seems obvious as a concept. But in the here and now, in the time frame left to our planet, what is feasible?

Here the left diverges. The faithful insist that Everything Will Be Okay. They play an emotional shell game of new technology, individual consumer choices, and hope as a moral duty. When all three shells turn up empty, the fall-back plan is an insistence in the belief that people can't really kill the planet. There will be bacteria if nothing else, they urge, as if that should give solace to the drowning bears and the vanished snails. Meanwhile, the facts tell a different story. Methane, a greenhouse gas twenty times more potent than carbon dioxide, is escaping from both land and sea where up until now it was sequestered by being frozen. This could lead to "a catastrophic warming of the earth."[1] *Catastrophic* meaning a planet too hot for life—any life, all life. Kiss your mustard seed of bacteria good-bye: yes, we can kill the planet.

It's a bankrupt approach regardless. Try this. Pretend that I have a knife and you don't. Pretend I slice off one of your fingers, then another, then a third. When you object—and you will object, with all your might—I tell you that I'm not going to kill you, just *change* you. Joint by joint, I continue to disarticulate someone still alive, who will very soon be dead. When you protest for your life, I tell you that you're not actually going to die, as there will surely be some bacteria remaining. Does that work for you?

One would hope that a looming mass extinction would compel us to seek something beyond emotional solace wrapped in pseudospiritual platitudes. But strategies for action are an affront to the faithful, who need to believe in individual action. This faith is really just liberalism writ large. One of the cardinal differences between liberals—those who insist that Everything Will Be Okay—and the truly radical is in their conception of the basic unit of society. This split is a continental divide. Liberals believe that a society is made up of individuals. Individualism is so sacrosanct that, in this view, being identified as a member of a group or class is an insult. But for radicals, society is made up of classes (economic ones in Marx's original version) or any groups or castes. In the radical's understanding, being a member of a group is not an affront. Far from it; identifying with a group is the first step toward political consciousness and ultimately effective political action.

But classical liberalism was the founding ideology of the US, and the values of classical liberalism—for better and for worse—have dispersed around the globe. The ideology of classical liberalism developed against the hegemony of theocracy. The king and church had all the economic, political, and ideological power. In bringing that power down, classic liberalism helped usher in the radical analysis and political movements that followed. But the ideology has limits, both historically and in its contemporary legacy.

The original founding fathers of the United States were not after a human rights utopia. They were merchant capitalists tired of the restrictions of the old order. The old world had a very clear hierarchy. This basic pattern is replicated in all the places that civilizations have arisen. There's God (sometimes singular, sometimes plural) at the top, who directly chooses both the king and the religious leaders. These can be one and the same or those functions can be split. Underneath them are the nobles, the priests, and the military. Again, sometimes these groups are folded into one, and sometimes they're discrete. Beneath them are the merchants, traders, and skilled craftsmen. The base of the pyramid contains the bulk of the population: people in slavery, serfdom, or various forms of indenture. And all of this is considered God's will, which makes resistance that much more difficult psychologically. Standing up to an abuser—whether an individual or a vast system of power—is never easy. Standing up to capital "G" God requires an entirely different level of courage, which may explain why this arrangement appears universally across civilizations and why it is so intransigent.

In the West, one of the first blows against the Divine Right of Kings was in 1215, when some of the landed aristocracy forced King John to sign Magna Carta. It required the king to renounce some privileges and to respect legal procedures. It established habeas corpus and due process. Most important was the principle it claimed: the king and the church are bound by the law, not above it, and citizens have rights against their government. Magna Carta plunged England into a civil war, the First Baron's War. Pope Innocent got involved as well, absolving the king from having to enforce Magna Carta—not because he'd been forced to sign it, but because it was blasphemous. Understand, it was a crime against God to suggest that people could question or make demands on the king.

The American Revolution can be seen as another Baron's revolt. This time it was the merchant-barons, the rising capitalist class, waging a rebellion against the king and the landed gentry of England. They wanted to take the king and the aristocrats out of the equation, so that the flow of power went God❯property owners. When they said "All men are created equal," they meant very specifically white men who owned property. That property included black people, white women, and more generally, the huge pool of laborers who were needed to turn this continent from a living landbase into private wealth. Less than 5 percent of the population could vote under the constitution as it was originally written. Under the rising Protestant ethic, amassing wealth was a sign of God's favor and God's grace. God was still operable, he'd just switched allegiance from the old inherited powers to the rising mercantile class.

This new class had a new set of priorities in the service of their God-given right to accumulate wealth. The West has had market economies for thousands of years; they are essential to feeding civilization. Goods have to be traded, first from the countryside, then from the colonies (and there are always colonies), to fill the ever-growing needs of the bloated power base. (The Sahara Desert once fed the Roman Empire, which should tell you everything you need to know about civilization's hunger and its supporting ecosystem's ultimate fate.)

Those original market economies in the West, and, indeed, around the world, were nestled inside a moral economy informed by community networks of care, concern, and responsibilities. Property owners and moneylenders were restricted by community norms and the influence of extralegal leaders like elders, healers, and religious officers. This social world was held together by personal bonds of affection and mutual obligation. These were precisely the bonds that the rising capitalist class needed to destroy. Their concept of freedom meant freedom from those obligations and responsibilities. In their schema, individuals were free from traditional moral and community values, as well as from the king and landed gentry, to pursue their own financial interests. What held this social world together wasn't bonds of affection and obligation, but impersonal contracts—and impersonal contracts favored the rich, the employers, the landlords, the owners, and the creditors while dispossessing the poor, the employees, the tenants, the slaves, and the debtors.

In 1776, half the immigrants to America were indentured servants. Three out of four people in Pennsylvania, Maryland, and Virginia were or had been indentured, 20 percent of the population were slaves, and 10 percent of the population owned half the wealth. George Washington was the wealthiest man in America.

Groups of people don't endure oppression without some of them fighting back. This is true everywhere, no matter what. There were huge and fertile populist movements in America at that time, with visions for a true democracy that have yet to be equaled. For instance, the commoners seized control of the Pennsylvania statehouse and wrote the following into their constitution: "An enormous portion of property vested in a few individuals is dangerous to the rights and destructive of the common happiness of mankind; and therefore every free state hath a right by its laws to discourage the possession of such property."

And here are a few other facts you probably didn't learn in public school. Between 1675 and 1700, militant confrontations brought down governments in Massachusetts, New York, Maryland, Virginia, and North Carolina. By 1760 there had been eighteen rebellions aimed at overthrowing colonial governments, six black rebellions, and forty major riots. "Freedom from all foreign or domestic oligarchy!" was a slogan of the common people. "Domestic" referred to George Washington and his friends, the merchant-barons. People knew who their enemies were—most of them had been literally *owned* by the rich. Contrast their slogan to the following quote from John Jay, the president of the First Continental Congress and the first Chief Justice of the Supreme Court: "The people who own the country ought to govern it." In fact, common soldiers mounted multiple attacks against the headquarters of the Continental Congress in Philadelphia. Nobody was taken in by the government that the merchant-barons were proposing.

What the merchant-barons wanted was a centralized national government with the ability to coercively suppress internal dissent movements, regulate trade, protect private property, and subsidize infrastructure that would drive the economy. What they ultimately wanted was to gut a vast, living continent and turn it into wealth, and they didn't want anyone to get in their way. That's the trajectory this culture has been on for 10,000 years, since the beginning of agriculture. The only thing that has changed is who gets to benefit from that gutting.

We need to understand the contradictory legacy of liberalism to understand the left today. Any political idea that can bring down theocracy, monarchy, and religious fundamentalism is worth considering, but any ideology that impedes a radical transformation of other equally violent systems of power needs to be rigorously examined and ultimately rejected.

Classical liberalism values the sovereignty of the individual, and asserts that economic freedom and property rights are essential to that sovereignty. John Locke, called the Father of Liberalism, made the argument that the individual instead of the community was the foundation of society. He believed that government existed by the consent of the governed, not by divine right. But the reason government is necessary is to defend private property, to keep people from stealing from each other. This idea appealed to the wealthy for an obvious reason: they wanted to keep their wealth. From the perspective of the poor, things look decidedly different. The rich are able to accumulate wealth by

Liberalism vs. Radicalism

LIBERAL	RADICAL
Individualism ■ basic social unit is individual ■ person is distinct from social group	*Group or Class* ■ basic social unit is group ■ person is socially constructed ■ active and critical embrace of group
Idealism ■ attitudes are sources and solutions for oppression ■ thinking as prime mover of social life ■ rational argument/education is engine of social change	*Materialism* ■ concrete systems of power are sources and solutions of oppression ■ thoughts and ideas are only one part of social life ■ organized political resistance compels social change
Naturalism ■ body exists independently of society/mind ■ gender/race as physical body	*Constructivism* ■ reality is socially constructed ■ gender/race are socially real categories, but biology is ideology
Voluntarism ■ social life comprised of autonomous, intentional, self-willed actions	*Social Determinism* ■ social life is comprised of a complex political determinism ■ the oppressed do not make or control conditions ■ but "with forms of power forged from powerlessness, conditions are resisted"[2]
Moralism ■ rightness means conforming behavior to rules that are abstractly right or wrong ■ equality before the law	*Feminist Jurisprudence* ■ abstract moralism works in the interests of power ■ material equality ■ while powerlessness is the problem, redistribution of power as currently defined is not its ultimate solution

taking the labor of the poor and by turning the commons into privately owned commodities; therefore, defending the accumulation of wealth in a system that has no other moral constraints is in effect defending theft, not protecting against it.

Classical liberalism from Locke forward has a contradiction at its center. It believes in human sovereignty as a natural or inalienable right, but only against the power of a monarchy or other civic tyranny. By loosening the ethical constraints that had existed on the wealthy, classical liberalism turned the powerless over to the economically powerful, simply swapping the monarchs for the merchant-barons. Adam Smith's *The Wealth of Nations*, published in 1776, provided the ethical justification for unbridled capitalism. As previously discussed, the pursuit of wealth for its own sake had been considered a sin and such pursuit had been constrained by a whole series of societal institutions. But Smith argued that the "Invisible Hand" of the market would provide what society needed; any government interference would be detrimental.

According to classical liberalism, government needs to refrain from any participation in the economic realm, beyond the enforcement of contracts. Classical liberalism's commitment to civil rights was based on a similar idea of what are termed "negative freedoms." The government must not interfere in arenas like speech and religion in order to guarantee liberty to individual citizens. The Bill of Rights is essentially a list of negative freedoms. In the real world, what negative freedoms mean is: if you have the power, you get to keep it. If you own the press or have the money to access it, you're free to "say" whatever you like. If you can't access it, well, the government can't interfere. The vast majority of citizens thus have no right to be heard in any way that is socially meaningful. This is how classical liberalism increased the rights of the powerful against the rights of the dispossessed.

In 1880, the growing monopolies of the big trusts (corporations) showed the inevitable end point of laissez-faire economics. Reformers saw that the government was the only institution that could break the economic stranglehold of the big trusts. Liberal thinkers started to abandon the classical commitment to laissez-faire economics, while they remained committed to individualism and the liberal concept of civil rights.

The big split between liberals and the true left came in the 1940s: as liberals took up an anti-Communist position, the actual leftists were purged from liberalism, especially from labor unions and the New Deal coalition. From the beginnings of classical liberalism, liberals have embraced capitalism. Indeed, classical liberalism was foundational to a capitalist economy. Hence, unlike in Europe, there is no real left in the US, as a true left starts with the rejection of capitalism. There is no political party in the US that represents a critique of capitalism. Congress is essentially filled with two wings of the Capitalist Party.

After the disaster of the Great Depression, liberalism shifted to the idea of government intervention to regulate business in order to assure competition and to enforce safety and labor standards. This was an attempt to make capitalism work, not to dismantle it. This approach is very different from state socialism, in which the state owns (not regulates) the means of production (and which has produced its own environmental and human rights disasters).

This modern version of liberalism is called social liberalism. It maintained its commitment to civil rights, especially as negative freedoms, and a capitalist system guided by government supports and regulates.

At this moment, the liberal basis of most progressive movements is impeding our ability, individually and collectively, to take action. The individualism of liberalism, and of American society generally, renders too many of us unable to think clearly about our dire situation. Individual action is not an effective response to power because human society is political; by definition it is built from groups, not from individuals. That is not to say that individual acts of physical and intellectual courage can't spearhead movements. But Rosa Parks didn't end segregation on the Montgomery, Alabama, bus system. Rosa Parks plus the stalwart determination and strategic savvy of the entire black community did.

Liberalism also diverges from a radical analysis on the question of the nature of social reality. Liberalism is idealist. This is the belief that reality is a mental activity. Oppression, therefore, consists of attitudes and ideas, and social change happens through rational argument and education. Materialism, in contrast, is the understanding that society is organized by concrete systems of power, not by thoughts and ideas, and that the solution to oppression is to take those systems apart brick by brick. This in no way implies that individuals are exempt from examining their privilege and behaving honorably. It does mean that antiracism workshops will never end racism: only political struggle to rearrange the fundamentals of power will.

There are three other key differences between liberals and radicals. Because liberalism erases power, it can only explain the subordinate position of oppressed groups through biology or some other claim to naturalism. A radical analysis of race understands that differences in skin tone are a continuum, not a distinction: race as biology doesn't exist. Writes Audrey Smedley in *Race in North America: Origin and Evolution of a Worldview,*

> Race originated as the imposition of an arbitrary value system on the facts of biological (phenotypic) variations in the human species. . . . The meanings had social value but no intrinsic relationship to the biological diversity itself. Race . . . was fabricated as an existential reality out of a combination of recognizable physical differences and some incontrovertible social facts: the conquest of indigenous peoples, their domination and exploitation, and the importation of a vulnerable and controllable population from Africa to service the insatiable greed of some European entrepreneurs. The physical differences were a major tool by which the dominant whites constructed and maintained social barriers and economic inequalities; that is, they consciously sought to create social stratification based on these visible differences.[3]

Her point is that race is about power, not physical differences. Racializing ideology was a tool of the English against the Irish and the Nazis against the Jews, groups that could not be distinguished by phenotypic differences—indeed, that was why the Jews were forced to wear yellow stars.

Conservatives actively embrace biological explanations for race and gender oppression. White liberals usually know better than to claim that people of color are naturally inferior, but without the systematic analysis of radicalism, they are stuck with vaguely uncomfortable notions that people of color are just . . . different, a difference that is often fetishized or sexualized, or that results in patronizing attitudes.

Gender is probably the ultimate example of power disguised as biology. There are sociobiological explanations for everything from male spending patterns to rape, all based on the idea that differences between men and women are biological, not, as radicals believe, socially created. This naturalizing of political categories makes them almost impossible to question; there's no point in challenging nature or four million years of evolution. It's as useless as confronting God, the right-wing bulwark of misogyny and social stratification.

The primary purpose of all this rationalization is to try to remove power from the equation. If God ordained slavery or rape, then this is what shall happen. Victimization becomes naturalized. When these forms of "naturalization" are shown to be self-serving rationalizations the fall-back position is often that the victimization somehow is a benefit to the victims. Today, many of capitalism's most vocal defenders argue that indigenous people and subsistence farmers *want* to "develop" (oddly enough, at the point of a gun); many men argue that women "want it" (oddly enough, at the point of a gun); foresters argue that forests (who existed on their own for thousands of years) benefit from their management.

With power removed from the equation, victimization looks voluntary, which erases the fact that it is, in fact, social subordination. What liberals don't understand is that 90 percent of oppression is consensual. As Florynce Kennedy wrote, "There can be no really pervasive system of oppression . . . without the consent of the oppressed."[4] This does *not* mean that it is our fault, that the system will crumble if we withdraw consent, or that the oppressed are responsible for their oppression. All it means is that the powerful—capitalists, white supremacists, colonialists, masculinists—can't stand over vast numbers of people twenty-four hours a day with guns. Luckily for them and depressingly for the rest of us, they don't have to.

People withstand oppression using three psychological methods: denial, accommodation, and consent. Anyone on the receiving end of domination learns early in life to stay in line or risk the consequences.

Those consequences only have to be applied once in a while to be effective: the traumatized psyche will then police itself. In the battered women's movement, it's generally acknowledged that one beating a year will keep a woman down.

While liberals consider it an insult to be identified with a class or group, they further believe that such an identity renders one a victim. I realize that identity is a complex experience. It's certainly possible to claim membership in an oppressed group but still hold a liberal perspective on one's experience. This was brought home to me while I was stuck watching television in a doctor's waiting room. The show was (supposedly) a comedy about people working in an office. One of the black characters found out that he might have been hired because of an affirmative action policy. He was so depressed and humiliated that he quit. Then the female manager found out that she also might have been ultimately advanced to her position because of affirmative action. She collapsed into depression as well. The emotional narrative was almost impossible for me to follow. Considering what men of color and all women are up against—violence, poverty, daily social derision—affirmative action is the least this society can do to rectify systematic injustice. But the fact that these middle-class professionals got where they were because of the successful strategy of social justice movements was self-evidently understood broadly by the audience to be an insult, rather than an instance of both individual and movement success.

Note that within this liberal mind-set it's not the actual material conditions that victimize—it's *naming* those unjust conditions in an attempt to do something about them that brings the charge of victimization. But radicals are not the victimizers. We are the people who believe that unjust systems can change—that the oppressed can have real agency and fight to gain control of the material conditions of their lives. We don't accept versions of God or nature that defend our domination, and we insist on naming the man behind the curtain, on analyzing who is doing what to whom as the first step to resistance.

The final difference between liberals and radicals is in their approaches to justice. Since power is rendered invisible in the liberal schema, justice is served by adhering to abstract principles. For instance, in the United States, First Amendment absolutism means that hate groups can actively recruit and organize since hate speech is perfectly legal. The principle of free speech outweighs the material reality of what hate groups do to real human people.

For the radicals, justice cannot be blind; concrete conditions must be recognized and addressed for anything to change. Domination will only be dismantled by taking away the rights of the powerful and redistributing social power to the rest of us. People sometimes say that we will know feminism has done its job when half the CEOs are women. That's not feminism; to quote Catharine MacKinnon, it's liberalism applied to women. Feminism will have won not when a few women get an equal piece of the oppression pie, served up in our sisters' sweat, but when all dominating hierarchies—including economic ones—are dismantled.

There is no better definition of oppression than Marilyn Frye's, from her book *The Politics of Reality*. She writes, "Oppression is a system of interrelated barriers and forces which reduce, immobilize and mold people who belong to a certain group, and effect their subordination to another group."[5] This is radicalism in one elegant sentence. Oppression is not an attitude, it's about systems of power. One of the harms of subordination is that it creates not only injustice, exploitation, and abuse, but also consent.

Subordination has also been defined for us. Andrea Dworkin lists its four elements:[6]

1. Hierarchy
Hierarchy means there is "a group on top and a group on the bottom." The "bottom" group has fewer rights, fewer resources, and is "held to be inferior."[7]

2. Objectification
"Objectification occurs when a human being, through social means, is made less than human, turned into a thing or commodity, bought and sold . . . those who can be used as if they are not fully human are no longer fully human in social terms."[8]

3. Submission
"In a condition of inferiority and objectification, submission is usually

essential for survival . . . The submission forced on inferior, objectified groups precisely by hierarchy and objectification is taken to be the proof of inherent inferiority and subhuman capacities."[9]

4. Violence

Committed by members of the group on top, violence is "systematic, endemic enough to be unremarkable and normative, usually taken as an implicit right of the one committing the violence."[10]

All four of these elements work together to create an almost hermetically sealed world, psychologically and politically, where oppression is as normal and necessary as air. Any show of resistance is met with a continuum that starts with derision and ends in violent force. Yet resistance happens, somehow. Despite everything, people will insist on their humanity.

Coming to a political consciousness is not a painless task. To overcome denial means facing the everyday, normative cruelty of a whole society, a society made up of millions of people who are participating in that cruelty, and if not directly, then as bystanders with benefits. A friend of mine who grew up in extreme poverty recalled becoming politicized during her first year in college, a year of anguish over the simple fact that "there were rich people and there were poor people, and there was a relationship between the two." You may have to face full-on the painful experiences you denied in order to survive, and even the humiliation of your own collusion. But knowledge of oppression starts from the bedrock that subordination is wrong and resistance is possible. The acquired skill of analysis can be psychologically and even spiritually freeing.

Once some understanding of oppression is gained, most people are called to action. There are four broad categories of action: legal remedies, direct action, withdrawal, and spirituality. These categories can overlap in ways that are helpful or even crucial to resistance movements; they can also be diversions that dead-end in despair. Crucial to our discussion, none of them are definitively liberal or radical as actions.

LEGAL REMEDIES

Most activist groups are centered around legal remedies to address specific harms. This is for a very good reason. As Catharine MacKinnon points out, "Law organizes power." Legislative initiatives and court challenges can run the gamut from useless pleading to potential structural change. It's too easy for radicals to dismiss this arena as inherently reformist. Much of it is, of course, and the main purpose of this book is to ask environmentalists to consider approaches beyond the usual legal response. But if we would like to organize power in an egalitarian distribution, we will need to grapple with the law. The trick is to do this as radicals, which means asking the questions: Does this initiative redistribute power, not just change who is at the top of the pyramid? Does it take away the rights of the oppressors and reestablish the rights of the dispossessed? Does it let people control more of the material conditions of their lives? Does it name and redress a specific harm? We can stand on the sidelines with a more-radical-than-thou attitude, but attitude will not help a single gasping salmon or incested girl child.

This is not a call to behave and ask nicely. I believe in breaking the law because the edifice is supported by a federal constitution that upholds a corrupt arrangement of power. It was written by white men who owned white women as chattel and black men and women as slaves, and those powerful men wrote it to protect their power. We have no moral obligation to respect it; quite the opposite. I also believe we will need to bring the whole edifice down or I wouldn't be a coauthor of this book. But there are legislative victories and court rulings—like the Civil Rights Act of 1964 and *Roe v. Wade*—that have changed people's lives in substantive ways, redirecting the flow of power toward justice. Further, a transition toward direct democracy built on a foundation of both human rights and human participation in the life of the planet is not conceptually difficult. Law is not just for liberals. The question is, what actions will get us from here to there? Neither sneering nor despairing has ever proven to be effective. It's easy for nothing to be radical enough, but an interior state of rage is also not enough. Structural change needs to happen. A radical analysis starts from that fact. How best to force that change is a strategic question.

DIRECT ACTION

Other activist groups bypass the legislative arena and focus on direct action. Sometimes this overlaps with a legal approach, such as civil disobedience to influence legislators and win specific goals. How many women chained themselves to the White House gate or endured the torture of force-feeding in Holloway Prison to win the right to vote? But actionists can also target other institutional arrangements of power, circumventing the law entirely. The Montgomery bus boycott is a good example of applying economic instead of political pressure. As with legal remedies, the goal of direct action can be liberal or radical.

No single action, whether "inside" or "outside" whatever system of power, is going to be definitive. A serious resistance movement understands that. Instead of closing off whole sectors of a power's organization, a successful movement aims at wherever power is vulnerable compared to the resources at hand. The "inside" and the "outside" actionists need to see themselves as working together toward that larger goal. Both are needed. Plenty of "outside" people do nothing effective their entire lives—indeed, a whole subculture of them declare that individual psychological change is a political strategy and attending personal growth workshops is "doing the work." You could not find a more liberal view. My point here is that "inside" and "outside" the identified system are not the bifurcation points of liberals and radicals.

A related mistake is in believing the most militant strategy to be the most radical. It isn't; it's only the most militant. I don't say this from a moral attachment to nonviolence. Derrick wrote 900 pages (in *Endgame*) to refute the pacifist arguments generally accepted across the left, and much of this current book is meant to inspire seriously militant action. But we need to examine calls for violence through a feminist lens critical of norms of masculinity. Many militant groups are an excuse for men to wallow in the cheap thrill of the male ego unleashed from social constraints through bigger and better firepower: real men use guns. Combined with ineffective strategic goals, and often rabidly masculinist behavioral norms, these groups can implode when the men start shooting each other. Michael Collins was killed by other Irish nationalists, Trotsky by Stalinist goons, and Malcolm X by other black Muslims. Leftist revolutions that used violence have often empowered a charismatic dictator and the next round of atrocities. Socialists and anarchists—many of whom believed in the Soviet Union as the utopian kingdom come—were stunned and appalled by the pact between Stalin and Hitler, and by the subsequent genocidal behavior of Joseph Stalin. Allowing violence to be directed by the wrong hands does nothing to bring down an oppressive system, and, indeed, reinscribes the system called patriarchy.

As Theodore Roszak points out, this strand of the male left has taken up "violence as self-actualization." Often tracing its roots to Franz Fanon's *The Wretched of the Earth*, or Jean Paul Sartre's introduction to that book, violence is not just considered as a potential tactic; it's urged as a psychological necessity for the manhood of the oppressed. "At this point," writes Roszak, "things do not simply become ugly; they become stupid. Suddenly the measure of conviction is the efficiency with which one can get into a fistfight with the nearest cop at hand."[11]

This approach is actually no different than that of the workshop hoppers; the goal is a satisfactory internal emotional state (and not a particularly liberatory one) rather than an egalitarian society or the resistance movement needed to get us there.

The misogynist entitlement of men on the left was what led to the resurgence of feminism in the 1960s. Women learned to think politically in the civil rights movement, the student movement, and the peace movement, and then applied that analysis to their own situation. The behavior of their male comrades was no different from that of men of the establishment—"no less foul, no less repressive, and no less unliberated," as three Students for a Democratic Society (SDS) veterans put it.[12] This was true across the racial spectrum. Former Weatherwoman Cathy Wilkerson said many women dropped out of the antiwar movement altogether because of the sexism: "You couldn't penetrate the left. It was just like a stone wall."[13] Writes historian Jeremy Varon, "As part of its infamous 'smash monogamy' campaign, Weatherman mandated the splitting apart of couples, whose affection was deemed impermissibly 'possessive' or even 'selfish'; the forced rotation of sex partners, determined largely by the leadership for reasons both political and, it is alleged, crudely 'personal' (the charge is that some male leaders essentially shuttled particular women between collectives in order to sleep with them); and even eruptions of group sex in which taboos broke down in variously uncomfortable and exhilarating scenes of libidinal confusion."[14]

Exhilarating for whom? is the question, answered by Varon's understated observation that "life in the collectives could be especially difficult for women . . . and also invited the sexual exploitation of female members."[15] Weather Underground collectives were "psychologically harsh environments [that] rewarded assertive and even aggressive personalities, while chewing up those less confident or able to defend themselves."[16] Even the women's cadres were "driven by a coerced machismo" that, not surprisingly, "encouraged neither true autonomy nor solidarity among the women."[17]

Underground newspapers like the *Free Press*, the *Berkeley Barb*, and *Rat* made money from ads that both used imagery of objectified women and sold actual women as sexual commodities. As early as 1969, women at the Underground Press Syndicate Conference proposed a resolution that "papers should stop accepting commercial advertising that uses women's bodies to sell records and other products, and advertisements for sex, since the use of sex as a commodity especially oppresses women."[18] Eventually "a particularly violent and pornography-filled issue of *Rat*, with articles trivializing women's liberation, so enraged the women on the magazine's staff" that they joined in coalition with other feminist groups and took over the magazine. Robin Morgan was a member of the editorial coup. Her foundational article, "Good-Bye To All That," was published in the new *Rat*, an article filled with justified feelings of rage and betrayal. The New Left looked just like the Old Patriarchy, a problem that has only increased on the left as it has embraced pornography as freedom. *Freedom for whom?*, *To do what?*, and *To whom?* are the dirty little questions that leftist men refused to face. The fact that an entire class of women was kept in conditions of abuse and servitude utterly contradicted any claim the left could make to defending universal human rights.

The leaders of the Black Power movement provided similar examples. Eldridge Cleaver wrote openly of raping black women as "practice" for raping white women.[19] He was eventually arrested and jailed for both.

Huey Newton, cofounder of the Black Panthers and its Minister of Defense, raped numerous women with the backup of his armed thugs. He is quoted as saying, "There are two kinds of rape. In one version, you simply take a woman's body. In the other, you not only take her body, you try to make her enjoy being raped."[20] He was arrested for embezzling money from the Black Panther's education and nutrition program, and he was convicted of embezzling money from a Panther school, probably to fund his drug habit. Newton was also tried twice for the murder of a seventeen-year-old prostituted girl, Kathleen Smith. Malcolm X wasn't much better. He was a batterer and a pimp with a hateful attitude to lesbian women before converting to Islam. Afterward, he instituted his male supremacist ideology in the guidelines for black Muslim family life, which, like all fundamentalist religions, gave men the ultimate ideological reassurance that dominating women was God's plan.

It is important to note that at the time, and continuing to the present day, there were and are men and women of all races who rejected this behavior as exploitative and unacceptable. The radical Puerto Rican group, the Young Lords, stands as a great example. Originally, the group had an all-male leadership and a point in their platform that stated, "Machismo must be revolutionary." Iris Morales remembers,

> Men in leadership were abusing their authority and women recruits would come in and the men would be sleeping around with them. They'd be sleeping with two and three women, of course, they were mucho machos and thought this was really cool. They pulled out their list to compare who had the most conquests, and we were outraged, the women were *outraged*.[21]

The women began meeting without men in their own caucus and came up with a list of demands. These included promoting women to leadership positions, child care at meetings, and including women in the defense ministry. They found support among the more progressive men because "they understood that without women you can't have a revolution."[22] Over an amazing six months, all ten of their demands were met, even the adoption of the slogan "Abajo con Machismo!" (Down with Machismo!). Feminism was taken so seriously that "almost every single central committee member was demoted for male chauvinism and they had to change their way of being, even the chairman of the organization."[23] The men even started their own caucus to discuss issues of machismo. This transformation was documented in Morales's film ¡Palante Siempre Palante![24]

Morales also speaks of "the sad story of the movement," a story replayed into heartbreak across so many movements. "There were one or two women who shunned us altogether. And they later emerge on the backs of the movement we had fought for. This is an important lesson because not every woman is my sister and not every Puerto Rican is my sister."[25] Solidarity with each other is such a precious commodity, often harder to come by than public courage against the oppressor. Attacking each other is doing his work for him.

Similarly, Norm R. Allen Jr. coined the phrase "Reactionary Black Nationalism" to describe the "bigotry, intolerance, hatred, sexism [and] homophobia" that he urged the black community to reject.[26] Mark Anthony Neal's New Black Man stands as an engaging template of moral agency and community building in the face of both oppression (he's African American) and privilege (he's also heterosexual and a man).

Even in this short discussion, the complexity of the issue of violence becomes apparent. It's understandable that people who care about justice want to reject violence; many of us are survivors of it, and we know all too well the entitled psychology of the men who used it against us. And whatever our personal experiences, we can all see that the violence of imperialism, racism, and misogyny has created useless destruction and trauma over endless, exhausting millennia. There are good reasons that many thoughtful people embrace a nonviolent ethic.

"Violence" is a broad category and we need to be clear what we're talking about so that we can talk about it as a movement. I would urge the following distinctions: the violence of hierarchy vs. the violence of self-defense, violence against people vs. violence against property, and the violence as self-actualization vs. the violence for political resistance. It is difficult to find someone who is against all of these. When clarified in context, the abstract concept of "violence" breaks down into distinct and concrete actions that need to be judged on their own merits. It may be that in the end some people will still reject all categories of violence; that is a prerogative we all have as moral agents. But solidarity is still possible, and is indeed a necessity given the seriousness of the situation and the lateness of the hour. Wherever you personally fall on the issue of violence, it is vital to understand and accept its potential usefulness in achieving our collective radical and feminist goals.

Violence of Hierarchy vs. Violence of Self-Defense

The violence of hierarchy is the violence that the powerful use against the dispossessed to keep them subordinated. As an example, the violence committed for wealth is socially invisible or committed at enough of a distance that its beneficiaries don't have to be aware of it. This type of violence has defined every imperialist war in the history of the US that has been fought to get access to "natural resources" for corporations to turn into the cheap consumer goods that form the basis of the American way of life. People who fight back to defend themselves and their land are killed. No one much notices. The powerful have armies, courts, prisons, and taxation on their side. They also own the global media, thus controlling not just the information but the entire discourse. The privileged have the "comforts or elegancies" (as one defender of slavery put it) to which they feel God, more or less, has entitled them, and the luxury to remain ignorant.[27] The entire structure of global capitalism runs on violence (Violence: The Other Fossil Fuel?). The violence used by the powerful to keep their hierarchy in place is one manifestation that we can probably agree is wrong.

In contrast stands the violence of self-defense, a range of actions taken up by people being hurt by an aggressor. Everyone has the right to defend her or his life or person against an attacker. Many leftists extend this concept of self-defense to the right to collective defense as a people. For example, many political activists supported the Sandinistas in Nicaragua, even taking personal risks in solidarity work like building schools and harvesting coffee. Indeed some people refuse to call this collective self-defense "violence," defining violence as only those brutal acts that support hierarchy. I believe it is more honest to call this violence, and accept that not all violence is equal, or equally bad.

Violence against Property vs. Violence against People

Again, some people reject that violence is the correct word to describe property destruction. Because physical objects cannot feel pain, they argue, tools like spray paint and accelerants can't be considered weapons and their use is not violent. I think the distinction between sensate beings and insensate objects is crucial. So is property destruction violent or nonviolent? This question is both pragmatic—we do need to call it *something*—and experiential. Destroying property can be done without harming a single sentient being and with great effect to stop an unjust system. Can anyone really argue against the French resistance blowing up railroad tracks and bridges to stop the Nazis?

But violence against property can also be an act meant to intimidate. This is the source of the unease that many progressives and radicals may feel toward property destruction. If you have been a person so threatened, you know how effective it is. Indeed, if violence against property were an ineffective approach to instilling fear and compliance, no one would ever use it. Burning a cross on someone's lawn is meant to traumatize and terrorize. So is smashing all the dinner plates to the floor. A friend who survived a right-wing terrorist attack on the building where she worked was later hospitalized with severe PTSD (posttraumatic stress disorder). Property destruction can have a crippling effect on sentient beings.

Whatever we decide to call property destruction, we need to weigh the consequences and strategic benefits and make our decisions from there. Again, "violence" is not a bad word, only a descriptive one. Obviously, many more people can accept an attack against a window, a wall, or an empty building than can accept violence against a person, and that's as it should be. But wherever you stand personally on this issue, basic respect for each other and for our movement as a whole demands that we acknowledge the distinction between people and property when we discuss violence.

Violence as Self-Actualization vs. Violence for Political Resistance

Male socialization is basic training for life in a military hierarchy. The psychology of masculinity is the psychology required of soldiers, demanding control, emotional distance, and a willingness and ability to dominate. The subject of that domination is a negative reference group, an "Other" that is objectified as subhuman. In patriarchy, the first group that boys learn to despise is girls. Franz Fanon quotes (uncritically, of course) a young Algerian militant who repeatedly chanted, "I am not a coward, I am not a woman, I am not a traitor."[28] No insult is worse than some version of "girl," usually a part of female anatomy warped into hate speech.

With male entitlement comes a violation imperative: men become men by breaking boundaries, whether it's the sexual boundaries of women, the cultural boundaries of other peoples, the physical boundaries of other nations, the genetic boundaries of species, or the biological boundaries of ecosystems. For the entitled psyche, the only reason "No" exists is because it's a sexual thrill to force past it. As Robin Morgan poignantly describes the situation of Tamil women,

> To the women, the guerillas *and* the army bring disaster. They complain that both sets of men steal, loot, and molest women and girls. They hate the government army for doing this, but they're terrified as well of the insurgent forces ostensibly fighting to free them. Of their own Tamil men, one says wearily, "If the boys come back, we will have the same experience all over again. We want to be left in peace."[29]

Eldridge Cleaver announced, "We shall have our manhood or the earth will be leveled by our attempts to gain it." This is a lose-lose proposition for the planet, of course, and for the women and children who stand in the way of such masculine necessity. Or as the Vietnamese say, when the elephants fight, it's the grass that suffers.

As we can see from these examples, whether from a feminist understanding or from a peace perspective, the concern that taking up violence could potentially be individually and culturally dangerous is a valid one. Many soldiers are permanently marked by war. Homeless shelters are peopled by vets too traumatized to function. Life-threatening situations leave scars, as do both committing and surviving atrocities.

But violence is a broad category of action; it can be wielded destructively or wisely. We can decide when property destruction is acceptable, against which physical targets, and with what risks to civilians. We can decide whether direct violence against people is appropriate. We can build a resistance movement and a supporting culture in which atrocities are always unacceptable; in which penalties for committing them are swift and severe; in which violence is not glorified as a concept but instead understood as a specific set of actions that we may have to take up, but that we will also set down to return to our communities. Those are lines we can inscribe in our culture of resistance. That culture will have to include a feminist critique of masculinity, a good grounding in the basics of abuse dynamics, and an understanding of posttraumatic stress disorder. We will have to have behavioral norms that shun abusers instead of empowering them, support networks for prisoners, aid for combatants struggling with PTSD, and an agreement that anyone who has a history of violent or abusive behavior needs to be kept far away from serious underground action. Underground groups should do an "emotional background check" on potential recruits. Like substance abuse, personal or relational violence should disqualify that recruit. First and foremost, we need a movement made of people of character where abusers have no place. Second, the attitudes that create an abuser are at their most basic level about entitlement. A recruit with that personality structure will almost certainly cause problems when the actionists need sacrifice, discipline, and dependability. Men who are that entitled are able to justify almost any action. If they're comfortable committing atrocities against their intimates and families, it will be all too easy for them to behave badly when armed or otherwise in a position of power, committing rape, torture, or theft. We need our combatants to be of impeccable character for our public image, for the efficacy of our underground cells, and for the new society we're trying to build. "Ours is not a war for robbery, not to satisfy our passions, it is a struggle for freedom," Nat Turner told his recruits, who committed no atrocities and stole only the supplies that they needed.

Only people with a distaste for violence should be allowed to use it. Empowering psychopaths or reinscribing the dominating masculinity of global patriarchy are mistakes we must avoid.

A very simple question to ask as we collectively and individually consider serious actions like property destruction is, is this action tactically sound? Does it advance our goal of saving the planet? Or does it simply answer an emotional need to do something, to feel something? I have been at demonstrations where young men smashed windows of mom and pop grocery stores and set fire to random cars in the neighborhood. This is essentially violence as a form of self-expression—for a very entitled self. Such random acts of destruction against people who are not the enemy have no place in our strategy or in our culture. It's especially the job of men to educate other men about our collective rejection of masculinist violence.

WITHDRAWAL

Another response to conditions of oppression is withdrawal. Withdrawal encompasses a vast range of possible actions. On one end of the spectrum are acts of personal detachment or refusal carried out by alienated individuals. Entire social enclaves—the inheritors of the Bohemian tradition—are filled with such people. Their goal is not to make broad-based social or political change, but to live "authentically." We can see the potential problem with this strategy in some synonyms for the word "withdrawal": abandonment, abdication, disengagement, marooning, resignation, retirement.

On the other end of the spectrum is withdrawal used as a political tactic, targeting specific economic, political, or social practices or institutions. As with legal remedies and direct action, this can be a radical—and successful—attempt to win liberty. It can also dead-end into political irrelevance and horizontal hostility. Horizontal hostility, a phrase coined by Florynce Kennedy in 1970,[30] describes the destruction that happens when oppressed groups fight amongst themselves instead of fighting back against the powerful (Figure 3-1). It's a predictable behavior, and one against which we must guard. A strategy of withdrawal risks exacerbating this tendency for the obvious reason that if you close off the possibility of fighting up the pyramid of hierarchy, the only people left to fight are each other.

The main difference between withdrawal as a successful strategy and withdrawal as a failed strategy is whether the withdrawal is linked to political resistance or instead seen as adequate in itself. This difference often hinges exactly on the distinction between the liberal and the radical. Remember that liberalism is idealist; it conceptualizes society as made up of ideas, not material institutions. Therefore, a strategy of simply withdrawing loyalty from the dominant system, of individual psychological, intellectual, or cultural positioning, is believed by liberals to be revolutionary. While issues of identification and loyalty are crucial to building the class consciousness needed for a resistance movement, this alone is not enough. The withdrawal has got to go beyond the intellectual, the emotional, and the psychological to include a goal of actually winning justice. "Worlds within worlds" may give solace, but ultimately they change nothing. We need to guard against these impulses, as seductive as they are. The idea that all we have to do is turn our attention to ourselves and our chosen community is appealing, but such actions will never be enough. Divorced from a larger goal of liberty and a strategy of direct confrontations with power, "withdrawn" communities end up irrelevant at best, and unpleasant places toxic with personal criticisms and cultlike elements at worst.

Often, the "withdrawalists" set withdrawal and direct confrontations with power in opposition to each other as strategies, rather than seeing the former as a necessary element for the latter. But living in a rarified bubble-world of the converted is a poor substitute for freedom—and such a world will certainly not save the planet. The distinction between a merely alternative culture and a culture of resistance is so important that we are devoting an entire chapter to it.

Figure 3-1: Horizontal Hostility

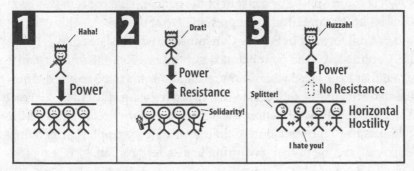

For now, a positive example for study is the American Revolution. The colonists' original strategy was one of withdrawal, which employed:

- identificational withdrawal and the subsequent creation of new personal loci of loyalties to the American colonies as opposed to the British crown;
- economic withdrawal and boycotts of everything from tea to wool;
- cultural withdrawal and the valuing of American art, products, and sensibilities;
- political withdrawal, built around the colonial court system and state- and colony-wide congresses for governance.

All of these forms of withdrawal came together in a culture of resistance that created, encouraged, and supported the revolution. People began to conceive of themselves as citizens of their state and ultimately of those states united. They also took on new political identities as patriots, as "Sons of Liberty," rather than sons and daughters of England.

These politicized self-definitions merged with cultural and economic withdrawal. The United States is singular as an ex-British colony that is a nation of coffee drinkers, not tea drinkers. This is a direct result of the colonial resistance to the tax on tea, still mythologized in the Boston Tea Party. No patriot drank tea, and the Sons of Liberty were willing to take the necessary measures to make sure no one else did either.

Some background history of the era may be necessary to the discussion. The British Constitution granted that taxation on British subjects could only be by consent of the people. That consent was seen to dwell in Parliament as the representative of the people. This concept was carried forward in the US Constitution, which states that only the US Congress has the power to tax, not the president. Samuel Adams wrote that to be taxed without representation was to be reduced "from the character of free Subjects to the miserable state of tributary Slaves." The insult of British taxation was felt all the more keenly because the colonists had representation in their own state assemblies, which they believed were the proper governing bodies for taxation.

Local uprisings—what would now be considered mob violence—were common throughout England and across the colonies because there was no police force in the eighteenth century. Since the Middle Ages, the government depended on institutions like "hue and cry," where lawbreakers would be apprehended by the community at large. By the eighteenth century, the preferred method was the *posse commitatus*, in which the magistrate or sheriff was empowered to call up as many able-bodied men as might be needed. The next line of defense was the militia. Explains historian Pauline Maier, "Both the *posse* and the militia drew upon local men, including many of the same persons who participated in extralegal uprisings. This meant that insurrections could naturally assume the manner of a lawful institution, as insurgents acted by habit with relative restraint and responsibility."[31]

What it also meant was that if the population at large was sympathetic to a cause or directly involved in a disturbance, the local magistrate was left "virtually helpless."[32] This happened repeatedly throughout the period leading up to the Revolutionary War as a groundswell of people felt their rights outraged by British policies.

The Stamp Act was in many ways the beginning of organized resistance. The act was passed by Parliament in 1765 to help pay for the Seven Years' War. Most official documents, like court records and land grants, and printed materials, like broadsheets and newspapers, had to carry a stamp, and the stamps cost money. The act was despised throughout the colonies, and colonial legislatures sent letters of protest back to England. But more important was the Stamp Act Congress. This was the first collective colony-wide effort to make common cause against Britain. Local groups opposed to the Stamp Act also created committees of correspondence, a network of activists that spanned the thirteen colonies. These committees proved crucial in providing the political infrastructure required to form the revolutionary movement that followed. According to Richard Bushman, "The network of activists meant that revolutionary language by 1773 was sounding in virtually every adult ear in Massachusetts, and that there was a fluid continuum of discourse joining the Boston press and town meeting and the talk in meetings and taverns through the Province."[33]

The Stamp Act was never enforced because of the resistance efforts of the common people. Those efforts largely took the form of property destruction and threats of bodily harm. The stamps required distributors, an official person responsible for their sale. Those officials were the leverage point, the easily identified target to stop the dreaded stamps. Street protests swelled in Boston, and then quickly spread to neighboring colonies. The distributors were hanged, burned, and/or beheaded in effigy. The mob then moved on to the distributor's house, which would be evicted of its residents and then looted or pulled to pieces. Often the distributor would be forced to resign from the duty publicly. As a result, no one could be found who would take up the job. According to Maier, "The solution was infectious. Without distributors the Stamp Act could not go into effect, so the coercion of stampmen seemed rational, even efficient."[34] The Massachusetts stamp distributor resigned on August 15th, 1765. On August 29th, Rhode Island's followed suit, and the strategy proved so successful that the rest fell in line with alacrity. The last distributor was from Georgia, and he had to be sent from England. On reaching the US, resigning was his first and only official act. By March 1766, the Stamp Act was repealed because it was simply unenforceable.

Boycotts against British goods were strengthened into a formal agreement called the Continental Association. The Association, as it was known, wrote a fourteen-point document, the Articles of Association, which was a pact between the colonies to resist British infringement on colonial rights. Its main goal was a broad-scale boycott. To quote from the document, "a non-importation, non-consumption, and non-exportation agreement, faithfully adhered to, will prove the most speedy, effectual, and peaceable measure." The ban on tea proved especially so.

The fascinating point is that the Association had no power of enforcement. Unlike the Crown, they could not arrest, fine, or jail offenders. Offenders could only be named and shamed in print and socially ostracized as "the enemies of American liberty." According to Walter H. Conser, et al.,

> If colonial merchants violated popular sentiments by continuing to import boycotted goods, people not only refused to buy from them but also to talk with them, to sit with them in

church, or to sell them goods of any kind. At times, colonial activists conducted regular business in violation of British law by using documents without required tax stamps, by settling legal disputes without courts, and by sending protest petitions to England without the permission of royal governors. They organized and served on local, county, and province committees designed to extend, support, and enforce resistance. In 1774 and 1775, many of these bodies assumed governmental powers on their own initiative, acting as extralegal authorities with powers greater than the remnants of colonial government.[35]

The Association tried to address the economic hardship that the colonies were sure to endure because of the boycott. Toward that end, they sought to "encourage frugality, economy, and industry, and promote agriculture, arts and the manufactures of this country, especially that of wool." Some provincial conventions thought through the economic implications and tried to encourage the manufacture of the following: "woolens, cottons, flannel, blankets, rugs, hosiery, coarse cloths, all sorts of dyes, flax, hemp, salt, saltpeter, gunpowder, nails, wire, steel, paper, glass, copper products, and malt liquors." Massachusetts added "tin plates, firearms, and buttons."[36] Conser, et al., explain that

The real work of the resistance was often carried on in villages and towns, in the country as well as the city, by forgotten patriots. These now nameless men and women were the people who spun, wove, and wore homespun cloth, who united in the boycott of British goods, and who encouraged their neighbors to join them and stand firm. Many came together in crowd actions and mass meetings to protest and served on, or supported, local resistance committees. They refused to obey the statutes and officers of the British Crown, which so short a time before had been the law of the land. It was these various acts of resistance and noncooperation that struck most openly at the authority of the Crown.[37]

Patriots also refused to quarter troops, published newspapers without the required stamps, and continued to run government bodies that the British declared dissolved.

The situation escalated in Massachusetts. With the Massachusetts Government Act, Parliament essentially wrested control of both governing bodies and the judiciary from the citizens. The first provision declared that judges were to be appointed by the governor, himself appointed by the Crown, instead of by the council, which had been under the control of the people. This was to take effect August 1, 1774. What happened instead has been called the first American Revolution.[38]

The patriots of Pittsfield, Massachusetts, the seat of Berkshire County, "proposed a new and more direct method for opposing British policy: Why not close down the courts? Since the weight of governmental authority was experienced most directly and frequently through the judicial system, closing the courts would effectively bring the Massachusetts Government Act to a halt."[39] The Pittsfield Committee of Correspondence circulated the strategy. Boston soon replied, "We acknowledge ourselves deeply indebted to your wisdom. . . . Nothing in our opinion could be better concerted than the measures come into by your County to prevent the Court's sitting."[40]

The Berkshire County Court never opened again until the United States was an independent nation. On August 16, 1,500 unarmed patriots—farmers, artisans, small business owners—took over the courthouse. As one witness described it, "The Sheriff commanded them to make way for the court; but they gave him to understand that they knew no court or any other establishment than the ancient laws and usages of their country, & to none other would they submit or give way on any terms."[41] The scene was repeated throughout the state. Anyone who had agreed to officiate as judge or magistrate was liable to face social shunning and intimidation at church, at home, and on the street by crowds that reached into the thousands, until they resigned, often in public and in print. Most of these encounters were restrained and even respectful. According to historian Ray Raphael,

These citizens took special care to distance themselves from any intimations that they might be a "mob." In their view, they acted like model citizens. The crowd conducted all its business according to strict democratic principles: ad hoc delegates were elected to conduct negotiations, while all decisions were put to a vote of the entire body. There were no "leaders" empowered to issue orders from above.[42]

Indeed, the crowds were so orderly some of them voted on whether to raise a cheer on the Sabbath. Wrote observer Abigail Adams, "It being Sunday evening it passed in the negative."[43]

Their strategy of withdrawal—economic, political, and identificational—created a true culture of resistance that successfully supported acts of further resistance. Writes Raphael, "While a group of renowned lawyers, merchants, and slave-owning planters were meeting as a Continental Congress in Philadelphia to consider whether or not they should challenge British rule, the plain farmers and artisans of Massachusetts, guarding their liberties jealously and voting at every turn, wrested control from the most powerful empire on earth."[44] By the time the shot heard 'round the world was fired, the Crown had already lost control of the colony. The Red Coats' march to Lexington was a last-ditch effort to gain control of the weapons.

Now, contrast the colonial revolutionary movement to the current strategy proposed by many of the leaders of the radical environmental movement. There is much to learn from these people, some of whom are also kind and caring individuals, and all of whom are courageous in their insistence on telling the truth to a public virulent with ignorance. We agree on basic values of justice, compassion, and sustainability, on the horrors wrought by human entitlement, and on the fact that both a reduction of human population and the end of industrial civilization are inevitable.

Where we disagree is on the idea of resistance. Daniel Quinn, for instance, explains in a very accessible way why civilization is unsustainable and based on exploitation. He is very clear that we humans are in for a very ugly time in the next few decades, and for the 200 species we are driving extinct every day there is no time left. The main strategy he proposes, however, is withdrawal, which he calls "walking away." To where? Well, there's no actual place that he has in mind, but rather a state of mind. This would be like the Massachusetts patriots deciding they could have freedom in their heads while actual freedom from unjust taxation, corrupt courts, nondemocratic government, billeted soldiers, press gangs, and economic exploitation weren't important or even achievable. The people of colonial America withdrew, but their withdrawal went well beyond a reframing of their intellectual and emotional loyalties. They engaged in acts of direct confrontation with power, to withdraw from the economic and political institutions that created their subordination. In the end, their withdrawal was so successful that it resulted in a war, though some historians argue that independence could have been won with the continued nonviolent techniques used to such great effect in Massachusetts.[45]

Quinn is worth quoting because his viewpoint is widely reflected across much of the left:

> Because revolution in our culture has always represented an attack on hierarchy, it has always meant upheaval—literally a heaving up from below. But upheaval has no role to play in moving beyond civilization. If the plane is in trouble, you don't shoot the pilot, you grab a parachute and jump. To overthrow the hierarchy is pointless; we just want to leave it behind.[46]

The metaphor of a plane in trouble is a bad fit to the situation the planet is facing. A more apt comparison would be a maniac with his finger two inches and closing above the red button. Would anyone really argue that "walking away" would be the order of the day?

To reframe the airplane image to the current crisis, the planet has to be included. Yet Quinn writes the planet out of the equation:

> When we talk about saving the world, what world are we talking about? Not the globe itself, obviously. But also not the biological world—the world of life. The world of life, strangely enough, is not in danger (though thousands and perhaps even millions of species are). Even at our worst and most destructive, we would be unable to render this planet lifeless. At

present it's estimated that as many as two hundred species a day are becoming extinct, thanks to us. If we continue to kill off our neighbors at this rate, there will inevitably come a day when one of those two hundred species is our own. . . . Saving the world can only mean one thing: saving the world as a human habitat.[47]

First, humans *can* render this planet lifeless. A nuclear war could do it. So could the "methane burp" released by the melting of the Arctic tundra; our planet could soon be too hot to support life.

But second, and more importantly, *why aren't those 200 species a day worth fighting for?* From the tiny snails building their perfect homes of logarithmic spirals to the great bears majestic with maternal rage, why don't the lives of these creatures provoke a ferocious tenderness of protection and solidarity? Why are they only valued as human "habitat"?

I have heard variations on this position repeated everywhere: we can't kill the planet; species loss is regrettable but inevitable; the best we can do is learn about permaculture so that me and mine might have some food when the crash arrives. I find this position morally reprehensible at a level that can't be argued, only mourned. Surely somewhere in the human heart empathy, loyalty, and love are still alive. What is the meaning otherwise of that heart—or is a pump for oxygen all we have left of ourselves?

Pretend instead that Quinn's plane is stocked with nuclear weapons—enough to kill every living creature on the planet—and the pilot intends to use them. Killing the pilot then becomes the urgent moral necessity of this thought experiment.

We have examples from recent events. The people on board the fourth plane in the September 11 attacks realized that the plane was intended as a weapon. They were dead anyway; their duty became to bring that plane down before it could be used to hurt anyone else. *That* is the situation we are in, on a massive scale, and life on Earth is at stake, 200 species at a time. Parachuting out to save only ourselves should not be the goal of a political movement worth the name, even if there were a safe place to which parachuting was possible.

Quinn's only other strategy is education about the nature of civilization: "Teach a hundred people what you've learned here and urge each of *them* to teach a hundred."[48] As we have already seen, this is a deeply liberal understanding of social change. Certainly radicals believe in the strategic necessity of education, but that education is toward a goal of transforming material conditions of socially sanctioned subordination to material conditions of justice. This book, for instance, is an attempt at education, but it's ultimately a call for direct confrontations with power. Quinn continues, "I know that nothing changes unless people's *minds* change first. You can't change a society by passing new laws—unless people see the *necessity* for new laws."[49] This statement is ignorant to the point of being bizarre. From the Thirteenth Amendment, to the Civil Rights Act of 1964, to antistalking, antirape, and sexual harassment laws, to the Clean Water Act, laws have profoundly changed society by forcing people to change their behavior, and providing for consequences when they don't. Further, leaving laws out of the picture entirely, Georg Elser nearly stopped World War II all by himself. He did this neither by educating nor by changing laws, but by attempting to assassinate Hitler. He tried to change material conditions, not hearts and minds, and very nearly saved tens of millions of lives.

A related concept is the "lifeboat" idea, proposed by Richard Heinberg. Heinberg has probably done more than anyone to raise awareness about peak oil and resource depletion. His work is cogent, compelling, and compassionate. Where we differ is on the necessity of resistance. He proposes the "lifeboat" as an option for action, which he defines as "the path of community solidarity and preservation."[50] This would include learning basic survival skills for food production and other necessities; preserving scientific, historic, and cultural knowledge; and (re)developing social norms for democratic decision making. These tasks are all necessary, and indeed make up a large part of our concept of a culture of resistance, as well as a great deal of our hope for the best-case scenarios. But as with Quinn, it's not enough. These activities have to be linked to both theoretical and public defense of resistance, and material support for actionists. To return to colonial Massachusetts as an example, the farmers already had basic survival skills, were inheritors of the knowledge of their time, and had strong local democracies in place. None of this alone stopped the British from subjugating them. That required resistance. But Heinberg doesn't believe that resistance to industrial culture is possible or advisable. He writes, "Efforts to try to bring industrialism to ruin prematurely seem to be pointless and wrongheaded; ruin will come soon enough on its own. Better to invest time and effort in personal and community preparedness."[51] I don't know why he thinks saving our relations—our parents and grandparents of plants and mycorrhizae, our cousins and siblings of birds and beasts—is pointless or wrongheaded. What indeed, in the whole history of human endeavor, could have more value than saving life itself? And ruin has already come to the Western black rhino and the Carolina parakeet. How many others have joined them in the forever of extinction since you started reading this book?

We can also contrast this fatalistic attitude with that of members of the German resistance to Hitler. After the Allied invasion of France, members of the resistance considered whether to call off their attempts to stop the Nazis; the war was lost, and the regime would be destroyed in any case. Yet they decided to risk their lives, and hundreds were tortured for their actions. They took those risks because, as Henning von Tresckow said, "Every day, we [the Germans] are assassinating nearly 16,000 additional victims." This is not so much math as a grim moral equation, and the resistance chose to try and save those lives.

Note well what von Tresckow also said, "How will future history judge the German people, if not even a handful of men had the courage to put an end to that criminal?" Future history will judge us just as surely, if anything that could be called a future survives our lack of courage against this criminal culture.

I will be the first to admit that we are up against a system of vast power, global in scale, with no sympathetic population upon which to draw for either combatants or support. Still, if illiterate farmers armed only with pitchforks could face off against the most powerful empire that had ever existed—and win—surely we can aim higher than a goal of simply creating really great gardens.

SPIRITUALITY

A withdrawalist stance is often a mixture of liberalism and despair. Liberalism can only offer individual solutions; despair threatens no solution. Systems of oppression like capitalism and patriarchy can feel overwhelming in their scale and sadism. The promise of withdrawal—that a strategy of personal change can compound into political change—is understandably appealing, especially because it often comes with a set of directives that, if not always easy, are at least simple. Indeed, a whole life—including an identity—can be built around these actions. Thus, it is an answer to despair, but it's an answer that relies on faith, not on strategy—which is to say it's an emotional solution and not a material one.

We have got to think past our emotional needs. Faith-based solutions can't stand up to intellectual scrutiny. When questioned, the adherents feel threatened and must retreat to the protection of repeatable platitudes and the reassuring company of like-minded others. This is the stance taken across much of the progressive community. And currently, there is a whole subculture of withdrawalists who have achieved true millenarianism.

Millenarianism is "any religious movement that predicts the collapse of the world order as we know it, with its replacement by the millennium, or period of justice, equality, salvation, etc. Millenarian movements are thought to be an extreme example of the use of religion as a 'way out' or reaction to social stress and its resulting anomie."[52]

The worst historic case of millenarianism was the cattle-killing cult of the Xhosa people. The Xhosa are a cattle-herding people who were living in eastern South Africa when the Dutch arrived in the mid-1600s. Their first encounter with Europeans was in the early 1700s. The century that followed was filled with the predictable displacement, resistance, and war. Along with those stressors, a lung disease spread through the Xhosa's cattle in 1854, leaving people even more vulnerable.

In April of 1856, a fifteen-year-old girl named Nongqawuse had a spiritual vision in which she was told that the Xhosa should kill their cattle, raze their crops, dump their food stores, and destroy their garden and kitchen tools. If these things were done, the dead would return;

sickness and old age would disappear; food would spring from the earth; fat, fertile cattle would materialize; and "the whole community will rise from the dead" to drive out the British.[53]

The story of the prophecy spread quickly. The Xhosa chief, Sarhili, was converted, and ordered the killing of the cattle. As the prophecy picked up speed, other people began to have visions, seeing the dead rising from the sea or hanging in the air. This encouraged more destruction of food stores and cattle. So many cattle were killed that the carrion birds couldn't keep up and the carcasses rotted. In total, 400,000 beasts were slaughtered. The Xhosa built bigger and better corrals for the promised new cattle and giant skin bags for their milk in preparation.

The first prophesized deadline came and went with no fulfillment. The date was moved. Still nothing. With that much psychological investment, the people's response was predictable: the problem was with the unbelievers. The few cattle left to provide for immediate needs had to be killed. They were, and still no miracle happened. Mass starvation ensued, with the attendant atrocities and cannibalism. People ate animal food, they ate grass, they ate their own children. The believers never gave up their belief, they simply blamed the skeptics. Between starvation and attendant diseases, the population collapsed from 105,000 to 26,000. Many of the survivors were forced to migrate. One hundred fifty years of imperialism couldn't destroy the Xhosa, but two years of millenarian fever nearly did.

From a different continent comes a related example, the Righteous Harmony Society Movement, also known as the Boxer Rebellion. They were a secret religious society in northern China who believed that a combination of martial arts, diet, and prayer would give them the power to fly and protection against bullets and swords. They also believed that an army of heavenly "spirit soldiers" would drive out foreigners. Bad flooding and drought conditions had created both hardship and starvation for farmers and refugees, and desperate people make good converts. The overarching context, of course, was British imperialism and the escalating exploitation and humiliation of the Opium Wars, forced trade, and the loss of Hong Kong.

The Righteous Harmony Society (RHS) members were able to scapegoat both Chinese Christians and European Christian missionaries for the famine. The scapegoating culminated in the Taiyuan Massacre, in which the Boxers killed over 18,000 Chinese Christians. In June of 1900, Righteous Harmony Society fighters massed in Beijing to lay siege to foreign embassies. The siege of the Legation Quarter resulted ultimately in the arrival of an international force (six European nations plus Russia and Japan) of over 20,000 troops, called the Eight-Nation Alliance, which ended the siege, occupied Beijing, and forced the Qing court to make reparations. The soldiers of most of the eight nations behaved abominably, looting and raping with the encouragement of their commanders; once again, it was the grass that suffered.

The point, for our purposes, is that the members of the RHS were not assisted by "spirit soldiers," couldn't fly, and had no immunity to bullets. Thousands of people wanted to believe it, but believing only brought useless atrocities and their own deaths.

Millenarian cults spring up with regularity even among people not enduring the stress of displacement and genocide. The Millerites, for instance, believed in the Second Coming of Christ and set the date a number of times. It built into a mania. People didn't plant crops, they broke up their furniture, and they gave away their valuables. Alas, what followed was called the Great Disappointment—Christ didn't arrive. The Seventh Day Adventists, not disappointed enough, grew out of the Millerites. They believe that Christ's coming is imminent but wisely refrain from picking a date. Jehovah's Witnesses, another descendant of the Millerites, picked a succession of dates: 1874, 1878, 1881, 1910, 1914, 1920, 1925 . . . World War II was interpreted as Armageddon to the point that people put off dental work and lived with the pain, so strong was their belief that they would soon be taken to heaven. Writes one former member, "To this day, I associate the fragrance of cloves [used for tooth pain] with the imminence of disaster."[54]

In the 1970s, the *Watchtower*, the legal organization of the Jehovah's Witnesses, began predicting that 1975 would be the year. One family member of a Jehovah's Witness observed her brother and his family giving away their belongings and scaring their young children with instructions on where to hide if they heard screaming. When the Second Coming didn't come, her brother was hospitalized with suicidal depression.[55]

Leon Festinger, with colleagues Henry W. Riecken and Stanley Schachter, developed the concept of "cognitive dissonance" to explain the behavior of people who continue to believe in millenarian sects even after the predicted catastrophe/utopia doesn't come to pass. Their book, *When Prophecies Fail*, is a psychological examination of the cult that sprang up around Marian Ketch, a woman who claimed to be in communication with aliens. Said aliens predicted a world cataclysm and offered survival in exchange for belief. When the prediction didn't come true, the believers clung more tightly to their belief system, as is common among disappointed believers. Festinger suggested the phrase "cognitive dissonance" to explain this phenomenon. When people try to believe two contradictory things, the resultant discomfort has to be resolved. The more strongly the beliefs are held, the more imperative the resolution becomes. That resolution often comes by actively proselytizing. Explains Festinger,

> The dissonance is too important and though they may try to hide it, even from themselves, the believers still know that the prediction was false and all their preparations were in vain. The dissonance cannot be eliminated completely by denying or rationalizing the disconfirmation. But there is a way in which the remaining dissonance can be reduced. If more and more people can be persuaded that the system of belief is correct, then clearly it must, after all, be correct. Consider the extreme case: if everyone in the whole world believed something there would be no question at all as to the validity of this belief. It is for this reason that we observe the increase in proselytizing following disconfirmation. If the proselytizing proves successful, then by gathering more adherents and effectively surrounding himself with supporters, the believer reduces dissonance to the point where he can live with it.[56]

This is a common psychological process, and one that we would do well to name and intervene against as it starts to take hold in our communities.

When the Black Plague devastated Europe, no one had ever seen a microbe. They can be forgiven for believing that Doomsday was at hand. We don't have the same excuse. We know what is causing mass extinctions and catastrophic climate change. Yes, these systems are massive, hegemonic, and well-armed. But unlike *Yersinia pestis*, they are at least visible to the naked eye. This is probably the reason that the millenarian leanings on the left tend not toward explanation, but resolution. We will be saved, though not by the Second Coming. Cosmic forces, often linked to indigenous myths, will appear. The Age of Aquarius faded, to be revived by the Harmonic Convergence in 1987, when eight planets—and all the self-proclaimed druids—lined up according with the Mayan calendar, presaging some Vague New World of the usual peace, light, and consciousness. Except nothing happened. Up next were the three syllables that we never have to speak again: Y2K. I had friends who were furious with me for not stockpiling food or water. One woman on the fringes of my social network gave away all her belongings and her cat, so sure was she that her demise drew near. Again, nothing happened. But never fear: the end of the Mayan calendar (once again) in 2012 clearly spells the end of the world.

So far, New Age millenarianism hasn't generally resulted in more personal trauma than a pantry full of MREs—military meals, ready to eat. But on a broader scale, the spiritual approach of the alternative culture is damaging to our movement. Instead of guiding people to face the hard reality of oppression and environmental destruction, and giving them the emotional and spiritual support to wage a resistance struggle, it offers a range of other-worldly events and characters who—*deus trans machina?*—will save the planet.

An example from life. In the middle of a perfectly reasonable dinner conversation about global warming, one of the other guests earnestly tried to reassure us that "the beings [Beings?] from the Akashic Plane would never let that happen."

The sudden silence was broken by a man with a long gray beard. "I don't want to save the whales," he said. "That's just karma." Was he missing a segue or some synapses? Never mind a conscience.

Another man joined in, in that smooth, soft voice that is supposed to signify spiritual attainment. "We were meant to be the consciousness for the earth. This was our childhood stage. If the human race doesn't grow into clear intention, the Holy Ones will step in."

The original speaker led the conversation into the inevitable cultural train wreck. "Yes! Haven't you heard the Native American prophecies?"

Millenarian beliefs can be a destructive force across communities, and they can also detour vulnerable people into a dead end. Thankfully, very few people have ever heard of Akashic Beings. But what if the dinner party participants provided the only context I knew, the only place to bring my biophilic despair? The stories and myths of a culture provide the matrix for the possible, and only extraordinary individuals are able to break out of their surrounding context.

There is a final level of complication. A claim of access to divine guidance is not one that can be proven. Mystical visions are both a compelling and an individual experience. Both of those characteristics render the realm of the mystic potentially dangerous to the mystic and the people in her community. The compelling, suprareal quality of religious visions produces intractable loyalty in the visionary, a loyalty that can lend seductive charisma to anyone. And if the source of the vision is illness rather than a friendly cosmos, the result can be disaster. The individual nature of the visions means that the received guidance can't be verified, only experienced. But more people having the same vision does not actually confirm its veracity: all it confirms is that the human brain is capable of producing ecstatic states. Religious mania is common among schizophrenics, for instance, but mental illness has never yet proved a sound basis for a political strategy. The mystical vision thus contains a contradiction in that people must apply rationality to an inherently nonrational experience.

Three examples from the same movement point to both the promise and pitfalls of mysticism: John Brown, Nat Turner, and Harriet Tubman. Brown was the leader of the Harpers Ferry Raid. He and his followers attacked the US arsenal at Harpers Ferry in an attempt to arm enslaved people and inspire an uprising. Brown believed that God told him to undertake the raid and that he would be given divine protection. But if God promised him protection, God lied, as the raid ended in disaster. Likewise Nat Turner, the enslaved man who inspired Turner's Rebellion, was a fiery preacher nicknamed "The Prophet" by his followers. He heard voices, had visions, and claimed to be the second coming of Jesus sent to end slavery. Some historians think he was in fact schizophrenic. Believing that an eclipse of the sun was a sign from God, he led seventy-five others in an insurrection that, like the Harpers Ferry Raid, ended in disaster. Turner was caught and hanged, and 200 other blacks were beaten, tortured, and murdered by white mobs. Laws were passed across the South prohibiting the education of both enslaved and free blacks and instituting other reductions of civil rights.

In counterpoint is the example of Harriet Tubman. Tubman received a severe head injury as a teenager when she tried to protect an escaping enslaved man. The injury gave her lifelong visions, seizures, and hypersomnia. She claimed the dreams and visions were from God. Some historians suggest she had temporal lobe epilepsy. No one can argue with her incredible success. She spent eleven years as a conductor on the Underground Railroad, guiding over 300 people to freedom, including her disabled father whom she hauled through the swamps on a jerry-rigged, hand-built cart. She was never caught and she never lost a single passenger. She also served as a scout and a spy in the Civil War, leading the spectacular Combahee River Raid, which liberated 700 enslaved people. When in need of guidance, she would lie down and fall unconscious for ten minutes, which she called "consulting with God." Her visions, even when counterintuitive, never led her astray.

There is a role for our spiritual longings and for the strength that a true spiritual practice can bring to social movements. There may even be guidance from other realms, but tread carefully: no one has yet developed a simple checklist to distinguish mysticism, desperation, and mental illness. And we need to learn from history. Despite all the suffering of genocide and depression over centuries, no spirit warriors have ever appeared to save the day. That's N-E-V-E-R. No special garments have stopped bullets except Kevlar, a gift from the Pentagon. I think we can all agree that they're not the Holy Ones. No amount of prayer can stop the harrow of oppression, and no special diet can produce special powers. The only miracle we're going to get is us.

◙ ◙ ◙

The four main categories of action discussed here—legal remedies, direct action, withdrawal, and spirituality—can be taken up by either liberals or radicals. What defines all four of these categories as liberal or radical is how they are used. It's the ultimate goal that will dictate their strategic use, and it's the goal that's either liberal or radical.

The main point of this chapter is that because of the historic dominance of liberalism, we've been handed a framework that truncates actions that could otherwise be effective. All four of these categories of action could play a role in dismantling civilization and creating a just and sustainable culture, but only if their strengths and liabilities are understood and acknowledged. That understanding will only come if we accept the insights of radicalism.

Remember that liberalism is a combination of idealism with individualism. For liberals, social reality is comprised of individuals, and it's essentially an intellectual event. Oppression is not about concrete systems of power to liberals, but about ideas and attitudes. Hence, education and moral suasion are the order of the day.

This has stranded the left with tactics that range from ineffectual to ridiculous. Nobody cares if we light candles to stop global warming; asking nicely will not help. This kind of pleading also keeps us forever trapped in a posture of dependent children. If we're good—compliant, quiet, well-behaved—if we follow the rules—someone in authority will listen and care. Meanwhile, power couldn't care less. Power will only care when it is threatened. And none of the strategies currently acceptable on the left contain any threat, precisely because liberalism deeply misunderstands the nature of power.

Consider the array of "political actions" we are offered. First we have the legal strategies, the usual petitions, demonstrations, and lawsuits aimed at protecting what shreds of the world the system will allow. People have dedicated their lives to saving a species, a river, a place, someone or something that they are brave enough to love and that they love enough to protect. I am in no way insulting their commitment or sneering at their passion. But it isn't working. The planet is dying. We do what we can; the planet keeps dying. We know the planet is dying but what else *can* we do? The avenues open to us, the petitions, the lawsuits, don't challenge the basic processes of civilization, the destructive and extractive activities on which this way of life depends. That is the insight from which activists are kept, not just by power and its endless propaganda, but also by the subculture of the left.

Direct action, even nonviolent direct action, has also been derailed by liberalism. I was born the same year as the Civil Rights Act of 1964. Twenty-five years later, I watched on TV as the people of Berlin pulled down that wall. Nonviolence is a form of resistance that works but *it needs to be understood* if it's to be used effectively.

Gene Sharp is the foremost scholar on nonviolent action. His three volume *The Politics of Nonviolent Action* should be required reading for all activists as a basic primer on the nature of political struggle. He starts with the insight that

> It is widely assumed that all social and political behavior must be clearly either violent or nonviolent. This simple dualism leads only to serious distortions of reality, however, one of the main ones being that some people call "nonviolent" anything they regard as good, and "violent" anything they dislike. A second gross distortion occurs when people totally erroneously equate cringing passivity with nonviolent action because in neither case is there the use of physical violence. Careful consideration of actual responses to social and political conflict requires that all responses to conflict situations be initially divided into those of *action* and those of *inaction*, and not divided according to their violence or lack of violence. In such a division nonviolent action assumes its correct place as *one* type of *active* response.[57]

Nonviolent direct action is a form of struggle which uses political, economic, or social leverage in an attempt to coerce the structures of power to change, up to and including complete abdication. Sharp continues,

> Several writers have pointed to the general similarities of nonviolent action to military war. Nonviolent action is a means of combat, as is war. It involves the matching of forces and the waging of "battles," requires wise strategy and tactics, and

demands of its "soldiers" courage, discipline, and sacrifice. This view of nonviolent action as a technique of active combat is diametrically opposed to the popular assumption that, at its strongest, nonviolent action relies on rational persuasion of the opponent, and that more commonly it consists simply of passive submission.[58]

If you are someone who embraces a nonviolent ethic, then you need to understand how the technique of nonviolent direct action works if you are going to employ it successfully. A radical analysis will lead you to the conclusion that justice will only be won by a struggle; oppression is not a mistake; and nice, reasoned requests will not make it stop. In the words of Frederick Douglass, who well knew, "Power concedes nothing without a demand; it never has and never will." Once we understand that, the activist's task becomes one of simple strategy: power must be forced, so how best to apply that force?

The left has often operated on the smug or sentimental belief that nonviolence works only by personal, moral example. It doesn't. Having said that, there is a moral high ground that has historically been useful for nonviolent struggles. When actionists stick to nonviolence while being attacked by the police or military, there is often an upswell of sympathy amongst the general public. Sharp calls this phenomenon a form of "political jujitsu." If you are building a mass movement, then nonviolent discipline is a good technique to employ for this reason alone. But we cannot lose sight of the nature of power and the nature of the struggle that is required to change it. Against power, only force will work. Progressives have repeatedly refused to understand that, from the abolitionists who thought that a pending spiritual transformation would end slavery, to Gandhi writing a letter to Hitler asking him to stop (and then being shocked when it didn't work), to both whites and blacks in the civil rights movement who thought lunch counter sit-ins were too confrontational.

Right now, the culture of most of the left has declared any action but "nonviolence" off-limits for discussion. I put nonviolence in quotes because by and large the people who have embraced such nonviolence don't actually understand the technique of nonviolent direct action. The correct name for them is pacifists, people who for moral or spiritual reasons have an "opposition to war or violence as a means of resolving disputes." Of course, by that definition I'm a pacifist, as I'm against war and I also think violence is a bad way to settle disputes. But it isn't disputes I'm concerned with here; it's global systems of oppression, especially the arrangement called civilization, which is right now devouring the world. Meanwhile, I've heard the proponents of so-called nonviolence declare that speaking in anything besides "I-statements" is violent. Fine; I feel that that is ridiculous.

A personal commitment to the rejection of violence can be an honorable and thoughtful act. But if this commitment leads to an inability to face the realities of systems of power—their inherent violence, their intransigence, their sociopathic destruction of anyone and anything in their way—and what is involved in changing those systems, then the wholesale embrace of such pacifism will only impede our ability to win justice and save what's left of our planet.

Systems of power are not swayed by moral exhortation. They don't care how well-behaved you are, how much you believe in the power of healing, or how much you want the inner child of perpetrators and CEOs to feel the love they supposedly never got. Their inner children are sociopathic. And out in the real world, they will turn fire hoses and German shepherds on your actual children. Nonviolent actionists have been gunned down in cold blood, tortured, thrown in jail to rot. Any quick perusal of the history of political struggle will yield the harsh truth, the lesson learned from Bloody Sunday to Tiananmen Square: nonviolence does not work by persuasion, nor does it offer protection, and the left needs to give up its maudlin belief in both. Those are not the reasons to employ it.

Nonviolence works by facing the ruthless reality of oppression, identifying its linchpins, and using direct action to interrupt the flow of power and hopefully dislodge some portion of its foundation. Instead of weapons, the technique uses people, usually large numbers of people willing to have direct confrontations with power, which means they risk getting killed. The sooner the left faces the reality of that danger, the better prepared we will be to make strategic and tactical decisions, individually and collectively.

Forms of withdrawalism are another popular offer from the left. This especially includes individual, personal "growth." One American Buddhist writes, "What I do for peace and justice is split wood."[59] To declare this political action is a level of narcissism that is insane. You are not the world. And guess what? How you feel will not change the world, no matter how much wood you chop and how peaceful you feel while chopping it.

Hyperindividualism renders this method useless. Withdrawal has to happen on a much larger scale to be effective: we need to think institutionally, not personally, which is the exact point of divergence between liberals and radicals. Alternative institutions like local food networks, communal child care, nonindustrial schooling, direct democracy, and community-based policing and justice are essential to both a culture of resistance and to postcarbon survival. Replacing one consumer choice with another is an act with almost no impact. Indeed, the choices themselves are often useless: ethanol has a net energy loss, and a solar panel may use more energy in its production than it will save in its use. But again, the individualism of liberalism obstructs our ability to use withdrawal as a serious political strategy.

We are encouraged to make lifestyle choices ranging from diet to "green weddings" to suburban sprawl ecovillages that use up slightly fewer resources while still using up plenty.[60] Again, these are essentially a withdrawalist approach. None of these challenge the systems of power that are actively dismembering our planet. Remember, there are no individual solutions to political problems, not ever. At best, these attempts are well-meaning, if misguided. At worst, they hijack the very real concern and despair of anyone who's even half awake, offering a deeply delusional sense of hope.

Spirituality, the last category of action we discussed, has played a strong role in many social change movements: the black churches have been called the cradle of the civil rights movement; Liberation Theology has been central to prodemocracy struggles in Latin America; and Christian missionaries helped end slavery and the caste system in Kerala, India, leaving a human rights legacy that still holds today. But spirituality plays a role in resistance by offering the exact opposite of the American Buddhist quoted above. First, it lends a moral-mythic framework for facing down power as in the Jews' flight from enslavement in Egypt or Jesus's throwing the moneylenders from the Temple. In contrast, the hyperindividualism of "inner peace" as a final goal offers nothing but moral and political disengagement. Second, a spirituality of resistance provides a connection to something way bigger than ourselves. Whatever you want to call it—the Great Mystery, the Goddess, a Higher Power—that source can lead us out of our personal pain, loss, and exhaustion, and lend us the courage and strength to fight for justice. The key words here are "way bigger than ourselves." This is not to say that our personal suffering should not be addressed—indeed, conditions like depression, addiction, and PTSD can be life-threatening and people in our communities that are afflicted need our compassion and help. But a spiritual system worth the name must ultimately lead us out, not in, both because it offers an experience of love or grace beyond our personal pain and because it connects us to the wider world—human, planetary, and cosmic—that must call us to action.

A serious strategy to save this planet has to consider every possible course of action. To state it clearly once more: *our planet is dying.* There could not be a greater call to responsibility than stopping the destruction of all life. A heartfelt belief in human goodness is not a political strategy. Neither is our spiritual growth or our moral purity. We all need to decide for ourselves what actions we can and cannot take, and as in all things that matter, "No" is absolute. That should be a given. There is room—indeed there is a necessity—for every level of engagement in this project. But it is long past time to stop playing make believe about the threats to our planet, solutions to those threats, and about the courage and sacrifice that will be required to bring the system down.

◻ ◻ ◻

So can it be done? Can industrial civilization be stopped? Theoretically, it's not that difficult. Industrialization is dependent on very fragile infrastructure. It requires vast quantities of fossil fuels, which come from relatively few places, enter through a small number of centralized ports and processing facilities, and then have to be transported out along vulnerable supply lines, including the highway system. Industrial civilization is utterly dependent on electricity, and the electric grid is a million fragile miles long. The system is also dependent on the Internet; globalization would not be possible without it to organize and transfer both information and capital. And finally there is capital itself, which flows every day through twenty major stock markets—a finite number indeed.

Any of the above could be targeted in a multitude of ways. Serious nonviolent actionists could blockade the ports, the processing facilities, the stock exchanges, the main highways outside New York, Washington, DC, Chicago. There are only sixteen main bridges into Manhattan. A flow of bodies would be necessary to keep the system at a standstill day after day, bodies provided by people willing to face the consequences. Ask yourself if you have that many people. No? Now ask yourself how long it would take to get that many people, how much political education, how much consciousness-raising against the sweet, numbing dream of conformity and cheap consumer goods? How much can you count on that slow build of courage when the planet is losing species and gaining heat every minute?

The human race as a whole could do with an honest assessment of the destruction inherent in civilization and in our resultant swollen numbers. We could make a series of difficult decisions, reorganize our societies economically, politically, spiritually, and sexually, and restore the monocultures of asphalt and agriculture to living, biotic communities inside which our species could take its humble place once more. Instead, China and India are hurtling into industrialization as fast as the coal can be mined, and the United States' entitlement to 4,000 pounds of steel for every citizen plus the gas to move it continues unabated. We're not on the edge of the "Great Turning," but on the brink of destruction.

In a similar vein, industrialization could be brought down by nonviolent direct action—but will it, when most environmentalists refuse to understand the basic nature of political power and hence the principles by which the strategy works? More importantly, do we have the sheer numbers of people that would be required? And how many species have gone extinct since you opened this book? I need hope to be backed up by more than a fundamentalist insistence on it: I need proof, actual evidence that either the bulk of humanity will willingly give up civilization, right now, or that enough of us are willing to risk our lives to bring it down to make nonviolent interventions feasible.

Reality tells me differently. That means we face a decision, individually and as a resistance movement. Because a small number of people could directly target that infrastructure; a few more, willing to persist, could potentially bring it down.

◙ ◙ ◙

Q: I believe in the hundredth monkey story, in which one monkey learned a new skill, and taught it to another, and another until when a critical mass of monkeys—say, one hundred—had learned this skill, suddenly all the monkeys knew the skill, even on other islands. If enough minds are changed, won't civilization transform itself into something sustainable?

Derrick Jensen: First, the hundredth monkey story is not true. It is a story made up by some New Agers. It is stupid to base a strategy for saving the planet on a fictional story. If we're going to base our strategy on the hundredth monkey, why don't we just base it on Santa Claus bringing us a sustainable culture for Christmas?

And, no, civilization will not transform itself into something sustainable. That's not physically possible. Civilization is functionally unsustainable. And the fact that ideas like the hundredth monkey are spoken of quite often in public discourse lets us know the extreme distance that we have to go to make the sort of changes that are necessary. The fact that people are still talking about this level of detachment from physical reality is evidence itself that there will not be a voluntary transformation.

No, the momentum is too fierce. What we need to do is stop this culture before it kills the planet. And I can't speak for you, but I'm not going to rely on a fictional hundredth monkey to do the work for me when I can do the work myself.

Chapter 4

Culture of Resistance
by Lierre Keith

Tell me, what is it you plan to do
with your one wild and precious life?
—Mary Oliver, poet

The culture of the left needs a serious overhaul. At our best and bravest moments, we are the people who believe in a just world; who fight the power with all the courage and commitment that women and men can possess; who refuse to be bought or beaten into submission, and refuse equally to sell each other out. The history of struggles for justice is inspiring, ennobling even, and it should encourage us to redouble our efforts now when the entire world is at stake. Instead, our leadership is leading us astray. There are historic reasons for the misdirection of many of our movements, and we would do well to understand those reasons before it's too late.[1]

The history of misdirection starts in the Middle Ages when various alternative sects arose across Europe, some more strictly religious, some more politically utopian. The Adamites, for instance, originated in North Africa in the second century, and the last of the Neo-Adamites were forcibly suppressed in Bohemia in 1849.[2] They wanted to achieve a state of primeval innocence from sin. They practiced nudism and ecstatic rituals of rebirth in caves, rejected marriage, and held property communally. Groups such as the Diggers (True Levelers) were more political. They argued for an egalitarian social structure based on small agrarian communities that embraced ecological principles. Writes one historian, "They contended that if only the common people of England would form themselves into self-supporting communes, there would be no place in such a society for the ruling classes."[3]

Not all dissenting groups had a political agenda. Many alternative sects rejected material accumulation and social status but lacked any clear political analysis or egalitarian program. Such subcultures have repeatedly arisen across Europe, coalescing around a common constellation of themes:

113

- A critique of the dogma, hierarchy, and corruption of organized religion;
- A rejection of the moral decay of urban life and a belief in the superiority of rural life;
- A romantic or even sentimental appeal to the past: Eden, the Golden Age, pre-Norman England;
- A millenialist bent;
- A spiritual practice based on mysticism; a direct rather than mediated experience of the sacred. Sometimes this is inside a Christian framework; other examples involve rejection of Christianity. Often the spiritual practices include ecstatic and altered states;
- Pantheism and nature worship, often concurrent with ecological principles, and leading to the formation of agrarian communities;
- Rejection of marriage. Sometimes sects practice celibacy; others embrace polygamy, free love, or group marriage.

Within these dissenting groups, there has long been a tension between identifying the larger society as corrupt and naming it unjust. This tension has been present for over 1,000 years. Groups that critique society as degenerate or immoral have mainly responded by withdrawing from society. They want to make heaven on Earth in the here and now, abandoning the outside world. "In the world but not of it," the Shakers said. Many of these groups were and are deeply pacifistic, in part because the outside world and all things political are seen as corrupting, and in part for strongly held moral reasons. "Corruption groups" are not always leftist or progressive. Indeed, many right-wing and reactionary elements have formed sects and founded communities. In these groups, the sin in urban or modern life is hedonism, not hierarchy. In fact, these groups tend to enforce strict hierarchy: older men over younger men, men over women. Often they have a charismatic leader and the millenialist bent is quite marked.

"Justice groups," on the other hand, name society as inequitable rather than corrupt, and usually see organized religion as one more hierarchy that needs to be dismantled. They express broad political goals such as land reform, pluralistic democracy, and equality between the sexes. These more politically oriented spiritual groups walk the tension between withdrawal and engagement. They attempt to create communities that support a daily spiritual practice, allow for the withdrawal of material participation in unjust systems of power, and encourage political activism to bring their New Jerusalem into being. Contemporary groups like the Catholic Workers are attempts at such a project.

This perennial trend of critique and utopian vision was bolstered by Romanticism, a cultural and artistic movement that began in the latter half of the eighteenth century in Western Europe. It was at least partly a reaction against the Age of Enlightenment, which valued rationality and science. The image of the Enlightenment was the machine, with the living cosmos reduced to clockwork. As the industrial revolution gained strength, rural lifeways were destroyed while urban areas swelled with suffering and squalor. Blake's dark, Satanic mills destroyed rivers, the commons of wetlands and forests fell to the highest bidder, and coal dust was so thick in London that the era could easily be deemed the Age of Tuberculosis. In Germany, the Rhine and the Elbe were killed by dye works and other industrial processes. And along with natural communities, human communities were devastated as well.

Romanticism revolved around three main themes: longing for the past, upholding nature as pure and authentic, and idealizing the heroic and alienated individual. Germany, where elements of an older pagan folk culture still carried on, was in many ways the center of the Romantic movement.

How much of this Teutonic nature worship was really drawn from surviving pre-Christian elements, and how much was simply a Romantic recreation—the Renaissance Faire of the nineteenth century—is beyond the scope of this book. Suffice it to say, there were enough cultural elements for the Romantics to build on.

In 1774, German writer Goethe penned the novel *The Sorrows of Young Werther*, the story of a young man who visits an enchanting peasant village, falls in love with an unattainable young woman, and suffers to the point of committing suicide. The book struck an over-sensitive nerve, and, overnight, young men across Europe began modeling themselves on the protagonist, a depressive and passionate artist. Add to this the supernatural and occult elements of Edgar Allan Poe's work, and, by the nineteenth century, the Romantics of that day resembled modern Goths. A friend of mine likes to say that history is same characters, different costumes—and in this case the costumes haven't even changed much.[4]

Another current of Romanticism that eventually influenced our current situation was bolstered by philosopher Jean Jacques Rosseau,[5] who described a "state of nature" in which humans lived before society developed. He was not the creator of the image of the noble savage—that dubious honor falls to John Dryden, in his 1672 play *The Conquest of Granada*. Rousseau did, however, popularize one of the core components that would coalesce into the cliché, arguing that there was a fundamental rupture between human nature and human society. The concept of such a divide is deeply problematical, as by definition it leaves cultures that aren't civilizations out of the circle of human society. Whether the argument is for the bloodthirsty savage or the noble savage, the underlying concept of a "state of nature" places hunter-gatherers, horticulturalists, nomadic pastoralists, and even some agriculturalists outside the most basic human activity of creating culture. All culture is a human undertaking: there are no humans living in a "state of nature."[6] With the idea of a state of nature, vastly different societies are collapsed into an image of the "primitive," which exists unchanging outside of history and human endeavor.

Indeed, one offshoot of Romanticism was an artistic movement called Primitivism that inspired its own music, literature, and art. Romanticism in general and Primitivism in particular saw European culture as overly rational and repressive of natural impulses. So-called primitive cultures, in contrast, were cast as emotional, innocent and childlike, sexually uninhibited, and at one with the natural world. The Romantics embraced the belief that "primitives" were naturally peaceful; the Primitivists tended to believe in their proclivity to violence. Either cliché could be made to work because the entire image is a construct bearing no relation to the vast variety of forms that indigenous human cultures have taken. Culture is a series of choices—political choices made by a social animal with moral agency. Both the noble savage and the bloodthirsty savage are objectifying, condescending, and racist constructs.

Romanticism tapped into some very legitimate grievances. Urbanism is alienating and isolating. Industrialization destroys communities, both human and biotic. The conformist demands of hierarchical societies leave our emotional lives inauthentic and numb, and a culture that hates the animality of our bodies drives us into exile from our only homes. The realization that none of these conditions are inherent to human existence or to human society can be a profound relief. Further, the existence of cultures that respect the earth, that give children kindness instead of public school, that share food and joy in equal measure, that might even have mystical technologies of ecstasy, can serve as both an inspiration and as evidence of the crimes committed against our hearts, our culture, and our planet. But the places where Romanticism failed still haunt the culture of the left today and must serve as a warning if we are to build a culture of resistance that can support a true resistance movement.

In Germany, the combination of Romanticism and nationalism created an upswell of interest in myths. They spurred a widespread longing for an ancient or even primordial connection with the German landscape. Youth are the perennially disaffected and rebellious, and German youth in the late nineteenth century coalesced into their own counterculture. They were called *Wandervogel* or wandering spirits. They rejected the rigid moral code and work ethic of their bourgeois parents, romanticized the image of the peasant, and wandered the countryside with guitars and rough-spun tunics. The *Wandervogel* started with urban teachers taking their students for hikes in the country as part of the *Lebensreform* (life reform) movement. This social movement emphasized physical fitness and natural health, experimenting with a range of alternative modalities like homeopathy, natural food, herbalism, and meditation. The *Lebensreform* created its own clinics, schools, and intentional communities, all variations on a theme of reestablishing a connection with nature. The short hikes became weekends; the weekends became a lifestyle. The *Wandervogel* embraced the natural in opposition to the artificial: rural over urban, emotion over rationality, sunshine and diet over medicine, spontaneity over control. The youth set up "nests" and "antihomes" in their towns and occupied abandoned castles in the forests. The *Wandervogel* was the origin of the youth hostel movement. They sang folk songs; experimented with fasting, raw foods, and vegetarianism; and embraced ecological ideas—all before the year 1900. They were the anarchist vegan squatters of the age.

Environmental ideas were a fundamental part of these movements. Nature as a spiritual source was fundamental to the Romantics and a guiding principle of *Lebensreform*. Adolph Just and Benedict Lust were a pair of doctors who wrote a foundational *Lebensreform* text, *Return to Nature*, in 1896. In it, they decried,

> Man in his misguidance has powerfully interfered with nature. He has devastated the forests, and thereby even changed the atmospheric conditions and the climate. Some species of plants and animals have become entirely extinct through man, although they were essential in the economy of Nature. Everywhere the purity of the air is affected by smoke and the like, and the rivers are defiled. These and other things are serious encroachments upon Nature, which men nowadays entirely overlook but which are of the greatest importance, and at once show their evil effect not only upon plants but upon animals as well, the latter not having the endurance and power of resistance of man.[7]

Alternative communities soon sprang up all over Europe. The small village of Ascona, Switzerland, became a countercultural center between 1900 and 1920. Experiments involved "surrealism, modern dance, dada, Paganism, feminism, pacifism, psychoanalysis and nature cure."[8] Some of the figures who passed through Ascona included Carl Jung, Isadora Duncan, Mikhail Bakunin, Peter Kropotkin, Vladimir Lenin, Leon Trotsky, and an alcoholic Herman Hesse seeking a cure. Clearly, social change—indeed, revolution—was one of the ideas on the table at Ascona. This chaos of alternative spiritual, cultural, and political trends began to make its way to the US. On August 20, 1903, for instance, an anarchist newspaper in San Francisco published a long article describing the experiments underway at Ascona.

As we will see, the connections between the *Lebensreform, Wander-vogel* youth, and the 1960s counterculture in the US are startlingly direct. German Eduard Baltzer wrote a lengthy explication of *naturliche lebensweise* (natural lifestyle) and founded a vegetarian community. Baltzer-inspired painter Karl Wihelm Diefenbach, who also started a number of alternative communities and workshops dedicated to religion, art, and science, all based on *Lebensreform* ideas. Artists Gusto Graser and Fidus pretty well created the artistic style of the German counterculture in the late nineteenth and early twentieth centuries. Viewers of their work would be forgiven for thinking that their paintings of psychedelic colors, swirling floraforms, and naked bodies embracing were album covers circa 1968. Fidus even used the iconic peace sign in his art.

Graser was a teacher and mentor to Herman Hesse, who was taken up by the Beatniks. *Siddhartha* and *Steppenwolf* were written in the 1920s but sold by the millions in the US in the 1960s. Declares one historian, "Legitimate history will always recount Hesse as the most important link between the European counter-culture of his [Hesse's] youth and their latter-day descendants in America."[9]

Along with a few million other Europeans, some of the proponents of the *Wandervogel* and *Lebensreform* movements immigrated to the United States at the beginning of the twentieth century. The most famous of these *Lebensreform* immigrants was Dr. Benjamin Lust, deemed the Father of Naturopathy, quoted previously. Write Gordon Kennedy and Kody Ryan, "Everything from massage, herbology, raw foods, anti-vivisection and hydro-therapy to Eastern influences like Ayurveda and Yoga found their way to an American audience through Lust."[10] In *Return To Nature*, he railed against water and air pollution, vivisection, vaccination, meat, smoking, alcohol, coffee, and public schooling. Any of this sound a wee bit familiar? Gandhi, a fan, was inspired by Lust's principles to open a Nature Cure clinic in India.

The emphasis on sunshine and naturism led many of these *Leben-sreform* immigrants to move to warm, sunny California and Florida. Sun worship was embraced as equal parts ancient Teutonic religion, health-restoring palliative, and body acceptance. It was much easier to live outdoors and scrounge for food where the weather never dropped below freezing. Called Nature Boys, *naturemensch,* and modern primitives, they set up camp and began attracting followers and disciples. German immigrant Arnold Ehret, for instance, wrote a number of books on fasting, raw foods, and the health benefits of nude sunbathing, books that would become standard texts for the San Francisco hippies. Gypsy Boots was another direct link from the *Lebensreform* to the hippies. Born in San Francisco, he was a follower of German immigrant Maximillian Sikinger. After the usual fasting, hiking, yoga, and sleeping in caves, he opened his "Health Hut" in Los Angeles, which was surprisingly successful. He was also a paid performer at music festivals like Monterey and Newport in 1967 and 1968, appearing beside Jefferson Airplane, Jimi Hendrix, and the Grateful Dead. Carolyn Garcia, Jerry's wife, was apparently a big admirer. Boots was also in the cult film *Mondo Hollywood* with Frank Zappa.

The list of personal connections between the *Wandervogel* Nature Boys and the hippies is substantial, and makes for an unbroken line of cultural continuity. But before we turn to the 1960s, it's important to examine what happened to the *Lebensreform* and *Wandervogel* in Germany with the rise of Nazism.

This is not easy to do. Fin de siècle Germany was a tumult of change and ideas, pulling in all directions. There was a huge and politically powerful socialist party, the Sozialdemokratische Partei Deutschlands (Social Democratic Party of Germany), or SPD, which one historian called "the pride of the Second International."[11] In 1880, it garnered more votes than any other party in Germany, and, in 1912, it had more seats in Parliament than any other party. It helped usher in the first parliamentary democracy, including universal suffrage, and brought a shorter workday, legal workers' councils in industry, and a social safety net. To these serious activists, the *Wandervogel* and *Lebensreform,* especially "the more manifestly idiotic of these cults,"[12] were fringe movements. To state the obvious, the constituents of SPD were working-class and poor people concerned with survival and justice, while the *Lebensreform,* with their yoga, spiritualism, and dietary silliness, were almost entirely middle class.

Here we begin to see these utopian ideas take a sinister turn. The seeds of contradiction are easy to spot in the *völkisch* movement entry on Wikipedia, which states, "The *völkisch* movement is the German interpretation of the populist movement, with a romantic focus on folklore and the 'organic.' . . . In a narrow definition it can be used to designate only groups that consider human beings essentially preformed by blood, i.e. inherited character."

Immediately, there are problems. The *völkisch* is marked with a Nazi tag. One Wikipedian writes, "Personally I consider it offensive to claim that an ethnic definition of 'Folk' equals Nationalism and/or Racism." Another Wikipedian points out that the founders of the *völkisch* concept were leftist thinkers. Another argues, "With regard to its origins . . . the völkisch idea is wholeheartedly non-racist, and people like Landauer and Mühsam (the leading German anarchists of their time) represented a continuing current of völkisch anti-racism. It's understandable if the German page focuses on the racist version—a culture of guilt towards Romanticism seems to be one of Hitler's legacies—but these other aspects need to be looked at too."[13]

Who is correct? Culture, ethnicity, folklore, and nationalism are all strands that history has woven into the word. But *völk* does have a first philosopher, Johann Gottfried von Herder, who founded the whole idea of folklore, of a culture of the common people that should be valued, not despised. He urged Germans to take pride in their language, stories, art, and history. The populist appeal in his ideas—indeed, their necessity to any popular movement—may seem obvious to us 200 years later, but at the time this valuing of a people's culture was new and radical. His personal collection of folk poetry inspired a national hunger for folklore; the brothers Grimm were one direct result of Herder's work. He also argued that everyone from the king to the peasants belonged to the *völk*, a serious break with the ruling notion that only the nobility were the inheritors of culture and that that culture should emulate classical Greece. He believed that his conception of the *völk* would lead to democracy and was a supporter of the French Revolution.

Herder was very aware of where the extremes of nationalism could lead and argued for the full rights of Jews in Germany. He rejected racial concepts, saying that language and culture were the distinctions that mattered, not race, and asserted that humans were all one species. He wrote, "No nationality has been solely designated by God as the

chosen people of the earth; above all we must seek the truth and cultivate the garden of the common good."[14]

Another major proponent of leftist communitarianism was Gustav
Landauer, a Jewish German. He was one of the leading anarchists
through the Wilhelmine era until his death in 1919 when he was
arrested by the Freikorps and stoned to death. He was a mystic as well
as being a political writer and activist. His biographer, Eugene Lunn,
describes Landauer's ideas as a "synthesis of völkisch romanticism and
libertarian socialism," hence, "romantic socialism."[15] He was also a
pacifist, rejecting violence as a means to revolution both individually
and collectively. His belief was that the creation of libertarian communities would "gradually release men and women from their childlike
dependence upon authority," the state, organized religion, and other
forms of hierarchy.[16] His goal was to build "radically democratic, participatory communities."[17]

Landauer spoke to the leftist writers, artists, intellectuals, and youths
who felt alienated by modernity and urbanism and expressed a very
real need—emotional, political, and spiritual—for community renewal.
He had a full program for the revolutionary transformation of society.
Rural communes were the first practical step toward the end of capitalism and exploitation. These communities would form federations
and work together to create the infrastructure of a new society based
on egalitarian principles. It was an A to B plan that never lost sight of
the real conditions of oppression under which people were living. After
World War I, roughly one hundred communes were formed in Germany, and, of those, thirty were politically leftist, formed by anarchists
or communists. There was also a fledgling women's commune movement whose goal was an autonomous feminist culture, similar to the
contemporary lesbian land movement in the US.

Where did this utopian resistance movement go wrong? The
problem was that it was, as historian Peter Weindling puts it, "politically ambivalent."[18] Writes Weindling, "The outburst of utopian social
protest took contradictory artistic, Germanic volkish, or technocratic
directions."[19] Some of these directions, unhitched from a framework
of social justice, were harnessed by the right, and ultimately incorpo-
rated into Nazi ideology. *Lebensreform* activities like hiking and eating

whole-grain bread were seen as strengthening the political body and were promoted by the Nazis. "A racial concept of health was central to National Socialism," writes Weindling. Meanwhile, Jews, gays and lesbians, the mentally ill, and anarchists were seen as "diseases" that weakened the Germanic race as a whole.

Ecological ideas were likewise embraced by the Nazis. The health and fitness of the German people—a primary fixation of Nazi culture—depended on their connection to the health of the land, a connection that was both physical and spiritual. The Nazis were a peculiar combination of the Romantic and the Modern, and the backward-looking traditionalist and the futuristic technotopians were both attracted to their ideology. The Nazi program was as much science as it was emotionality. Writes historian David Blackborn,

> National socialism managed to reconcile, at least theoretically, two powerful and conflicting impulses of the later nineteenth century, and to benefit from each. One was the infatuation with the modern and the technocratic, where there is evident continuity from Wilhelmine Germany to Nazi eugenicists and *Autobahn* builders; the other was the "cultural revolt" against modernity and machine-civilization, pressed into use by the Nazis as part of their appeal to educated élites and provincial philistines alike.[20]

Let's look at another activist of the time, one who was political. Erich Mühsam, a German Jewish anarchist, was a writer, poet, dramatist, and cabaret performer. He was a leading radical thinker and agitator during the Weimar Republic, and won international acclaim for his dramatic work satirizing Hitler. He had a keen interest in combining anarchism with theology and communal living, and spent time in the alternative community of Ascona. Along with many leftists, he was arrested by the Nazis and sent to concentration camps in Sonnenburg, Brandenburg, and finally Oranienburg. Intellectuals around the world protested and demanded Mühsam's release, to no avail. When his wife Zenzl was allowed to visit him, his face was so bruised she didn't recognize him. The guards beat and tortured him for seventeen months. They made him dig his own grave. They broke his teeth and burned a swastika into his scalp. Yet when they tried to make him sing the Nazi anthem, he would sing the International instead. At his last torture session, they smashed in his skull and then killed him by lethal injection. They finished by hanging his body in a latrine.

The intransigent aimlessness and anemic narcissism of so much of the contemporary counterculture had no place beside the unassailable courage and sheer stamina of this man. Sifting through this material, I will admit to a certain amount of despair: between the feckless and the fascist, will there ever be any hope for this movement? The existence of Erich Mühsam is an answer to embrace. Likewise, reading history backwards, so that Nazis are preordained in the *völkish* idea, is insulting to the inheritors of this idea who resisted Fascism with Mühsam's fortitude. There were German leftists who fought for radical democracy and justice, not despite their communitarianism, but with it.

Our contemporary environmental movement has much to learn from this history. Janet Biehl and Peter Staudenmaier, in their book *Ecofascism: Lessons from the German Experience*,[21] explore the idea that fascism or other reactionary politics are "perhaps the unavoidable trajectory of any movement which acknowledges and opposes social and ecological problems but does not recognize their systemic roots or actively resist the political and economic structures which generate them. Eschewing societal transformation in favor of personal change, an ostensibly apolitical disaffection can, in times of crisis, yield barbaric results."[22]

The contemporary alterna-culture won't result in anything more sinister than silliness; fascism in the US is most likely to come from actual right-wing ideologues mobilizing the resentments of the disaffected and economically stretched mainstream, not from New Age workshop hoppers. And friends of Mary Jane aren't known for their virulence against anything besides regular bathing. German immigrants brought the *Lebensreform* and *Wandervogel* to the US, and it didn't seed a fascist movement here. None of this leads inexorably to fascism. But we need to take seriously the history of how ideas which we think of as innately progressive, like ecology and animal rights, became intertwined with a fascist movement.

An alternative culture built around the project of an individualistic and interior experience, whether spiritual or psychological, *cannot create a resistance movement*, no matter how many societal conventions it trespasses. Indeed, the *Wandervogel* manifesto stated, "We regard with contempt all who call us political,"[23] and their most repeated motto was "Our lack of purpose is our strength." But as Laqueur points out,

> Lack of interest in public affairs is not civic virtue, and . . . an inability to think in political categories does not prevent people from getting involved in political disaster . . . The Wandervogel . . . completely failed. They did not prepare their members for active citizenship. . . . Both the socialist youth and the Catholics had firmer ground under their feet; each had a set of values to which they adhered. But in the education of the free youth movement there was a dangerous vacuum all too ready to be filled by moral relativisim and nihilism.[24]

We are facing another disaster, and if we fail there will be no future to learn from our mistakes. That same "lack of interest"—often a stance of smug alienation—is killing our last chance of resistance. We are not preparing a movement for active citizenship and all that implies—the commitment, courage, and sacrifice that real resistance demands. There is no firm moral ground under the feet of those who can only counsel withdrawal and personal comfort in the face of atrocity. And the current *Wandervogel* end in nihilism as well, repeating that it's over, we can do nothing, the human race has run its course and the bacteria will inherit the earth. The parallels are exact. And the outcome?

The *Wandervogel* marched off to World War I, where they "perished in Flanders and Verdun."[25] Of those who returned from the war, a small, vocal minority became communists. A larger group embraced right-wing protofascist groups. But the largest segment was apolitical and apathetic. "This was no accidental development," writes Laqueur.[26]

The living world is now perishing in its own Flanders and Verdun, a bloody, senseless pile of daily species. Today there are still wood thrushes, small brown angels of the deep woods. Today there are northern leopard frogs, but only barely. There may not be Burmese star tortoises, with their shells like golden poinsettias; the last time anyone looked—for 400 hours with trained dogs—they only found five. If the largest segment of us remains apolitical and apathetic, they will all surely die.

◪ ◪ ◪

This is the history woven through the contemporary alternative culture. It takes strands of the Romantics, the *Wandervogel*, and the *Lebensreform*, winds through the Beatniks and the hippies, and splits into a series of subcultures with different emphases, from self-help and twelve-step believers to New Age spiritual shoppers. There is a set of accumulated ideas and behavioral norms that are barely articulated and yet hold sway across the left. It is my goal here to fully examine these currents so we may collectively decide which are useful and which are detrimental to the culture of resistance.

For the purposes of this discussion, I've set "alternative culture" against "oppositional culture," knowing full well that real life is rarely lived in such stark terms. Many of these norms and behaviors form a continuum along which participants move with relative ease.

In my own experience, these conflicting currents have at times merged into a train wreck of the absurd and the brave, often in the same evening. The righteous vegan dinner of even more righteously shoplifted ingredients, followed by a daring attack on the fence at the military base, which included both spray painting and fervent Wicca-esque chanting—in case our energy really *could* bring it down—rounded out with a debrief by Talking Stick which became a foray into that happy land where polyamory meets untreated bipolar disorder (medication being a tool of The Man), a group meltdown of such operatic proportions that the neighbors called the police.

Ah, youth.

I was socialized into some of these cultural concepts and practices as a teenager. I know my way around a mosh pit, a womyn's circle, and a chakra cleansing. I embraced much of the alternative culture for reasons that are understandable. At sixteen, fighting authority felt like life and death survival, and all hierarchy was self-evidently domination.

Alternative vs. Oppositional Culture

ALTERNATIVE CULTURE	OPPOSITIONAL CULTURE
Apathetic-to-hostile to concept of political engagement	Consciously embraces resistance
Change seen in psychological and cultural terms	Change seen in economic and political terms
Individual consciousness is the target	Concrete institutions are targeted
Adolescent values of youth movement	Adult values of discernment, responsibility
All authority is rejected out of hand	Legitimate authority is accepted and cultivated
Rejection of moral judgment	Strong moral code based on universal human rights
Attacks on conventions ■ all boundaries fair game ■ shock value	Attacks on power structures
Alienated individual valorized	Loyalty and solidarity valued
Goal is to feel intense, "authentic," unmediated emotions	Goals are adult concerns: guide the community, socialize the young, enforce norms, participate in larger project of righting the world
A politics of emotion in which feeling states outweigh effective strategy or tactics	A politics of community that values responsibility, mutual aid, work ethic—dependent on self-regulation of mature adults
Politics is *who you are*	Politics is *what you do*
Human relations are corrupted in the act of political resistance; only right consciousness can prevail	Human relations are corrupted by systems of power and oppression; justice must prevail even if it takes generations
Generalized withdrawal as strategy	Withdraw loyalty from systems of oppression and the oppressors but active engagement to stop injustice
Moral vigor of youth cut off from action ■ horizontal hostility ■ questions of in-group/out-group	Idealism tempered by experience
Cultural appropriation	■ Cultural reclamation and protection (oppressed group) ■ Cultural respect, political solidarity (allies)

Meanwhile, all around me, in quite varied venues, people said that personal change *was* political change—or even insisted that it was the *only* sphere where change was possible. I knew there was something wrong with that, but arguing with the New Age branch led to defeat by spiritual smugness and Gandhian clichés. The fact that I have a degenerative disease was always used as evidence against me by these people. Arguing with the militant, political branch (Did it really matter if someone ate her pizza with "liquid meat," aka cheese? Was I really a sell-out if I saw my family on Christmas?) led to accusations of a lack of true commitment. With very little cross generational guidance and the absence of a real culture of resistance, I was left accepting some of these arguments despite internal misgivings.

Way too many potential activists, lacking neither courage nor commitment, are lost in the same confusion. It's in the hope that we are collectively capable of something better that I offer these criticisms.

This focus on individual change is a hallmark of liberalism. It comes in a few different flavors, different enough that their proponents don't recognize that they are all in the same category. But underneath the surface differences, the commonality of individualism puts all of these subgroups on a continuum. It starts with the virulently antipolitical dwellers in workshop culture; only individuals (i.e., themselves) are a worthy project and only individuals can change. The continuum moves toward more social consciousness to include people who identify oppression as real but still earnestly believe in liberal solutions, mainly education, psychological change, and "personal example." It ends at the far extreme where personal lifestyle becomes personal purity and identity itself is declared a political act. These people often have a compelling radical analysis of oppression, hard won and fiercely defended. This would include such divergent groups as vegans, lesbian separatists, and anarchist rewilders. They would all feel deeply insulted to be called liberals. But if the only solutions proposed encompass nothing larger than personal action—and indeed political resistance is rejected as "participation" in an oppressive system—then the program is ultimately liberal, and doomed to fail, despite the clarity of the analysis and the dedication of its adherents.

The defining characteristic of an oppositional culture, on the other hand, is that it consciously claims to be the cradle of resistance. Where the alternative culture exists to create personal change, the oppositional culture exists to nurture a serious movement for political transformation of the institutions that control society. It understands that concrete systems of power have to be dismantled, and that such a project will require tremendous courage, commitment, risk, and potential loss of life. In the words of Andrea Dworkin,

> Now, when I talk about a resistance, I am talking about an organized political resistance. I'm not just talking about something that comes and something that goes. I'm not talking about a feeling. I'm not talking about having in your heart the way things should be and going through a regular day having good, decent, wonderful ideas in your heart. I'm talking about when you put your body and your mind on the line and commit yourself to years of struggle in order to change the society in which you live. This does not mean just changing the men whom you know so that their manners will get better—although that wouldn't be bad either. . . . But that's not what a political resistance is. A political resistance goes on day and night, under cover and over ground, where people can see it and where people can't. It is passed from generation to generation. It is taught. It is encouraged. It is celebrated. It is smart. It is savvy. It is committed. And someday it will win. It will win.[27]

As you can see there is a split to the root between the Romantics and the resistance, a split that's been present for centuries. They both start with a rejection of some part of the established social order, but they identify their enemy differently, and from that difference they head in opposite directions. Again, this difference often forms a continuum in many people's lived experience, as they move from yoga class to the food co-op to a meeting about shutting down the local nuclear power plant. But we need to understand the differences between the two poles of the continuum, even if the middle is often murky. Those differences have been obscured by two victories of liberalism: the conflation of personal change with political change, and the broad rejection of real resistance. But a merely alternative culture is not a culture of resistance, and we need clarity about how they are different.

For the alternative culture—the inheritors of the Romantic movement—the enemy is a constraining set of values and conventions, usually cast as bourgeois. Their solution is to "create an alternative world within Western society" based on "exaggerated individualism."[28] The Bohemians, for instance, were direct descendants of the Romantic movement. The Bohemian ethos has been defined by "transgression, excess, sexual outrage, eccentric behavior, outrageous appearance, nostalgia and poverty."[29] They emphasized the artist as rebel, a concept that would have been incomprehensible in the premodern era when both artists and artisans had an accepted place in the social hierarchy. The industrial age upset that order, and the displaced artist was recast as a rebel. But this rebellion was organized around an internal feeling state. Stephen Spender wrote in his appropriately titled memoir *World Within World*, "I pitied the unemployed, deplored social injustice, wished for peace, and held socialist views. These views were emotional."[30] Elizabeth Wilson correctly names Bohemia as "a retreat from politics."[31] She writes, "In 1838, Delphine de Girardin commented on the way in which the best-known writers and artists were free to spend their time at balls and dances because they had taken up a stance of 'internal migration.' They had turned their back on politics, a strategy similar to the 'internal exile' of East European dissidents after 1945."[32]

The heroization of the individual, in whatever admixture of suffering and alienation, forms the basis of the Romantic hostility to the political sphere. The other two tendencies follow in different trajectories from that individualism. First is a valuing of emotional intensity that rejects self-reflection, rationality, and investigation. For instance, Rosseau wrote, "For us, to exist is to feel; and our sensibility is incontestably more important than reason."[33] Second is a belief that the polis, the political life of society, is yet another stultifying system for the romantic hero to reject:

> Romantics . . . rejected the possibility of effecting change
> through politics. The Romantics were skeptical about merely

organizational reform, about the effects of simply rearranging
a society's institutions. . . . The Romantics revolted not in the
name of equality or to effect economic change but to enable
the development of the 'inner man.' In this sense, they were
opposed to the bourgeoisie *and* the radicals. Bourgeois conventions were rejected because they were shallow and artificial,
and the radical's program of social and economic change was
rejected because it did nothing to free the human spirit.[34]

The Beatniks were inheritors of this tradition. Their main project
was to "reject . . . the conformity and materialism of the middle class,"
mostly through experimentation with drugs and sex, and to lay claim to
both emotion and art as unmediated and transcendent.[35] But the Beatniks were a small social phenomenon. They didn't blossom into the
hippies until the demographics of both the baby boom and the middle
class provided the necessary alienated youth in the 1960s.

◙ ◙ ◙

The youth origin of the alternative culture is crucial to understanding it.
As previously discussed, the *Wandervogel* was a youth movement. In fact,
in 1911, "there were more Germans in their late teens than there would
ever be again in the twentieth century."[36] The seeds of that original youth
culture were transplanted to the US, where they lay dormant until a similar critical mass of young people reached adolescence. The alternative
culture as we know it is largely a product of the adolescent brain.

Because the brain of an adolescent is the same size as an adult brain,
scientists once concluded that it was fully developed sometime around
puberty. But with new technology like MRI and PET scans, we can literally see that the adolescent brain is very much "a work in progress."[37]

To begin with, the prefrontal cortex (PFC) isn't utilized in an adolescent brain to the extent that it will be in an adult brain. David Walsh,
in his book on the adolescent brain, *Why Do They Act That Way?*, calls
the PFC "the brain's conscience." According to Walsh, it is "responsible
for planning ahead, considering consequences, and managing emotional states."[38] As well, a person's ability to judge time is not fully

developed until age twenty-one. Adolescents literally cannot understand cause and effect or long-term consequences the way an adult can.

The PFC is the "executive center of the brain."[39] When impulses fire from other areas of the brain, the PFC's job is to control them. But because this region is still under construction for adolescents, they lack impulse control. Delayed gratification is not exactly the gift of that age group, who are also routinely associated with rudeness, irresponsibility, and laziness. All of this is a function of an underactive PFC. The "laziness" is compounded by a few other brain development processes. The ventral striatal circuit is responsible for motivation and it goes inactive during adolescence. As well, the adolescent brain undergoes a huge shift in sleep patterns. The amount of sleep and the timing of the sleep cycle are both affected. Much of the process is complicated and still under scrutiny. Fifty different neurotransmitters and hormones may be involved.[40] Two things are certain: teens need more sleep, and they often can't fall asleep at night. Forced to conform to an industrial regimentation of time, they're often dead tired during the day, a tiredness based on their biology, not their moral failings.

Myelination is crucial to brain development. Myelin is a form of fat that protects and insulates our axons, which are the cablelike structures in the neurons. Myelination is the process whereby the neurons build up that protective fat. Without it, the electrical impulses are impeded in their travel along their axons—by a factor of a hundred. Unprotected axons are also vulnerable to electrical interference from nearby axons. A generation ago, scientists thought that myelination was complete by age seven, but nothing could be further from the truth. The myelination process is not only incomplete for adolescents, in some areas of the brain it "increases by 100 percent."[41] One of the areas responsible for emotional regulation undergoes myelination during adolescence, which, according to David Walsh, "accounts for the lightening quick flashes of anger" that are the hallmark of youth.[42]

Hormonal fluctuations are another factor that can create an amplification of emotional intensity, leading to the risk taking, impulsive behavior, anger, and overall emotionality of the teen years.

Walsh is clear that while "it is not the teen's fault that his brain isn't fully under his control, it's his responsibility to get it under his control."[43] It's the role of parents and their stand-ins in the larger culture to provide the guidance, support, and structure to help young people toward adulthood. Without adults to supply expectations and consequences, the developing brain will never connect the neurons that need to be permanently linked at this stage of life. This has been an important task of functioning communities for thousands of years: to raise the next crop of adults.

There is a window of opportunity for every period of development in the brain. Walsh reports that neurologists have a saying: the neurons that fire together, wire together. This is true from infancy—where basic neurological patterns for functions like hearing and sight are laid down—on through adolescence, where our capacity for self-regulation, assessing consequences, and relational bonding are either cultivated into lifelong strengths or ignored to wither away.

Beyond the biology of the teen brain is the psychology of adolescence. Psychologist Erik Erikson says that the biggest task of those years is identity formation. It is the time when the question of *Who I Am* takes on an intensity and importance that will likely never be matched again.

And thank goodness. I remember my relationship with my high school best friend. We would see each other before first period, at lunch, and for shared classes. When we got home, we'd talk on the phone immediately—having been separated for all of forty-five minutes, there were crucial things to say. Then after dinner, we'd have to talk again. The next morning, it started all over in the five minutes at her locker before homeroom. Looking back I wonder: what in the world were we talking about? But that's the project of adolescence, self-revelation and exploration. It was all so new, so intense, so compelling. We talked about our feelings and then our feelings about our feelings and then our feelings about our . . . By the time I was twenty, it wasn't half so interesting. By the time I was thirty, it was boring. And past thirty-five, you couldn't pay me enough to have those kinds of conversations.

But this is where the counterculture—a product of adolescent biology and psychology—has been permanently stuck. The concerns of adolescence—its gifts and its shortcomings—are the framework for the alternative culture, and these community norms and habits have become accepted across the left in what Theodore Roszak calls a "progressive 'adolescentization' of dissenting thought and culture."[44] Its main project is the self, its exploration, and its expression, to the point where many adherents are actively hostile to political engagement. One common version of this is a concession that some kind of social change is necessary, but that the only thing we can change is ourselves. Thus injustice becomes an excuse for narcissism. As one former activist explained to sociologist Keith Melville,

> "I had done the political trip for awhile, but I got to the point where I couldn't just advocate social change, I had to live it. Change isn't something up there, out there, and it isn't a power trip. It's in here," he thumped his chest, and little puffs of dust exploded from his coveralls. "This is where I have to start if I want to change the whole fucking system."[45]

Timothy Leary, the high priest of Psychedelia, continuously urged the youth movement to "turn on, tune in, and drop out." He believed that the activists and the "psychedelic religious movement" were "completely incompatible."[46] John Lennon and John Hoyland debated the conflict between individual and social change in a public exchange of letters in 1968. Lennon argued by defending the lyrics to "Revolution."

> You say you'll change the constitution
> well, you know
> we all want to change your head.
> You tell me it's the institution,
> well, you know,
> you better free your mind instead.

To which Hoyland replied, "What makes you so sure that a lot of us haven't changed our heads in something like the way you recommend—and then found out it wasn't enough, because you simply cannot be turned on and happy when you know kids are being roasted to death in Vietnam?"[47]

The endless project of the self is fine for people who are fifteen, as long as they are surrounded by a larger community of adults who can provide the structure for the physical and psychological developments that need to happen to produce a mature individual. But anyone past adolescence should be assuming her or his role as an adult: to provide for the young and the vulnerable, and to sustain and guide the community as a whole. For a culture of resistance, these jobs are done with the understanding that resistance is primary in whatever tasks our talents call us to undertake. We are never delinked from the larger goal of creating a movement to fight for justice.

The legacy of the Romantics is especially prominent in the politics of emotion embraced by many different strands of the alternative culture. Emotions are understood as pure, unmediated by society, a society whose main offense is seen to be the suppression of these always-authentic feelings. The paramount emotional state varies—for the hippies and New Agers, it's love; for the punks, it's rage; and for the Goths, it's exquisite suffering—but the ultimate goal is to achieve the selected emotion and maintain it. Emotional states are not always clearly defined as a goal in these subcultures, but these efforts are accepted as the unexamined norms.

Under the influence of therapy and "personal growth" workshops, the expression of all emotions has achieved a status that approaches a human right. To tell someone you refuse to "process" or to suggest that a group stay focused on discussion and decision making is to provoke outrage. All appropriate sense of boundaries and discernment are considered not the hallmark of adults but conditioning that must be overcome. We must be willing and able to reveal the most intimate details of our personal histories with strangers, and the more intense and performative that sharing, the better.

This individualist stance was taken up as politics across the counterculture in the '60s. It found its zenith in Abbie Hoffman and the Yippies. The title of Hoffman's book, *Revolution for the Hell of It*, is just an update on "Our lack of purpose is our strength" and is about as useful for a political movement. Set aside the misogyny (Hoffman molesting flight attendants), homophobia ("the peace movement is fags"), and the excruciating right-on racism. It's the self-centered idiocy of this book that's unbearable. Yet it inspired a counterculture that still plagues the left today.

It's also hard to critique this book knowing that Hoffman was bipolar and committed suicide. The mental illness shrieks from the page.

> The Diggers left after we had talked the whole night. The SDS'ers slept all night very soundly. They had nothing to talk about in those wee morning hours when you rap on and on and a dialogue of non-verbal vibrations begins. You Relate!! You Plan!! You Think!! You Get Stoned!! You Feel!![48]

You need lithium and a caring support system.

The book is a scattershot of antiauthoritarian rants that claim intense emotion—usually brought on by staged drama—as the ultimate goal. Hoffman urged actions like this:

> Stand on a street corner with 500 leaflets and explode. . . . Recruit a person to read the leaflet aloud while all this distribution is going on. Run around tearing the leaflets, selling them, trading them. Rip one in half and give half to one person and half to another and tell them to make love. Do it all fast. Like slapstick movies. Make sure everyone has a good time. People love to laugh—it's a riot. Riot—that's an interesting word-game if you want to play it.[49]

This self-conscious display stands in stark contrast to a serious resistance movement. Comparing this behavior to the courage, spiritual depth, and personal dignity of Erich Mühsam or the rank and file in the civil rights movement, it's hard not to cringe.

The continuum between bipolar disorder and the adolescent brain is apparent in Hoffman: the lack of judgment, the runaway emotional intensity, the knee-jerk reaction against all constraint, the entitlement, even the sleeplessness, all tragically magnified by the manic states of his illness. A culture of youth without the guidance of adults will produce exactly what Hoffman envisioned. It will also be unable to recognize frank mental illness when it's costumed by a radical stance, or to help the people consumed by such illness. That help can only come from a stable, committed community. Ironically, building and maintaining such a community requires that some people embody the values that Hoffman and the youth culture rejected out of hand: responsibility, commitment, respect.

Beyond the personal tragedy lies the political tragedy that befell the left, as the drop-out culture diverted disaffected youth from building a serious resistance movement against real systems of oppression—racism, capitalism, patriarchy—and a culture of resistance that could support that movement. Instead, with the enemy identified as "middle-class hang-ups"—as anything that got in the way of any impulse—and liberation defined as an internal emotional state, the idealism and hard-won gains of the '60s collapsed into the "me" generation of the '70s.[50] And now all that's left is a vaguely liberal alterna-culture, identifiable by its meditation classes and under-cooked legumes, its obsession with its own psychology, and its New Age spiritual platitudes. Nothing bad will ever happen if you keep your mind, colon, and/or aura pure, which leaves believers in a very awkward position of having to blame the victim when disease, heartbreak, or smart bombs fall. This is in no way to erase those stalwart individuals who have never lost their commitment to a just world and continued to fight. It is to mourn with them a generational moment of promise that was squandered and has yet to come again.

◙ ◙ ◙

Radical groups have their own particular pitfalls. The first is in dealing with hierarchy, both conceptually and practically. The rejection of authority is another hallmark of adolescence, and this knee-jerk reactivity filters into many political groups. All hierarchy is a tool of The Man, the patriarchy, the Nazis. This approach leads to an insistence on consensus at any cost and often a constant metadiscussion of group power dynamics. It also unleashes "critiques" of anyone who achieves public acclaim or leadership status. These critiques are usually nothing more than jealousy camouflaged by political righteousness. "Bourgeois" is a perennial favorite, as well as whatever flavor of "sell-out" matches the group's criteria. It's often accompanied by a hyperanalysis of the victim's language use or personal lifestyle choices. There is a reason that the phrase "politically correct" was invented on the left.[51]

There's a name for this trashing. As noted, Florynce Kennedy called it "horizontal hostility."[52] And if it feels like junior high school by another name, that's because it is. It can reach a feeding frenzy of ugly gossip and character assassination. In more militant groups, it may take the form of paranoid accusations. In the worst instances of the groups that encourage macho posturing, it ends with men shooting each other. Ultimately, it's caused by fighting horizontally rather than vertically (see Figure 3-1, p. 85). If the only thing we can change is ourselves or if the best tactics for social change are lifestyle choices, then, indeed, examining and critiquing the minutiae of people's personal lives will be cast as righteous activity. And if you're not going to fight the people in power, the only people left to fight are each other. Writes Denise Thompson,

> Horizontal hostility can involve bullying into submission someone who is no more privileged in the hierarchy of male supremacist social relations than the bully herself. It can involve attempts to destroy the good reputation of someone who has no more access to the upper levels of power than the one who is spreading the scandal. It can involve holding someone responsible for one's own oppression, even though she too is oppressed. It can involve envious demands that another woman stop using her own abilities, because the success of someone no better placed than you yourself "makes" you feel inadequate and worthless. Or it can involve attempts to silence criticism by attacking the one perceived to be doing the criticising. In general terms, it involves misperceptions of the source of domination, locating it with women who are not behaving oppressively.[53]

This behavior leaves friendships, activist circles, and movements in shreds. The people subject to attack are often traumatized until they permanently withdraw. The bystanders may find the culture so unpleasant and even abusive that they leave as well. And many of the worst aggressors burn out on their own adrenaline, to drop out of the movement and into mainstream lives. In military conflicts, more soldiers may be killed by "friendly fire" than the enemy, an apt parallel to how radical groups often self-destruct.

To be viable, a serious movement needs a supportive culture. It takes time to witness the same behaviors coalescing into the destructive patterns that repeat across radical movements, to name them, and to learn to stop them. Successful cultures of resistance are able to develop healthy norms of behavior and corresponding processes to handle conflict. But a youth culture by definition doesn't have that cache of experience, and it never will.

A culture of resistance also needs the ability to think long-term. One study of student activists from the Berkeley Free Speech Movement interviewed participants five years after their sit-in. Many of them felt that the movement—and hence political action—was unsuccessful.[54] Five years? Try five generations. Movements for serious social change take a long time. But a youth movement will be forever delinked from generations.

Contrast the (mostly white) ex-protestors' attitude with the history of the Pullman porters, the black men who worked as sleeping car attendants on the railroad. The porters were both the generational and political link between slavery and the civil rights movement, accumulating income, self-respect, and the political experience they would need to wage the protracted struggle to end segregation. The very first Pullman porters were in fact formerly enslaved men. George Pullman hired them because they were people who, tragically, could act subserviently enough to make the white passengers happy. (When Pullman tried hiring black college kids from the North for summer jobs as porters, the results were often disastrous.) Yet the jobs offered two things in exchange for the subservience: economic stability (despite the gruesomely long hours) and a broadening outlook. Writes historian Larry Tye:

> The importance of education was drilled into porters on the sleepers, where they got an up-close look at America's elite that few black men were afforded, helping demystify the white race

at the same time it made its advantages seem even more unfair and enticing. That was why they worked so hard for tips, took on second jobs at home, and bore the indignities of the race-conscious sleeping cars. . . . It was an accepted wisdom that they turned out more college graduates than anyone else. And those kids, whether or not they made lists of the most famous, grew up believing they could do anything. The result . . . was that Pullman porters helped give birth to the African-American professional classes.[55]

The porters knew that in their own lives they would only get so far. But their children were raised to carry the struggle forward. The list of black luminaries with Pullman porters in their families is impressive, from John O'Bryant (San Francisco's first black mayor) to Florynce Kennedy to Justice Thurgood Marshall. Civil rights lawyer Elaine Jones, whose father worked as a porter to put his three kids through prestigious universities, has this to say: "All he expected in return was that we had a duty to succeed and give back. Dad said, 'I'm doing this so they can change things.' He won through us."[56]

One reason the civil rights struggle was successful was that there was a strong linkage between the generations, an unbroken line of determination, character, and courage, that kept the movement pushing onward as it accumulated political wisdom.

The gift of youth is its idealism and courage. That courage may veer into the foolhardy due to the young brain's inability to foresee consequences, but the courage of the young has been a prime force in social movements across history. For instance, Sylvia Pankhurst describes what happened when the suffragist Women's Social and Political Union (WSPU) embraced arson as a tactic:

> In July 1912, secret arson began to be organized under the direction of Christabel Pankhurst. When the policy was fully under way, certain officials of the Union were given, as their main work, the task of advising incendiaries, and arranging for the supply of such inflammable material, house-breaking tools, and other matters as they might require. A certain exceedingly

feminine-looking young lady was strolling about London, meeting militants in all sorts of public and unexpected places to arrange for perilous expeditions. Women, *most of them very young*, toiled through the night across unfamiliar country, carrying heavy cases of petrol and paraffin. Sometimes they failed, sometimes succeeded in setting fire to an untenanted building—all the better if it were the residence of a notability— or a church, or other place of historic interest.[57] (emphasis added)

Add to this that they performed these activities—including scaling buildings, climbing hedges, and running from the police—while wearing corsets and encumbered by pounds of skirting. It's overwhelmingly the young who are willing and able to undertake these kinds of physical risks.

A great example of a working relationship between youth and elders is portrayed in the film *Kanehsatake: 270 Years of Resistance*.[58] The movie documents the Oka crisis (mentioned in Chapter 6), in which Mohawk people protected their burial ground from being turned into a golf course. The conflict escalated as the defenders barricaded roads and the local police were replaced by the army. Alanis Obomsawin was behind the barricades, so her film is not a fictional replay, but actual footage of the events. Of note here is the number of times she captured the elders—with their fully functioning prefrontal cortexes—stepping between the youth and trouble, telling them to calm down and back away. Without the warriors, the blockade never would have happened; without the elders, it's likely there would have been a massacre.

Youth's moral fervor and intolerance of hypocrisy often results in either/or thinking and drawing too many lines in the sand, but serious movements need the steady supply of idealism that the young provide. The psychological task of middle age is to remember that idealism helps protect against the rough wear of disappointment. Adulthood also brings responsibilities that the young can't always understand. Having children, for instance, will put serious constraints on activism. Aging parents who need care and support cannot be abandoned. And then there's the activist's own basic survival needs, the demands of shelter, food, health care. The older people need the young to bring idealism and courage to the movement.

The women's suffrage movement started with a generation of women who asked nicely. In an age when women had no right to ask for anything, they did the best they could. The struggle, like that of the Pullman porters and the succeeding civil rights movement, was handed down to the next generation. Emmeline Pankhurst recalls a childhood of fund raisers to help newly freed blacks in the US, attending her first women's suffrage meeting at age fourteen, and bedtime stories from *Uncle Tom's Cabin*. She wrote,

> Those men and women are fortunate who are born at a time when a great struggle for human freedom is in progress. It is an added good fortune to have parents who take a personal part in the great movements of their time. . . . Young as I was—I could not have been older than five years—I knew perfectly well the meaning of the words "slavery" and "emancipation."[59]

Emmeline married Dr. Richard Pankhurst, who drafted the first women's suffrage bill and the Married Women's Property Act, which, when it passed in 1882, gave women control over their own wages and property. Up until then, women did not even own the clothes on their backs—men did. (The next time you buy your own shirt with your own money, remember to thank all Pankhursts great and small.) Emmeline and Richard's daughters, Sylvia and Christabel, were the third generation of Pankhursts born to be activists. It was in large part the infusion of their youthful idealism and courage that fueled the battle for women's suffrage. Emmeline wrote,

> All their lives they had been interested in women's suffrage. Christabel and Sylvia, as little girls, had cried to be taken to meetings. They had helped in our drawing-room meetings in every way that children can help. As they grew older we used to talk together about the suffrage, and I was sometimes rather frightened by their youthful confidence in the prospect, which they considered certain, of the success of the movement. One

day Christabel startled me with the remark: "How long you women have been trying for the vote. For my part, I mean to get it."

Was there, I reflected, any difference between trying for the vote and getting it? There is an old French proverb, "If youth could know; if age could do." It occurred to me that if the older suffrage workers could in some way join hands with the young, unwearied, and resourceful suffragists, the movement might wake up to new life and new possibilities. After that I and my daughters together sought a way to bring about that union of young and old which would find new methods, blaze new trails.[60]

Emmeline raised her girls in a serious culture of resistance. As a strategist, she wisely understood that the moment was ripe for the young to push the movement on to new tactics. Thus was formed the WSPU. "We resolved to . . . be satisfied with nothing but action on our question. 'Deeds, not Words' was to be our permanent motto."[61] Those deeds would run to harassing government officials, civil disobedience, hunger strikes, and arson. They would also be successful.

The transition from one generation to the next, and an increase in confrontational tactics, is rarely smooth. The older activists may try to obstruct the young. It often splits movements. When the WSPU embraced more militance, women who had been crucial to its founding had to leave the organization. Wrote Emmeline Pethick-Lawrence,

Mrs. Pankhurst met us with the announcement that she and Christabel had determined on a new kind of campaign. Henceforward she said there was to be a widespread attack upon public and private property, secretly carried out by Suffragettes who would not offer themselves for arrest, but wherever possible would make good their escape. As our minds had been moving in quite another direction, this project came as a shock to us both. We considered it sheer madness . . . Although we had been at one with Mrs. Pankhurst in her objective of women's political emancipation, and for six years had pursued the same path,

> there had always been an underlying difference between us that
> had not come into the open, mainly because of the close union
> of mind and purpose . . . we found ourselves for the first time
> in something that resembled a family quarrel.[62]

These are painful moments inside organizations and across movements. But it is more or less inevitable. The overall pattern is one we should be aware of so we can work with it rather than struggling against it. This transition is likely to be linked with the ethical issues around nonviolence. As with those disagreements, we have to find a way to build a serious movement despite our differences.

Building radical movements has been harder since the creation of a youth culture. Breaking the natural bonds (could there be a deeper bond than the cross generational one between mother and child?) between young and old means that the political wisdom never accumulates. It also means that the young are never socialized into a true culture of resistance. The values of a youth culture—an adolescent stance rejecting all constraints—prevent both the "culture" and the "resistance" from really developing. No culture can exist without community norms based on responsibility to each other and some accepted ways to enforce those norms. And the "resistance" will never amount to more than a few smashed windows, the low-hanging tactical fruit for an adolescent strategy of emotional intensity.

Currently there are young people emboldened by a desperate fearlessness, ready to take up militance. I get notes from them all the time; each one both revives and drains my hope. Because, though they burn for action, they have no guidance and no support. This is the deep irony of history: the countercultures of the Romantics, the *Wandervogel*, the hippies—created by youth—have stranded our young.

⊠ ⊠ ⊠

While the alternative culture "celebrates political disengagement," what it attacks are conventions, morals, and boundaries. It comes down to a simple question: Are we after shock value or justice? Is the problem a constraining set of values or an oppressive set of material conditions?

Remember that one of the cardinal points of liberalism is that reality is made up of values and ideas, not relationships of power and oppression. So not only is shock value an adolescent goal, it's also a liberal one.

This program of attacking boundaries rather than injustice has had serious consequences on the left, and to the extent that this attack has won, on popular culture as a whole. When men decide to be outlaw rebels, from Bohemians to Hell's Angels, one primary "freedom" they appropriate is women. The Marquis de Sade, who tortured women, girls, and boys—some of whom he kidnapped, some of whom he bought—was declared "the freest spirit that has yet existed" by Guillaume Apollinaire, the founder of the surrealist movement.[63] Women's physical and sexual boundaries are seen as just one more middle-class convention that men have a right to overcome on their way to freedom. Nowhere is this more apparent—and appalling—than in the way so many on the left have embraced pornography.

The triumph of the pornographers is a victory of power over justice, cruelty over empathy, and profits over human rights. I could make that statement about Walmart or McDonalds and progressives would eagerly agree. We all understand that Walmart destroys local economies, a relentless impoverishing of communities across the US that is now almost complete. It also depends on near-slave conditions for workers in China to produce the mountains of cheap crap that Walmart sells. And ultimately the endless growth model of capitalism is destroying the world. Nobody on the left claims that the cheap crap that Walmart produces equals freedom. Nobody defends Walmart by saying that the workers, American or Chinese, want to work there. Leftists understand that people do what they have to for survival, that any job is better than no job, and that minimum wage and no benefits are cause for a revolution, not a defense of those very conditions. Likewise McDonalds. No one defends what McDonalds does to animals, to the earth, to workers, to human health and human community by pointing out that the people standing over the boiling grease consented to sweat all day or that hog farmers voluntarily signed contracts that barely return a living. The issue does not turn on consent, but on the social impacts of injustice and hierarchy, on how corporations are essentially weapons of mass destruction. Focusing on the moment of individual choice will get us nowhere.

The problem is the material conditions that make going blind in a silicon chip factory in Taiwan the best option for some people. Those people are living beings. Leftists lay claim to human rights as our bedrock and our north star: we know that that Taiwanese woman is not different from us in any way that matters, and if going blind for pennies and no bathroom breaks was our best option, we would be in grim circumstances.

And the woman enduring two penises shoved up her anus? This is not an exaggeration or "focusing on the worst," as feminists are often accused of doing. "Double-anal" is now standard fare in gonzo porn, the porn made possible by the Internet, the porn with no pretense of a plot, the porn that men overwhelmingly prefer. That woman, just like the woman assembling computers, is likely to suffer permanent physical damage. In fact, the average woman in gonzo porn can only last three months before her body gives out, so punishing are the required sex acts. Anyone with a conscience instead of a hard-on would know that just by looking. If you spend a few minutes looking at it—not masturbating to it, but actually looking at it—you may have to agree with Robert Jensen that pornography is "what the end of the world looks like."

> By that I don't mean that pornography is going to bring about the end of the world; I don't have apocalyptic delusions. Nor do I mean that of all the social problems we face, pornography is the most threatening. Instead, I want to suggest that if we have the courage to look honestly at contemporary pornography, we get a glimpse—in a very visceral, powerful fashion—of the consequences of the oppressive systems in which we live. Pornography is what the end will look like if we don't reverse the pathological course that we are on in this patriarchal, white-supremacist, predatory corporate-capitalist society. . . . Imagine a world in which empathy, compassion, and solidarity—the things that make decent human society possible—are finally and completely overwhelmed by a self-centered, emotionally detached pleasure-seeking. Imagine those values playing out in a society structured by multiple hierarchies in which a domination/subordination dynamic

shapes most relationships and interaction. . . . [E]very year my sense of despair deepens over the direction in which pornography and our pornographic culture is heading. That despair is rooted not in the reality that lots of people can be cruel, or that some number of them knowingly take pleasure in that cruelty. Humans have always had to deal with that aspect of our psychology. But what happens when people can no longer see the cruelty, when the pleasure in cruelty has been so normalized that it is rendered invisible to so many? And what happens when for some considerable part of the male population of our society, that cruelty becomes a routine part of sexuality, defining the most intimate parts of our lives?[64]

All leftists need to do is connect the dots, the same way we do in every other instance of oppression. The material conditions that men as a class create (the word is *patriarchy*) mean that in the US battering is the most commonly committed violent crime: that's men beating up women. Men rape one in three women and sexually abuse one in four girls before the age of fourteen. The number one perpetrator of childhood sexual abuse is called "Dad." Andrea Dworkin, one of the bravest women of all time, understood that this was systematic, not personal. She saw that rape, battering, incest, prostitution, and reproductive exploitation all worked together to create a "barricade of sexual terrorism"[65] inside which àll women are forced to live. Our job as feminists and members of a culture of resistance is not to learn to eroticize those acts; our task is to bring that wall down.

In fact, the right and left together make a cozy little world that entombs women in conditions of subservience and violence. Critiquing male supremacist sexuality will bring charges of being a censor and a right-wing antifun prude. But seen from the perspective of women, the right and the left create a seamless hegemony.

Gail Dines writes, "When I critique McDonalds, no one calls me anti-food."[66] People understand that what is being critiqued is a set of unjust social relations—with economic, political, and ideological components—that create more of the same. McDonalds does not produce generic food. It manufactures an industrial capitalist product for profit.

The pornographers are no different. The pornographers have built a $100 billion a year industry, selling not just sex as a commodity, which would be horrible enough for our collective humanity, but sexual cruelty.[67] This is the deep heart of patriarchy, the place where leftists fear to tread: male supremacy takes acts of oppression and turns them into sex. Could there be a more powerful reward than orgasm?

And since it feels so visceral, such practices are defended (in the rare instance that a feminist is able to demand a defense) as "natural." Even when wrapped in racism, many on the left refuse to see the oppression in pornography. *Little Latina Sluts* or *Pimp My Black Teen* provoke not outrage, but sexual pleasure for the men consuming such material. A sexuality based on eroticizing dehumanization, domination, and hierarchy will gravitate to other hierarchies, and find a wealth of material in racism. What it will never do is build an egalitarian world of care and respect, the world that the left claims to want.

On a global scale, the naked female body—too thin to bear live young and often too young as well—is for sale everywhere, as the defining image of the age, and as a brutal reality: women and girls are now the number one product for sale on the global black market. Indeed, there are entire countries balancing their budgets on the sale of women.[68] Is slavery a human rights abuse or a sexual thrill? Of what use is a social change movement that can't decide?

We need to stake our claim as the people who care about freedom, not the freedom to abuse, exploit, and dehumanize, but freedom from being demeaned and violated, and from a cultural celebration of that violation.

This is the moral bankruptcy of a culture built on violation and its underlying entitlement. It's a slight variation on the Romantics, substituting sexual desire for emotion as the unmediated, natural, and privileged state. The sexual version is a direct inheritance of the Bohemians, who reveled in public displays of "transgression, excess, sexual outrage." Much of this ethic can be traced back to the Marquis de Sade, torturer of women and children. Yet he has been claimed as inspiration and foundation by writers such as "Baudelaire, Flaubert, Swinburne, Lautréamont, Dostoevski, Cocteau, and Apollinaire" as well as Camus and Barthes.[69] Wrote Camus, "Two centuries ahead of time . . . Sade extolled totalitarian societies in the name of unbridled freedom."[70] Sade also presents an early formulation of Nietzsche's will to power. His ethic ultimately provides "the erotic roots of fascism."[71]

Once more, it is time to choose. The warnings are out there, and it's time to listen. College students have 40 percent less empathy than they did twenty years ago.[72] If the left wants to mount a true resistance, a resistance against the power that breaks hearts and bones, rivers and species, it will have to hear—and, finally, know—this one brave sentence from poet Adrienne Rich: "Without tenderness, we are in hell."[73]

❋ ❋ ❋

The alternative culture of the '60s offered a generalized revolt against structure, responsibility, and morals. Being a youth culture, and following out of the Bohemian and the Beatniks, this was predictable. But a rejection of all structure and responsibility ends ultimately in atomized individuals motivated only by self interests, which looks rather exactly like capitalism's fabled Economic Man. And a flat out refusal of the concept of morality is the province of sociopaths. This is not a plan with a future.

Take the pull of the alternative culture across the left. Now add the ugliness and the authoritarianism of the right's "family values." It's no surprise that the left has ceded all claim to morality. But it's also a mistake. We have values, too. War is a moral issue. Poverty is a moral issue. Two hundred species driven extinct every day is a moral issue. Underneath every instance of injustice is a violation of what we know is right. Unrestricted personal license in a context that abandons morals to celebrate outrage will not inspire a movement for justice, nor will it build a culture worth living in. It will grant the powerful more entitlements—for instance, the rich will get richer, and the poor will be conceptually nonexistent, except as a resource. "If it feels good, do it" isn't even the province of adolescence; it's the morality of a toddler. For the entitled individual, in whatever version—*Homo economicus*, *Homo bohemicus*, or *Homo sadeus*—pleasure is reduced to cheap thrills, while the deepest human joys—intimacy, belonging, participation from community to cosmos—are impossible. This is because those joys depend on a realization that we need other people and other beings, ultimately a whole web of existence, all of whom deserve our protection and respect. In return we get rewards, rewards that can accrue into profound satisfaction: from the contented joy of communal well-being to the animal ecstasy of sex to the grace of participation in the mystery.

Currently, the right places the blame for the destruction of both family and community at the feet of liberalism. The real culprit, of course, is capitalism, especially the corporate and mass media versions. But as long as the left refuses to fight for our values *as values*—and to enact those values in our lives and our movements—the right will be partially correct. They will also have recruitment potential that we're squandering: people know that civic life and basic social norms have degenerated.

It is a triumph for capitalism that the right is winning the US culture war by pinning this decay of family and community on the left. But the right is willing to take a moral stance, even though the man behind the curtain isn't Sodom or Gomorrah, it's corporate capitalism. Meanwhile the left might identify capitalism as the problem, but by and large refuses a moral stance.

The US is dominated by corporate rule. The Democrats and Republicans are really the two wings of the Capitalist Party. Neither is going to critique the masters. It is up to us, the people who hold human rights and our living planet dear above all things, to speak the truth. We need to rise above individualism and live in the knowledge that we are the only people who are going to defend what is good in human possibility against the destructive overlapping power-grab of capitalism, patriarchy, and industrialization.

◙ ◙ ◙

We can begin by picking up the pieces of community and civic life in the US. People of my parent's generation are correct to mourn the loss of the community trust and participation that they once experienced. And as Robert Putnam makes clear in his book on the subject, *Bowling Alone*, social trust is linked to both civic and political participation in ways that are mutually reinforcing—or mutually reducing. My mother and her friends have the addresses of their state and federal congresspeople memorized. Twenty years behind them, I at least know their names. And the current college-aged generation? They explain earnestly how the government works: "The President tells Congress what to do, and Congress tells the Supreme Court what to do." In two generations, there goes every advance since Magna Carta.

We're getting stupider, crueler, and more depressed by the minute. Oliver James calls the values of the corporate media "Affluenza," likening it to a virus that spreads across societies. He points out that anxiety, depression, and addiction rise in direct proportion to the inequity in a country. The values required to institutionalize inequality are values that are destructive to human happiness and human community. Injustice requires reducing people—including ourselves—to "manipulable commodities."[74] James writes, "Intimacy is destroyed if you regard another person as an object to be manipulated to serve your ends, whether at work or at play. . . . This leaves you feeling lonely and craving emotional contact, vulnerable to depression."[75]

How did this happen? When did people stop caring? One insight of Marxist cultural theorists like Antonio Gramsci is that in order for oppression to function smoothly, ideology must be transferred from the oppressors to the oppressed. They can't stand over us all with guns twenty-four hours a day. This transfer must be consensual and actively embraced to work on a society-wide scale. If the dominant class can make the ideology pleasurable, so much the better. Nothing could have done the job better than the passivity-inducing, addictive, and isolating technologies of first television and then the Internet.

Corporations have managed to coerce a huge percentage of the population into abandoning the values and behaviors that make people happy—to act against our own interests by instilling in us a new mythos and a set of compulsive behaviors. There is no question that television and other mass media are addictive, leading to "habituation, desensitization, satiation, and an increasing level of arousal . . . required to maintain satisfaction."[76] Clearly, there is an intense short-term pleasure capturing people, because the long-term losses are tremendous. Literally thousands of studies have documented television's damage to children; indeed, a coalition of professional groups, including the American Medical Association and the American

Academy of Pediatrics, put out a joint report in 2000 declaring media violence a serious public health issue to children, with effects that are "measurable and long-lasting."[77] The American Academy of Pediatrics reports, "Extensive research evidence indicates that media violence can contribute to aggressive behavior, desensitization to violence, nightmares, and fear of being harmed."[78] The most chilling studies link television to teen depression, eating disorders, and suicide. If the destruction of our young isn't enough to get us to fight back, what will be? As a culture, we are actively handing over the young to be socialized by corporate America in a set of values that are essentially amoral. The average child will spend 2,000 hours with her parents and 40,000 hours with the mass media. Why even bother to have children?

If culture is a set of stories we collectively tell, the stories have now been reduced to the sound bites of profit, offered up in a tantalizing, addictive flash that barricades access to our selves, if not our souls. Writes Maggie Jackson, "The way we live is eroding our capacity for deep, sustained, perceptive attention—the building blocks of intimacy, wisdom, and cultural progress."[79] For the young, those barricades may be permanent. Children need to experience bonding or they will end up with personality disorders, living as narcissists, borderlines, and sociopaths. They must learn basic values like compassion, generosity, and duty to become functioning members of society. They must have brains that can learn, contemplate, and question in order to have both a rich internal life and to have something to offer as participants in a democracy. For the developing child, bonding, values, and expectations create neurologic patterns that last a lifetime. Their absence leaves voids that can never be filled. The brain gets one opportunity to build itself, and only one.

The job of a parent is to socialize the young. Until recently, parents and children were nestled inside a larger social system with the same basic values taught at home. Now, parents are being told to "protect" their kids from the culture at large—a task that cannot be done. Society is where we all live, unless you want to move to Antarctica. Even if you managed to keep the worst excesses of consumerist, violent, and misogynist elements out of your child's immediate environment, the child still has to leave the house. If the culture is so toxic that we can't entrust our children to it, we need to change the culture.

The values taught by the mass media encourage the worst in human beings. If people are objects, neither intimacy nor community are possible. If image is all we are, we will always need to be on display. Social invisibility is a kind of death to social creatures. We buy more and more, whether higher-status cars or lower-cut jeans, so that we can have a better shot at being noticed as the object du jour. People surrounded by a culture of mass images experience themselves and the world as depersonalized, distant, and fractured. This is the psychological profile of PTSD. Add to that the sexual objectification and degradation of those images, and you have girls presenting with PTSD symptoms with no history of abuse.[80] *The culture itself has become the perpetrator.*

Yes, we can try to inoculate ourselves and our children against the mass media, both its messages and its processes. But why should anyone need to be protected from the culture in which they live? And what good are all your heartfelt conversations and empowering feminist fairy tales when your girl child is surrounded by people who are not fans of *Gaia Girls*, but *Girls Gone Wild*?

As Pat Murphy bravely writes,

> Suggesting that media is in general harmful and should be eliminated (or a dramatic reduction in the time spent imbibing it) at first seems absurd. But it is no more absurd than suggesting the age of oil and other fossil fuels is over. Media, energy and corporate control have evolved together. We need different concepts and new world views to transition away from fossil fuels and its infrastructure of corporations (including those of the media).[81]

Again, the right does not have a monopoly on values. We can reject authoritarianism, conformity, social hierarchy, anti-intellectualism, and religious fundamentalism. We can defend equality, justice, compassion, intellectual engagement, civic responsibility, and even love against the corporate jihad. We have to.

❈ ❈ ❈

Past movements for social justice insisted on character in their recruits, in honor, loyalty, and integrity. The culture of resistance created by the Spanish Anarchists valued ethical personal behavior. Writes Murray Bookchin, "They were working men and women, *obrera consciente*, who abjured smoking and drinking, avoided brothels and the bloody bull ring, purged their talk of 'foul' language, and by their probity, dignity, respect for knowledge, and militancy, tried to set a moral example for their entire class."[82] We could do worse. The right will continue to successfully blame the left for the destruction of culture and community as long as the left can't or won't stand firmly in defense of our values.

This is probably the right time to defend the concept of a work ethic. The alternative culture of the '60s was in part a reaction against the conformity of the '50s and its obedience to authority. In 1959, my mother and her friends decided to start an underground newspaper at their school. Their first step? Asking permission from the principal. He said no. They dropped the idea. No wonder the '60s happened.

The alternative culture was based on the premise that essentially nobody had to do anything they didn't feel like doing. A major part of their rebellion was the rejection of a work ethic, always cast as Protestant. But taken to its logical end, this is the position of a parasite. The dropouts either got money from their parents, from friends who got it from parents, or from the state. Eventually, each life has to be supported with resources from somewhere. I have seen a few too many protests and alternative communities surviving on the Mooch Ethic. I have sat on couches that housed rats, eaten off dishes that gave me gastroenteritis, and learned (secondhand, thankfully) that an itchy butt at sundown means pinworms. I've watched incredible resources go to waste—houses fall to ruin, land repossessed—for refusal to do basic adult tasks like paying the taxes. I don't know which is worse: the general ethos's entitlement, or the stupidity; the smell of the outhouses, the unwashed bodies, or the marijuana.

The rebellion against a work ethic is another characteristic of youth culture. The ventral striatal circuit, which is the seat of motivation in the human brain, doesn't function well during adolescence, which is why teens are often accused of being lazy. This means that the norms of youth culture will gravitate toward structureless days with no expectations or goals. It also means that the youth culture and marijuana aren't a good match.

The war on drugs is appalling. It has a corrosive effect on communities of color especially and has also made it difficult for those with legitimate need to get pain relief from drugs like marijuana.[83] Medical cannabis is a legitimate treatment for a number of conditions, some of which, like autoimmune disorders, are life-threatening. People who need it should be able to get it, and society as a whole would probably be better off if cannabis was legalized.

But drugs and alcohol have been a terrible detriment to both activist cultures and oppressed communities. I have watched people that I love erode with addiction, a slow death I'm powerless to stop. I am very sympathetic to the straight-edge punks. It was obvious to me at age fourteen that there were two weapons I would need for the fight: a mind that could think and the heart of a warrior. Drugs would destroy the one and numb the other. I swore away from drugs and I've never regretted that decision.

Drug and alcohol addiction has had terrible effects on both oppressed communities and cultures of resistance. Such effects are broad and deep: the self-absorption, lack of motivation, and broken synapses create a population in semipermanent "couch lock." Drugs and alcohol will not help us when we need commitment, hard work, and sacrifice, which are the foundation of all cultures of resistance. Addicts have no place on the front lines of resistance because an addict will always put their addiction first. Always.

I came of age in a post-Stonewall lesbian community that recognized the role that alcohol had played in destroying gay and lesbian lives. Our events specifically avoided bars as venues, and were often labeled "chem-free." These were and are acts of communal self-care that were linked to survival and resistance. It was an important ethic, and it was understood and embraced. There are parallel calls for a chem-free ethic in some Native American activist groups, and for the same reason: drugs and alcohol have been damaging enough to name them genocidal. The radical left would do well to model itself on these recent examples and to consider an ethic of sobriety as both collective self-care and resistance. We need everyone's brain. If our goal is a serious movement, then we also need focus, dependability, and commitment. On the front lines, we need to know our comrades are rock solid. In our culture, we need a set of ethics and behavioral norms that can build a functioning community. Basic awareness of addiction—its symptoms, its treatment options—is important both to help the afflicted and to keep our groups safe and strong.

A related issue is the general lassitude caused by poor nutrition exacerbated by vegetarian and vegan diets. One investigator of alternative communities writes, ". . . for many of the rural groups, common activity is limited to part-time farming. In their permissive climate, there is often a debilitating, low-thyroid do-nothingness that looks like nothing so much as the reverse image of the compulsive busyness of their parents."[84]

The diet that holds sway across the left will produce that state exactly. A food ethic stripped of protein and fat may meet ideological needs, but it will not meet the biological needs of the human template. Our neurotransmitters—the brain chemicals that make us happy and calm—are made from amino acids; amino acids are protein. Serotonin, for instance, is produced from the amino acid tryptophan. We cannot produce tryptophan; we can only eat it. Likewise endorphins and catecholamines. We must eat protein to have brains that work. We need fat, too, and you'll notice that in nature, protein and fat come packaged together. In order for your neurotransmitters to actually transmit, dietary fat is crucial. This is why people on low-fat diets are twice as likely to suffer from depression or die from suicide or violent death. If you need more reason to eat real food, your sex hormones are all made from dietary cholesterol: please eat some. A steady diet of carbohydrates, on the other hand, will produce depressed, anxious, irritable people too exhausted to do much beyond attend to the psychodramas created by their blood sugar swings, which about sums up the emotional ambiance of my youth. And the author's inclusion of "low-thyroid" in his description is right on the mark. Soy is often the only acceptable protein on the menu. Besides its poor quality—plant protein comes wrapped in cellulose, which humans cannot digest—soy is a known goitrogen. In large enough quantities, like when eaten not as a condiment but as a protein source, it can suppress and even destroy the thyroid.

I've been to a few too many potlucks with brown rice, dumpster-dived mangoes, and the ubiquitous chips and hummus. I feel my grandmother's horror from the grave: why are we feeding each other poverty food? This is the only time I feel sorry for men, watching them repeatedly—and I mean four and five times—approach my pot of (pasture-raised) beef-and-leek chili for more. They're desperate. They may be getting enough bulk calories every day, but they're starving. Men tend to crave protein because their protein needs are higher—testosterone means men have more muscle than women, and muscle is built from protein. Women tend to crave fat because our bodies are designed to store fat for pregnancy and lactation.[85] The current anorexic beauty standards, besides being a very effective tool of patriarchy and capitalism, also point to a profound death wish embedded in this culture. Humans have been celebrating female fat—a veneration both aesthetic and spiritual—since we created art and religion. Our first two art projects reverenced the lives that made ours possible: the large ruminants we ate and the large women who birthed us.

We must stop hating the animals that we are. Only ideological fanatics (I was the most extreme version—vegan—for almost twenty years, so I'm allowed to say that) will be able to stick to such body-punishing fare for any length of time. Everyone else will "cheat" and feel guilty over moral or even spiritual failings without understanding why they failed. The answer is simple: we have paleolithic bodies, we need paleolithic food. If you're fighting evolution, you are not going to win. There is a reason you feel hungry without fat and protein, a reason for the exhaustion that aches in your muscles and surrounds you like fog, a reason for the gray weight of depression. A plant-based diet is not adequate for long-term maintenance and repair of the human brain or body, and it has been taking a heavy toll on the left for several generations.

◘ ◘ ◘

The final difference between the alternative culture and a culture of resistance is the issue of spirituality. Remember that the Romantic Movement, arising as it did in opposition to industrialization, upheld Nature as an ideal and mourned a lost "state of nature" for humans.

Emotions were privileged as unmediated and authentic. Nonindustrialized peoples were cast as living in that pure state of nature. The *Wandervogel* idealized medieval peasants, developing a penchant for tunics, folk music, and castles. Writes Keith Melville,

> Predictably, this attraction to the peasantry never developed into a firm alliance. For all their vague notions of solidarity with the folk, the German youths did not remain for long among the peasants, nor did they take up political issues on their behalf. What the peasants provided was both an example and a symbol which sharpened the German Youth Movement's dissent against the mainstream society, against modernity, the industrialized city, and "progress."[86]

When the subculture was transplanted to the US, there were no peasants on which the new Nature Boys could model themselves. Peasant blouses and folkwear patterns found a role, but the real exploitation was saved for Native Americans and African Americans. Primitivism, an offshoot of Romanticism, constructs an image of indigenous people as timeless and ahistoric. As I discussed in the beginning of the chapter, this stance denies the indigenous their humanity by ignoring that they, too, make culture. Primitivism sees the indigenous as childlike, sexually unfettered, and at one with the natural world. The indigenous could be either naturally peaceful or uninhibited in their violence, depending on the proclivities of the white viewer. Hence, Jack Kerouac could write:

> At lilac evening I walked with every muscle aching among the lights of 27th and Welton in the Denver colored section, wishing I were a Negro, feeling that the best the white world had offered was not enough ecstasy for me, not enough life, joy, kicks, darkness, music, not enough music, not enough night.[87]

He'd rather be black? Really? Would he rather have a better chance of going to jail than going to college? Would he rather have only one thirty-eighth the wealth of whites? Would he really want to face fire hoses and lynching for daring to struggle for the right to vote? This is Romanticism at its most offensive, a complete erasure of the painful realities that an oppressed community must endure in favor of the projections of the entitled. And depressingly, it's all too common across the alternative culture.

The appropriation of Native American religious practices has become so widespread that in 1993 elders issued a statement, "The Declaration of War Against Exploiters of Lakota Spirituality." The Declaration was unanimously passed by 500 representatives from forty Lakota tribes and bands. The statement could not be clearer: white people helping themselves to Native American religious practices is destructive enough to be called genocide by the Lakotas. The elders have spoken loud and clear and, indeed, even reaffirmed their statement. We should have learned this in kindergarten: don't take what's not yours. Other people's cultures are not a shopping mall from which the privileged get to pick and choose.

Americans are living on stolen land. The land belongs to people who are still, right now, trying to survive an ongoing genocide. Those people are not relics of some far distant, mythic natural state before history. They live here, and they are very much under assault. Native Americans have the highest alcoholism rate, highest suicide rate, poorest housing, and lowest life expectancy in the United States. From every direction, they're being pulled apart.

Let's learn from the mistakes of the *Wandervogel*. Their interest in peasants had nothing to do with the actual conditions of peasants, nor with the solidarity and loyalty that the rural poor could have used; it had everything to do with their own privileged desires. Judging from my many years of experience with the current alternative culture, nothing has changed. The people who adopt the sacred symbols or religious forms of Native Americans—the pipe ceremony, *inipi*—do it to fulfill their own perceived needs, even over the Native Americans' clear protests. These Euro-Americans may sometimes go a step further and try to claim their actions are somehow antiracist, a stunning reversal of reality. It doesn't matter how much people feel drawn to their own version of Native American spirituality or how much a sweat lodge (in all probability led by a plastic shaman) means to them. No perceived need outweighs the wishes of the culture's owners. They have said no. Respect starts in hearing no—in fact, it cannot exist without it. Just because something moves you deeply, or even speaks to a painful absence in your life, does not give you permission. As with the *Wandervogel*, the current alternative culture's approach is never a call for solidarity and political work with Native Americans. Instead, it's always about what white people want and feel they have a right to take. They want to have a sweat lodge "experience." They don't want to do the hard, often boring, work of reparation and justice. If, in doing that work, the elders invite you to participate in their religion, that's their call.

Many people have longings for a spiritual practice and a spiritual community. There aren't any obvious, honorable answers for Euro-Americans. The majority of radicals are repulsed by the authoritarian, militaristic misogyny of the Abrahmic religions. The leftist edges of those religions are where the radicals often congregate, and that's one option; you don't have to check your brain at the door, and you usually get a functioning community. But for many of us, the framework is still too alienating, and feels frankly unreformable. These religions have had centuries to prove what kind of culture they can create, and the results don't inspire confidence.

Next up are the pagans and the Goddess people. Unlike the Abrahmists, they often offer a vision of the cosmos that's a better fit for radicals. Some of them believe in a pantheon of supernaturals, and show an almost alarming degree of interest in the minutia of the believers' lives. Other pagans believe in an animist life force: everything is alive, sentient, and sacred. But if the theology is a better fit, the practice is where these religions often fall apart. They may be based on ancient images, but the spiritual practices of paganism are new, created by urban people in a modern context. The rituals often feel awkward, and even embarrassing. We shouldn't give up on the project; ultimately, we need a new cosmic story and religious practices that will keep people linked to it. But new practices don't have the depth of tradition or the functioning communities that develop over time.

In order to understand where the pagans have gone astray, it may be helpful to discuss the function of a spiritual tradition. Three elements that seem central are a connection to the divine, communal bonding, and reinforcement of the culture's ethic. What forms of the sacred are sought by the subculture, and by what paths does it intend to reach them? Obviously a community broad enough to encompass everything from crystal healing to "Celtic Wicca" will have a multitude of specific answers. But taken as a whole, the spiritual impulse has been rerouted to the realm of the psychological—the exact opposite of a religious experience.

By whatever name you wish to call it, the sacred is a realm beyond human description, what William James rightly describes as "ineffable." The religious experience is one of "overcoming all the usual barriers between the individual and the Absolute . . . In mystic states we both become one with the Absolute and we become aware of our oneness."[88] He describes this experience as one of "enlargement, union, and emancipation."[89] James offers a startlingly accurate description of that ineffable experience. But spiritual enlargement, union, and emancipation do not emerge from a focus on our psychology. We experience them when we leave the prisons of our personal pains and joys by connecting to that mystery that animates everything. The arrow—the spiritual journey—leads out, not in. But like everything else that might lend our lives strength and meaning, spiritual life—and the communities it both needs and creates—has been destroyed by the dictates of capitalism. The single-pointed focus on ourselves as some kind of project is not just predictably narcissistic, but at odds with every religion worth the name. The whole point of a spiritual practice is to experience something beyond our own needs, pains, and desires.

Ten years ago, I attended a weekend workshop called "The Great Goddess Returns." I was already leery of these events back then, but there was one scholar I wanted to hear. The description, in so many words, offered what many people long to find: support, community, empowerment, relief from pain and isolation, and connection to ourselves, each other, the cosmos. These are valid longings and I don't mean to dismiss anyone's struggle with loneliness, alienation, or trauma. My criticism is directed instead at the standard form of the faux solutions into which neopaganism has fallen.

Drumming from a CD thumped softly through the darkened room.

A hundred people were told to shut their eyes and imagine a journey back through time to an ancient foremother in a cave. I wasn't actually sure what the point was, but I didn't want to cultivate a spiritual Attitude Problem so early in the day, so I visualized. We were then handed a small piece of clay. No talking was allowed to break the sacrosanct if technological drumming. We were told to make something with the clay. Okay. It being March, and I being a gardener, I formed a peapod. Time ticked on. The drumming was more baffling than meaningful. And how long could it take people to mold a brownie size bit of clay? I kept waiting, the drumming kept drumming. Finally we were told to crumble up what we had made. All right. I smooshed up my peapod, and went back to waiting and my internal struggle against the demons of attitude. Boredom is annoying. It's also really boring. I didn't want to look around—everyone was hunched over with a gravitas that left me bewildered—but I was starting to feel confused on top of bored. Had I missed the part where they said, "Destroy your sculpture one mote at a time"? Finally, the rapture descended: further instructions. "Make your sculpture again," came the hushed voice. What? Why? I hadn't particularly wanted the first peapod. Did I have to make another one? Meanwhile, the drums banged on and on, emphasizing my growing ennui, and again, heads all around me bent to the work of clay like it was Day Six in the Garden. I reformed my peapod, which took about sixty seconds, then waited another eternity. I was ready to have a Serious Talk with whoever invented the drum.

Then the lights were slowly raised, a dawn to this long night of the bored soul. We were quietly divided into groups of ten and given the following instructions: "Talk about what you just experienced."

Talk about . . . what? I made a pea pod. I crushed a pea pod. I remade a pea pod. For dramatic tension, I tried not to get bored.

Luckily, I was the seventh person in the circle, which gave me time to recognize the pattern and understand the rules. Because everyone else already understood. Being dwellers in the Land of Psychological Ritual, they knew too well what was expected. First up was a woman in her fifties. I don't remember what she made with her clay. I do remember what she said. Crumbling up her sculpture brought her back to the worst loss of life, the death of her infant daughter. She cried over her clay, and she cried again while telling us, a group of complete strangers.

The next one up said it was her divorce, that crumbling the clay was the end of marriage. She cried, too.

For the third, the destruction of her clay was the destruction of her child-self when her brother raped her when she was five. She trembled, but didn't cry.

The fourth woman's clay was her struggle with cancer.

I had to stop paying attention right about then because I had to figure out what I was going to say.[90] But I was also reaching overload. Not because of the pain in these stories—after years as an activist against male violence, I have the emotional skills to handle secondary trauma—but because the pain in their stories deserved respect that this workshop culture actively destroyed. This was a performance of pain, a cheapening of grief and loss that I found repulsive. How authentic to their experiences could these women have been when their response was almost Pavlovian, with tears instead of saliva? Smoosh clay, feel grief. Not knowing the expectation—not having trained myself to produce emotion on demand—I felt very little, beyond annoyance, during the exercise, and a mixture of unease, pity, and repugnance during the "sharing circle." I had no business hearing such stories. We were strangers. I did not ask for their vulnerability nor did I deserve it. To be told the worst griefs of their lives was a violation both of the dignity such pain deserves and of the natural bonds of human community. This was not a factual disclosure—"I lost my first child when she was an infant"—but a full monty of grief. And it was wrong.

A true intimacy with ourselves and with others will die beneath that exposure. Intimacy requires a slow, cumulative build of safety between people who agree to a relationship, an ongoing connection of care and concern. The performance of pain is essentially a form of bonding over trauma, and people can get addicted to their endorphins. But whatever else it is, it's not a spiritual practice. It's not even good psychotherapy, divorced as it is from reflection and guidance. If you're going to explore the shaping of your past and its impact on the present, that's what friends are for, and probably what licensed professionals are for.

This "ritual" was, once more, a product of the adolescent brain and the alterna-culture of the '60s, which imprinted itself unbroken across the self-help workshop culture it stimulated. No amount of background drumming will turn self-obsession and emotional intensity into an experience of what Rudolph Otto named "the numinous." It will not build a functioning community. "Instant community" is a contradictory as "fast food," and about as nourishing.

I have done grocery shopping after someone's surgery, picked up a 2:00 am call to help keep a friend's first, bottomless drink at bay, and taken friends into my home to die. I've also celebrated everything from weddings to Harry Potter releases. True community requires time, respect, and participation; it means, most simply, caring for the people to whom we are committed. A performative ethic is ultimately about self-narration and narcissism, which are the opposite of a communal ethic, and its scripted intensity is an emotional sugar rush. Why would anyone try to make this a religious practice?

I have way too many examples of this ethos to leave me with much hope. Some of the worst instances still make me cringe (white people got invited to an inipi and all I got was this lousy embarrassment?). I've been included in indigenous rituals and watched the white neopagans and other alterna-culturites behave abominably. Pretend you got invited to a Catholic Mass: would you start rolling on the floor screaming for your mother as the Catholics approached the rail for communion? And would you later defend this behavior as a self-evidently necessary "catharsis," "discharge," or "release of power"? When did pop psychology get elevated to a universal component of religious practice? Meanwhile, do I even need to say, the traditional people would never behave that way either at their own or anyone else's sacred ceremonies. And they'd rather die than do it naked. Their dignity, the long stretches of quiet, the humility before the mystery, all build toward an active receptivity to the spiritual realm and whatever dwells there. The performative endorphin rush is a grasping at empty intensity that will never lead out of the self and into the all. Nor will it strengthen interpersonal bonds or reinforce the community's ethics, unless those ethics are a self-indulgent and increasingly pornified hedonism, in which case it's doomed to failure anyway.

So we're stuck with some primary human needs and, as yet, no way to fill them. Many of us have traveled a continuum of spiritual communities and practices and found that none of them fit. My attempts to name cultural appropriation in the alternative culture have been largely met with hostility. For me, grief has given way to acceptance. The forces misdirecting attempts to "indigenize" Euro-Americans and other settlers/immigrants have been in motion since the *Wandervogel*. I will not be able to find or create an authentic and honorable spiritual practice or community in my lifetime. All I can do is lay out the problems as I see them and perhaps some guidelines and hope that, over time, something better emerges. It will take generations, but it's not a project we can abandon.

Humans are hard-wired for spiritual ecstasy. We are hungry animals who need to be taught how to participate, respectfully and humbly, in the cycles of death and rebirth on which our lives depend. We're social creatures who need behavioral norms to form and guide us if our cultures are to be decent places to live. We're suffering individuals, faced with the human condition of loss and mortality, who will look for solace and grace. We also look for beauty. Soaring music produces an endorphin release in most people. And you don't even need to believe in anything beyond the physical plane to agree with most of the above.

Some white people say they want to "reindigenize," that they want a spiritual connection to the land where they live. That requires building a relationship to that place. That place is actually millions of creatures, the vast majority too small for us to see, all working together to create more life. Some of them create oxygen; many more create soil; some create habitat, like beavers making wetlands. To indigenize means offering friendship to all of them. That means getting to know them, their histories, their needs, their joys and sorrows. It means respecting their boundaries and committing to their care. It means learning to listen, which requires turning off the chatter and static of the self. Maybe then they will speak to you or even offer you help. All of them are under assault right now: every biome, each living community is being pulled to pieces, 200 species at a time. It's a thirty-year mystery to me how the neopagans can claim to worship the earth and, with few exceptions, be indifferent to fighting for it. There's a vague liberalism but no clarion call to action. That needs to change if this fledgling religion wants to make any reasonable claim to a moral framework that sacrilizes the earth. If the sacred doesn't deserve defense, then whatever will?

We once again have choices to make, as individuals and as a movement. If our task is to create a culture of resistance, then every element of it must support our political resistance and continually reinscribe our values into both our personal and communal behavior. A spirituality of resistance could be an important element. Practical techniques to connect people to the other beings with whom we share this land, to build back those relationships, could lend both strength and commitment to the fight. That spirituality could also, hopefully, guide us as we construct a way of life based on the values we hold dear, values like justice, compassion, and equity. It could reassert our place as humble participants in our human communities, our living communities, and in the cosmos. It could direct us in everything from socializing the young to our daily interactions to our material culture. That is one of the functions of religion, to frame the moral code of a given society inside a mythos that stretches from the individual to the cosmos. A moral code may inscribe obedience to authority throughout society or it may call us to fight injustice; we can find examples of both even in the same religious tradition. Religious condemnation of usury, for instance, kept the worst excesses of capitalism at bay for centuries in Europe, and still impedes capitalism in the Muslim world. Denying the working man his wages was a sin that cried out to heaven for justice. Of course, so was the sin of Sodom.

Neopaganism is still trying to find its moral bearings, along with its community norms and its religious practices. It may be a group too diverse to call a community. That emphasis on plurality gives it an intellectual and emotional flexibility that monotheists, especially the "No Gods Before Me" exclusivist Abrahmists, will always lack. This could emphasize the quality of personal conscience instead of personal entitlement, given the right context. Spending time on an Internet search led me to some interesting discussions. June was apparently declared Pagan Morals Month. I am not the only one who feels alienated by the bad behavior and community norms of much of the pagan movement. Families with children are at a loss; women are sick of men threatening and violating them at public rituals; and people have found that drinking and drumming around a fire produces a trance state that's ultimately pointless. There is a deep, unmet set of needs here that will not be answered any time soon.

The legacy of the *Wandervogel* is at its most hollow when applied to the deepest longings of which we're capable, that of the religious impulse. Stealing other people's symbols and ceremonies is selfish and immoral; adolescent self-absorption and drama cannot produce a resilient and lasting community; and entrainment in emotional intensity and display will never prepare us for the great and tender communion that brings us home. It is my hope that this discussion can help build something both more useful and more graceful. We need that new religion to help set the world right, and to nestle each human life in an unbroken circle of individual conscience and longing, communal bonding, connection to the multitude of members of this tribe called carbon, and finally our safe place in the mystery.

ⁿ ⁿ ⁿ

Resistance is a simple concept: power, unjust and immoral, is confronted and dismantled. The powerful are denied their right to hurt the less powerful. Domination is replaced by equity in a shift or substitution of institutions. That shift eventually forms new human relationships, both personally and across society.

Most of the population is never going to join an actual resistance. We're social creatures; by definition, it's hard to stand against the herd. Add to that how successful systems of oppression are at disabling the human capacity for resistance. As Andrea Dworkin said, "Feminism requires precisely what misogyny destroys in women: unimpeachable bravery in confronting male power."[91] The pool of potential resisters is going to be small. Conformity brings rewards and privileges; fighting back brings punishment and alienation. Most people are not psychologically suited to the requirements of resistance. The sooner we accept that, the better.

Personally, we can stop wasting time on conversations that will never produce anything but frustration. Politically, we can make better strategic decisions based on a more realistic assessment of our potential recruits. We all need to make our choices about personal risk. And there's a role for everyone. There are people who agree with the goals of a cause but for a variety of legitimate reasons can't undertake front-line or underground actions. Therefore, most recruits, by circumstance and by character, will be part of a culture of resistance.

Resistance movements require two things: loyalty and material support. Acquiring them are the two main tasks of the culture of resistance, although there may be others depending on the scope of the resistance at hand. Those others would include building alternative institutions for egalitarian, participatory governance; installing systems of justice for settling disputes; creating economic networks that can provide for basic survival needs apart from the injurious system; and socialization processes for both children and adults to reclaim and defend an indigenous culture under assault or create a new culture for those escaping the dominant culture. In real life, all these projects may not always be distinct, but instead form a reinforcing series of activities.

What ties them all together is an underlying set of values that include a self-conscious embrace of political resistance. This means first and foremost understanding what political resistance is and what it isn't. Without that understanding, all we will have is the same withdrawalist alternative culture, which will be content to coexist alongside injustice in all its horrors, no matter how repelled we are by those horrors. I don't know if it's a failure of courage or, as Adrienne Rich said, "the failure to want our freedom passionately enough,"[92] but it's a failure that haunts too many radical movements. Gene Sharp is worth quoting at length on this point (the people I call "withdrawalists" he calls "utopians").

Utopians are often especially sensitive to the evils of the world and, craving certainty, purity, and completeness, firmly reject the evils as totally as possible, wishing to avoid any compromises with them. Instead, utopians assert an alternative vision of the world which they would like to come into being. Their visionary belief may be labeled "religious" or "political"—it matters little for this discussion. They await a "new world" which is to come into being by an act of God, a change in the human spirit, by autonomous changes in economic conditions, or by a deep spontaneous social upheaval—all beyond deliberate human control. These believers are primarily concerned with espousing the "true" understanding of the evil and the principles by which people should live, gaining converts, living with the least possible compromise until the great change arrives. They may deliberately seek to establish ways of living and communities which exemplify their principles and which may inspire others to do likewise.[93]

The most serious weakness of this response to the problem of this world is not the broad vision, or the commitment of the people who believe in it. The weakness is that these believers have no effective way to reach the society of their dreams. Condemnation of social evil, espousal of an alternative order of life, a deep personal commitment, and an effort to live according to it, are all good and necessary, but unfortunately alone they do not transform human society and institutions. To do that, an instrumentally effective program of achievable steps for dealing with the evils of existing society and for creating an improved social order is required.[94]

One historian calls the Spanish Anarchist movement—a movement more serious than anything in the contemporary left by an order of magnitude—"secular millennialism."[95] Those two words could stand in for my entire youth. Among true radicals, there is a tremendous strain between a strategy of withdrawalist purity and a strategy based on an emotional need to act (which is essentially a compilation of undirected tactics). Neither of these are actually strategic; they're both usually stances based on a desperation and, with apologies to Rich, a collective failure to *plan* for our freedom passionately enough. Both tendencies can fall back on liberalism: we're going to change the world by "personal example." That example may be our permaculture garden and its attendant bike commute or it may be our brave, but useless, attacks on property at the bottom of the oppressor's food chain. We have these vague notions that these actions will inspire others, and then even more vaguely accumulate into a societal transformation or kickoff a spontaneous insurrection. But it will happen because it must: because either the "Great Turning" type narrative of progress says it must or the fires of our righteous rage will make it be so. But millennialism is a poor substitute for a real resistance movement. And given that victory is not, in fact, inevitable, we would be well advised to understand the basic principle of resistance: dislodging injustice requires organized, political resistance. Power will only change when it is forced to. Whether that force is applied violently or nonviolently is a discussion that comes later. Too many leftists refuse to face the nature of power, the nature of systematic oppression, and the nature of the social psychology we are up against.

For those of us who can't be active on the front lines—and this will be most of us—our job is to create a culture that will encourage and promote political resistance. The main tasks will be loyalty and material support.

Loyalty is sorely lacking across the left. First, and worst, is the out and out betrayal. Most of the Green Scare victims were turned in by former friends, in one case by an ex-husband. In any serious movement, snitches would be treated seriously. This is because snitching means that your people—your comrades, your friends—will be arrested, tortured, and killed. As a Deep Green Resistance movement becomes more serious, and hence the consequences in state repression become more serious, our collective response to snitching will be forced to keep pace. This will not be a fun moment; it is in fact likely to be permanently traumatizing. Our best hope is to instill the value of loyalty in our culture of resistance now, to stop snitching before it begins. Christabel Pankhurst wrote of the culture of the militant suffragettes:

> The spirit of the movement was wonderful. It was joyous and grave at the same time. Self seemed to be laid down as the women joined us. Loyalty, that greatest of virtues, was the keynote of the movement—first to the cause, then to those who were leading, and member to member. Courage came next, not simply physical courage, though so much of that was present, but still more the moral courage to endure ridicule and mis-

understandings and harsh criticism and ostracism. There was a touch of the "impersonal" in the movement that made for its strength and dignity. Humour characterised it, too, in that our militant women were like the British soldier who knows how to joke and smile amid his fighting and trials.[96]

Everything we need for a culture of resistance is in her description, and indeed played out across the suffrage movement. Over a thousand women endured "solitary confinement, hard labor, brutality, broken health and ultimately death."[97] Even more women committed acts of physical courage that didn't result in arrest, ranging from confrontations with police to stealth property destruction. And then there were the foot soldiers engaged in constant, daily tasks like fund raising, educating, public speaking, printing newspapers, door-to-door lobbying, organizing rallies, and prisoner support. All of these women supported their militant comrades. There were also between 500 and 600 non-militant women's suffrage societies across Britain, and it's interesting to note that the increase in militance by the WSPU resulted in a reinvigoration of those groups as well. Writes historian Midge MacKenzie, "The controversial tactics of the WSPU and their widespread news coverage revitalised the question of votes for women and the non-militant suffrage societies became stronger and more powerful."[98]

The first priority of their movement was *loyalty*, both to their cause and to those who were leading. Therein lies one of the major problems with modern radical groups. We tend to destroy our leaders with criticism, often personal and vicious. The antihierarchical stance of radicals leads to an adolescent reaction against anyone who rises to a public position. Writer after writer gets accused of "selling out," although not a single one can even make a living—let alone a killing—as a writer. This charge is also leveled at dedicated people who run small presses, bookstores, and, indeed, anyone with the temerity to actually get something done. It's a combination of petty jealousy and "rooster battling." Though the same attack-the-leader default is occasionally present in women's groups, the demands of masculinity make this way more of an issue for men. We must call it what it is when we see it happening. If the offenders refuse to stop, they should be shunned until their behavior improves. Attacking our leaders is painful and destructive to both individuals and movements. The younger members can't be expected to be able to identify and take a stand against this behavior; they don't have the life experience, and they're naturally inclined to be "combatants" at that stage of life. It is up to the middle-aged and older members to set the tone and behavioral expectations, to guide the community norms. People decades too old for this particular behavior publicly engage in it with glee. It's frustrating and heartbreaking, because no individual except the most resilient can survive those kinds of sustained personal attacks. And no one can hurt you like your own.

Let's look again to the WSPU and the militant suffrage movement. Their use of both civil disobedience and property destruction landed many women in jail. As any political prisoner might, they went on serious, sustained hunger strikes. The government was forced to release the strikers for fear of their deaths. The militants, of course, went right back to political action once they recovered. The government found itself in a quandary and decided to force-feed the next prisoners who went on hunger strikes. Make no mistake: this is a form of torture. It backfired; the courage of the prisoners against government torture made a very stark and easy choice for the general public in deciding who to sympathize with.

Some members of parliament (MPs) argued that prisoners should be left to die of starvation and that would be the end of it. But the Home Secretary had the pulse of the movement: "It has been said that not many women would die, but I think you would find that thirty, forty, or fifty would come up, one after another."[99] First up would have been Emmeline and her daughter Christabel Pankhurst, the beloved leaders of the movement. The government understood something that contemporary radicals tend to reject: movements without leaders are not movements, but random individuals incapable of waging a sustained campaign for justice. Emmeline Pankhurst had been to jail, gone on hunger strike, and been released too many times. The government wanted her broken or gone, without creating a dead martyr. The result was the Prisoners (Temporary Discharge for Ill Health) Act in 1913, quickly dubbed the Cat and Mouse Act. A hunger striker would be tortured until half-dead, then released to recover. The prisoner was given a ticket of leave. The act stated that once the prisoner was deemed fit, she (the act used the female pronoun) would be required to present herself for reimprisonment. Wrote Emmeline Pankhurst,

> Of course the act was, from its inception, treated by the Suffragettes with the utmost contempt. We had not the slightest intention of assisting Mr. McKenna in enforcing unjust sentences against soldiers in the army of freedom, and when the prison doors closed behind me I adopted the hunger-strike exactly as though I expected it to prove, as formerly, a means of gaining my liberty.[100]

Pankhurst was held for ten days, during which she went on a hunger strike. She writes of the day when, exhausted and half-conscious, she was about to be released under the Cat and Mouse Act:

> The Governor came to my cell and read me my license, which commanded me to return to Holloway in fifteen days, and meanwhile to observe all the obsequious terms as to informing the police of my movements. With what strength my hands remained I tore the document in strips and dropped it on the floor of the cell. "I have no intention," I said, "of obeying this infamous law. You release me knowing perfectly well that I shall never voluntarily return to any of your prisons."[101]

Other suffragettes showed humor as well as fortitude in the face of the Cat and Mouse Act. Annie Kenney, another of the movement's leaders, was released after three days of a hunger strike with a prison license to reappear. She escaped from her home, where she was recovering while being watched, under cover of night, and attended a suffrage meeting unannounced, where she auctioned her prison license to the highest bidder. She was immediately arrested.

Many of the hunger strikers released under the Cat and Mouse Act were protected in hiding by a network of loyal supporters who refused to give them up. And as a wonderful example of material support, one woman, Mrs. Brackenbury, gave the WSPU use of a large house in

London where hunger strikers could recover as long as needed with appropriate nursing care. The house was affectionately named "Mouse Castle."

The leadership was forced into another kind of cat and mouse game with the police when they were on medical leave. They were determined to do their job as leaders and equally determined to evade capture with its "slow judicial murder."[102] At public appearances, women bodily defended Emmeline Pankhurst against the police to try to keep her safe. She was rearrested on July 21, 1913, but released within three days as her health began to fail on a food and water fast. She was carried to the next public suffrage meeting, too ill to speak, but determined to appear. The auction of her prison license brought in £100, which she had promised the governor would be spent on "militant purposes."[103] She fled to the US to recuperate, and was arrested off the coast of Plymouth on her return, but not for lack of defenders. A pair of women in a powerboat attempted, unsuccessfully, to rescue her against *two warships*. On being taken into custody, she immediately began a hunger strike. All told, she was arrested six times under the Cat and Mouse Act, at the age of fifty-five.

An official bodyguard, trained in jujitsu, was organized specifically to protect Emmeline Pankhurst from the police. The police were so nervous about confrontations that in one instance a platform at Victoria Station was commandeered and the whole station surrounded by "battalions of police" to take Mrs. Pankhurst into custody. All this to keep Emmeline's defenders from protecting her against rearrest.

Her daughter, Sylvia Pankhurst, also faced potential rearrest and further torture, yet continued to speak in public. At one event, a "Women's May Day" in Victoria Park, twenty women chained themselves in a tight group around Sylvia, in an attempt to thwart the police. They were not successful—the police isolated the whole group and broke the locks with their truncheons—but even in the blurry photographs of the "chained guard," their expressions of stalwart loyalty shine through. An entire street in London's working-class East End was offered money by the police so that they could arrest Emmeline Pankhurst on her way to a rally; not a single household would take the money.

Real movements require leaders. Despite all the contempt that contemporary radicals heap on anyone who rises to a public position, leaders emerge. A collection of individuals, no matter how angry or inspired, will remain inchoate without language and ineffective without direction. Movements are easily destroyed by imprisoning or killing the leaders; that's why governments do it. The Spanish Anarchists never recovered from the assassination of their beloved Buenaventura Durruti, just as Dr. Martin Luther King Jr.'s death left a hole in the heart of the civil rights movement. Successful movements are always training new leaders because they recognize their critical functions. The British government hoped to break the suffrage movement on numerous occasions by arresting the leadership, but the women had a chain of command in place, from Emmeline and Christabel Pankhurst to Annie Kenney to Grace Roe and on down, with each woman preparing her replacement. We can reject the concept of leadership all we want, but that will not eradicate its necessity.

Of course, small-scale and aboveground groups should be democratic whenever possible, but that does not change the fact that leaders must emerge nor does it change the fact that underground groups engaged in coordinated or paramilitary activities require hierarchy. Combatants, especially, need leadership. Emmeline Pethick-Lawrence explained the nondemocratic structure of the WSPU: "The very fact that militant action involved individual sacrifice imposed heavy responsibilities upon the leaders of the campaign. Individuals who were ready to make the sacrifice that militancy entailed had to be sustained by the assurance of complete unity within the ranks."[104]

If we accept the reality of leadership, we can trade protection for expectation. Loyalty works both ways. Clarity of ideas, explication of goals, and personal courage can elevate an organizer, a teacher, a writer, or a minister to a leadership position. In exchange, those agreeing to be led have a right to expect sterling personal ethics, self-sacrifice, and the leaders' prioritizing of the movement. Charisma and status can be used in very ugly ways, and individuals who use power for personal gain or sexual exploits should, of course, be rejected from a leadership position. But a wholesale rejection of leadership means a movement will be stuck at a level of ineffective small groups. It may feel radical but it will change nothing.

Loyalty to each other, especially to frontline actionists who are taking serious risks, is just as important as loyalty to leaders. That loyalty requires those of us who work aboveground to declare our support for direct action at every opportunity. We need to use words like "resistance" and phrases like "culture of resistance"; we need to reject personal consumer choices as a solution and explain why to anyone who listens; and we need to defend whatever degree of militance we're comfortable with plus one.

Loyalty also implies material support. Time, money, and other resources are always needed by actionists. Think of what it took for the Underground Railroad to operate. Yes, it took leaders like Harriet Tubman to be militant, brave, and committed despite the serious consequences; we remember her name, and well we should. But it also took Quakers repeatedly opening their homes to fugitives and themselves to risk; it took people willing to sing a song about trains as they made their way through the slave quarters. It took communities of free blacks and white sympathizers who were willing to support escaped enslaved people with food, housing, and employment. Those people had to have absolute loyalty to the conductors and to the enslaved people risking their lives for freedom. It took people willing to say in front of family, friends, and congregations that the Underground Railroad was a fine and worthy project. We now accept the necessity and morality of the Underground Railroad, but at the time it was seen as a radical fringe project by many abolitionists. It was successful in part because activists published newspapers, arranged lectures, petitioned Congress, and even moved to Kansas to fight for the broad structural change that surrounded and explicated the Underground Railroad. All these people worked in concert, despite profound differences in ethics and strategies, taking what risks they could.

Quakers developed an ethic and a practice of community support and care very quickly: persecution has a tendency to encourage that. In Bristol, England, so many Quakers were arrested—3,000—that prisoners died from suffocation in jail. The children were largely left in charge of the community, and they heroically carried on with Meetings for Worship. Some of them were tortured in turn—flogged and put in stocks. The culture of resistance carried into the community at large;

some juries in Bristol refused to try Quakers. Other juries found them innocent and were punished by the judge, leading to the legal principle of a jury's immunity from prosecution, a principle we take for granted now, but one that had to be won. And when three Quakers were sentenced to transport to America and the West Indies, Bristolian sailors refused to carry them.

Loyalty and material support are also evident in a story about the militant antiwar group the Weather Underground. After describing a close call in which a WU cell in the San Francisco Bay area was almost apprehended, one member remembers:

> The narrow escape was due less to underground skills or harrowing Hollywood tricks than to the political context and the support the Weather Underground had. "While quick wits and fast maneuvers provide the most dramatic story," Gilbert says, "the basic reason for our escape was the anti-state political consciousness that prevailed in youth culture, which meant that information did not flow to the state but flowed to us," he says. Even people who weren't directly aware of or in contact with the Weather Underground tended to resist police questioning and spread the word when the FBI was around. That support, Gilbert says, is what thwarted the FBI.[105]

Communities that are used to taking care of each other have a much easier time mobilizing those existing networks into a culture of resistance. Such established networks could be called a *culture of survival*. For instance, the men who worked as Pullman porters often found themselves stranded in southern towns where walking on the wrong street could get them killed. There were black women along the rail lines who would open their homes to porters. Even if all they had for a bed was a mat in the pantry, at least it was safe and there'd be a meal to go with it. Usually these women had a husband or son who was also a porter, and they understood firsthand the literal terrors in being a black man in an unknown segregated town. These women later became the heart, soul, and foot soldiers of the civil rights movement.

The civil rights movement did not arise spontaneously. Explains historian Tye:

> In the heyday of its long battle for a union, the Brotherhood ran a series of labor conferences for the wider community, with as many as three thousand people attending. Each conference carried clear messages: blacks deserve full civil and economic rights, they need to be self-sufficient and free of white benefactors, and it is time to issue demands rather than make requests. The Brotherhood did not just precede the civil rights movement in Chicago; it planted the seeds for it. Rather than a conventional labor union, it had become a school for protest politics.[106]

It ran similar schools from Canada to Florida, creating a movement from building blocks of education, political consciousness, savvy strategy, and the existing culture of survival among blacks.

The Spanish Anarchists provide another example of a culture of survival that can be mobilized into a resistance. Murray Bookchin writes,

> It is essential to emphasize that Spanish anarchism was not merely a program embedded in a dense theoretical matrix. It was a way of life: partly, the life of the Spanish people as it was lived in the closely-knit villages of the countryside and the intense neighborhood life of the working class barrios; partly, too, the theoretical articulation of that life as projected by Bakunin's concepts of decentralization, mutual aid, and popular organs of self-management. . . . Spain had a long tradition of agrarian collectivism. . . . Spanish anarchism . . . sought out the precapitalist collectivist traditions of the village, nourished what was living and vital in them, evoked their revolutionary potentialities as liberatory modes of mutual aid and self-management, and deployed them to vitiate the obedience, hierarchical mentality, and authoritarian outlook fostered by the factory system. . . . The Spanish anarchists tried to use the pre-capitalist traditions of the peasantry and working class

against the assimilation of the workers' outlook to an authoritarian industrial rationality. . . . Their efforts were favored by the continuous fertilization of the Spanish proletariat by rural workers who renewed these traditions daily as they migrated to the cities. . . . Along the Mediterranean coastal cities of Spain, maybe workers retained a living memory of a non-capitalist culture—one in which each moment of life was not strictly regulated by the punch clock, the factory whistle, the foreman, the machine, the highly regulated workday, and the atomized world of the large city.[107]

The Spanish Anarchists were renowned for their loyalty. Workers' strikes in support of comrades in prisons were larger than strikes to demand better working conditions.

The radical environmental movement is largely white and well-assimilated into the noncommunity of the corporate-controlled, mass-media dominated, industrially produced culture of the contemporary United States and its colonies. Community has been destroyed to the point where we don't know the names of the people living twenty feet from us and communication has been reduced to keystrokes of consonants. Those of us from that world are not even starting from scratch; we're starting from negative. Hopefully, we can learn by example from comrades who come from more intact communities, from elders who remember a way of life organized around human needs instead of corporate profits, and from history. Necessity will have to reinvent us. Or, as Monique Wittig famously wrote, "Remember. Make an effort to remember. Or, failing that, invent."[108]

Perhaps we can take heart in the fact that resistance always has to be created. Every movement is faced with the task of nurturing the will to fight in the people at large and in potential recruits especially. People need a mythic matrix that includes a narrative of courage in the face of power, loyalty to comrades and cause, and the eventual triumph of good over evil. They need the emotional support of a functioning community that believes in resistance. And they need an intellectual atmosphere that encourages analysis, discussion, and the development of political consciousness.

One example from history is the life and accomplishments of Maud Gonne. Born in 1865 to an Anglo-Irish family, she took up the Irish struggle for independence early in life. Her first activism was with the Irish National Land League, a group that agitated on behalf of tenant farmers. Such farmers didn't own the land they worked, and in years of bad harvest, would be evicted for nonpayment of rent. Land ownership had been consolidated into fewer and fewer hands, causing widespread poverty and suffering—and, finally, resistance. Organized rent strikes led to what is known as the Land War.

The campaign to reduce rents and allow tenants to buy their land was ultimately successful, and it was won almost entirely by using nonviolent tactics. By 1914, the holdings of large landowners had been redistributed to small farmers. Many of these activists went on to fight for Irish independence. This was the context in which Gonne learned political activism. She was extremely active in Irish cultural activities. The period of the late nineteenth century was a cultural renaissance called the Gaelic Revival. Organizations like the Gaelic League and the Gaelic Athletic Association (GAA) sprang up to encourage Irish sports, literature, and, especially, the Gaelic language. Created in 1892, the National Literary Society was founded by Douglas Hyde and William Butler Yeats for the purpose of "de-Anglicizing the Irish people." The cultural activities, always contextualized within a framework of occupation, worked their magic, helping to lay the groundwork both emotionally and politically for resistance. In the first decade of the twentieth century, members of the Irish Republican Brotherhood rose to prominence in the Gaelic League and the GAA. All these forces resulted in the creation of Sinn Fein. Writes one historian, "The Nationalist movement of the early twentieth century was born out of the Gaelic revival of the late nineteenth century."[109]

Gonne played a prominent role in the Revival. She created Inghidhe na hÉireann (Daughters of Ireland) for women and girls to pursue Irish language, drama, and literature. She was also an active member of the Celtic Literary Society and the National Players Movement. She founded *L'Irlande Libre*, a journal dedicated to the Irish struggle. She also found time to illustrate books of Celtic folklore.

But the cultural work was not an end in itself for Gonne or for the movement as a whole. Her husband, John MacBride, took part in the Easter Rising and was executed for it. Gonne did time in Holloway Prison and after her release, she worked tirelessly for political prisoners. Some of these men had been in jail ten years without a single visit from anyone. Gonne was arrested again for smuggling supplies into Mountjoy Prison. She went on a hunger strike for thirty-one days, which nearly killed her, but she and the other strikers won some basic rights for prisoners. When the Republic Courts of Justice were organized to supersede the British courts, Gonne was elected and served as a judge. She also helped with the Irish White Cross, providing material relief to families in need after the War of Independence. She was nicknamed the Irish Joan of Arc, and it's not hard to see why.

The Irish struggle didn't set its culture and its resistance against each other. Instead, the Irish understood that each was necessary for the other. Gonne's life stands as an example of the entire continuum of cultural work to serious direct action.

Gonne also produced a son, Seán MacBride, whose CV is at least as impressive as hers. At the age of fifteen, he signed on with the Irish Volunteers, and fought in the Irish War of Independence. He was against the Anglo-Irish Treaty, and was arrested a number of times by the Irish Free State. He was personal secretary to Eamon de Valera, one of the leaders of the Easter Rising, and served as both director of intelligence of the Irish Republican Army (IRA) and chief of staff. As a lawyer, he defended many IRA political prisoners throughout his career. In 1948, he was appointed to a cabinet position in the government as Minister of External Affairs. He played a decisive role in the repeal of the External Relations Act and the Declaration of the Republic, in which Ireland declared its independence from the British Commonwealth. During his tenure as minister, the European Convention on Human Rights was drafted, and he was a driving force behind its signing. He was also the main reason that Ireland refrained from joining NATO. He was a cofounder of Amnesty International; he drafted the first constitution of Ghana and the constitution of the Organization of African Unity; served as the UN's High Commissioner for Refugees as well as the High Commissioner for Human Rights; and was appointed chairman of UNESCO. He lobbied the International

Court of Justice to declare nuclear weapons illegal. He won the Nobel Peace Prize. The judges said he had "mobilized the conscience of the world in the fight against injustice." He never stopped. If most of us achieved even one of these things, we could probably die content. By all accounts he remained humble to the end.

This is a culture of resistance that worked. It created profound material changes in the organization of power by producing activists of courage and stamina across generations. We can look at political struggles through history and find similar patterns of activity. The actionists that we remember were formed by their context, by a culture of resistance. That culture forms actionists' characters around a core set of values like courage, loyalty, and a commitment to justice. It gives them the intellectual tools needed for political consciousness. It weaves them into a social network of comradeship and belonging. And it encourages them in their acts of resistance by providing money, supplies, lawyers, and prisoner support. As the struggle takes shape, the people in the aboveground take on the tasks of building alternative institutions, ranging from schools to militias, institutions that will be needed when the oppressive system is brought down.

The environmental movement has made a choice, a choice we're asking each reader to reevaluate against industrial culture's relentless assault on our planet. The collective decision to date has been to reject the possibility of a serious resistance movement. That conclusion has been fostered by many cultural forces, some of which, as we have seen, go back centuries. Religious movements, from both East and West, have long declared the world a place of suffering and corruption, with withdrawal and personal salvation as the corrective activities. Classical liberalism, with its individualism and idealism, has also been a continuous drain of confusion and obstruction. The contemporary alternative culture, with its roots in the *Lebensreform, Wandervogel,* and Bohemian movements, has for over a hundred years been pulled between poles of confronting power and breaking boundaries, of fighting for justice on the one hand and displays of adolescent intensity on the other.

This is the moment when we have to decide: does a world exist outside ourselves and is that world worth fighting for? Another 200 species went extinct today. They were my kin. They were yours, too. If we know them as such, why aren't we fighting to save them with everything we've got?

◙ ◙ ◙

You will find an answer to that question amongst some very earnest people, people who know that industrial civilization is killing the planet and who may hold deep wells of grief and despair in their hearts. They will try to convince you that political resistance is neither possible nor advisable. They have coalesced around the ideas of permaculture, simple living, and Transition Towns. What follows are the main arguments that the authors of this book have heard repeatedly from the Permaculture Wing of the environmental movement.

"The human race is now in its adolescent phase. We have to grow up."

If this is true, then current destruction is inevitable, a natural part of the "life cycle" of humans as a species. Some people even claim that human destruction is part of the *earth's* life cycle. We could spend hours trying to puzzle out the psychological needs motivating people to create such a narrative, but does it matter? Some of them have morally collapsed from despair, and to quote Isak Dinesen, "All suffering is bearable if it is seen as part of a story." Others are too attached to their comfortable lives to want them disrupted even though they can intellectually admit to the destruction embodied in their computer chips and housing suburbs. The third group are simply cowards: if human destructiveness is natural and inevitable, then it can't be fought and they don't have to risk anything. But the current destruction is not a developmental stage. The idea is offensive and condescending to all the cultures that have come before. Were they the "children" that led inevitably to glorious us?

And there are plenty of examples of cultures that didn't destroy the living communities in which they participated. It's only a few that have gone psychopathic. There's nothing inevitable about any human culture. In fact, this argument doesn't even work on its own merits: faced with an abusive or psychopathic adolescent, the first order of the day is still to stop him. In any case, this argument is pathologically narcissistic: the world is being murdered so we can learn some lessons? Only in an utterly insane culture could such an idea be conceptualized, much less given voice.

"The only way to change things is to change people's hearts and minds individually."

This is liberalism condensed to one sentence, and we have covered it previously. Movements for social change must have a program of popular political education. Successful movements get very good at it. But the point isn't to change people one at a time; it's to create a movement that can alter or abolish the institutions that organize power.

"Our assaults on [fill in the blank: empire, industrial civilization, patriarchy] won't work unless we change the culture of endless destruction and consumption. The question is really one of how to change culture."

This is a neat liberal trick that elides the nature of power, which is both sadistic and systematic. Imagine if blacks in the segregated South decided that changing "the culture" of segregation or "the hearts and minds of whites" was a workable strategy; they'd still be sitting at the back of the bus. What they attacked instead were some key instances of segregation in public accommodations. The Montgomery bus boycott was brilliant because blacks had economic leverage, and it was that economic power that brought down segregation on the buses. While the frontline activists were risking their lives at lunch counter sit-ins and registering voters, other activists were rallying for laws that would outlaw segregation and shift the balance of power. Hence the Civil Rights Act of 1964. And guess what? The culture changed. So did hearts and minds. A whole generation of now middle-aged people have never had to drink from a "colored" water fountain. One of them is even president. In parallel is a generation of white people whose psychology of entitlement and institutionalized ability to dehumanize blacks has been curtailed. That's because structural change toward justice affects hearts and minds and does it on a broad scale. That's why liberalism, with its focus on individual consciousness, will never change the world.

Even if your passion is to do that "cultural" work, you need to be thinking in terms of institutional power and how our movement is going to attack it. And we all need to stand with the people willing to take the biggest risks. But we haven't got a chance in hell without facing the facts. It is possible that the Prime Minister of Monsanto or the Crown Prince of Porn will have a spiritual epiphany, but is it probable? A one in a billion chance is not a solid base on which to build a political strategy.

"We can't stop them."

This is the Om of the alternative wing. There can be understandable personal reasons for believing in the invincibility of an oppressive system. And there are certainly reasons that those in power want us to see them as invincible. Abusive systems, from the most simple to the most sophisticated, from the familial to the social and political, work best when the victims and bystanders police themselves. And one of the best ways to get victims and bystanders to police themselves is for those victims and bystanders to internalize the notion that the abusers are invincible. Even better is to get the victims and bystanders to proselytize about the abusers' "invincibility" to anyone who threatens to break up the stable abuser-victim-bystander triad.

But those who believe in the invincibility of perpetrators and their systems are wrong. Systems of power are created by humans and can be stopped by humans. The people in power are never supernatural or immortal, and they can be brought down. People with a lot fewer resources collectively than *any single one of us* in rich countries have fought back against systems of domination, and won. There is no reason we can't do the same.

But resistance starts by believing in it, not by talking ourselves out if it. And certainly not by trying to talk others out of it.

History provides many examples of successful resistance. So do current events. Right now, the Movement for the Emancipation of the Niger Delta (MEND) has disabled 30 percent of the oil industry's production in Nigeria, and the industry is considering *pulling out altogether.* If we had one hundredth of their courage and commitment to their land and community, we could do the same thing here. We have vastly more resources at our disposal, and the best we can come up with is, what, compost piles? The world is being killed and environmentalists think that riding bikes is some sort of answer?

"Because we feel such a strong need to fight, is it better to fight battles that we chronically lose just to fight?"

Leaving aside the fact that the environmental movement has never fought militantly in the US, and taking "fight" to be the more general idea of resistance, this is a question worth asking. Why *are* environmentalists content to use the same strategies when they are clearly not working (for example, attempting to "change the culture" through discourse or example: that was tried once or twice or a million times by indigenous peoples)? Why not talk about what really needs to happen to save this planet? Burning fossil fuels has to stop. This is not negotiable. You cannot negotiate with physical reality. It's real.

Next, the infrastructure is vulnerable, as any reasonably informed member of a resistance movement—or any competent military strategist or historian—could tell us. Why not do what needs doing? Why are we not even discussing a serious strategy to save this planet?

A real culture of resistance would see that activities like biological remediation, the creation of local food networks, and teaching people self-sufficiency skills are part of a larger struggle to *actually save the planet*. Those activities should not be at odds with political resistance; they should be nestled inside each other in mutually nourishing and encouraging ways. Instead, the lifestylists take every opportunity to shut down discussion about action, actively discouraging a resistance movement from forming.

"We need to question some basic assumptions about how the culture has taught us to fight. We need to think outside the cultural box."

We agree. And three of those basic assumptions are that (a) resistance is futile; (b) the most meaningful resistance today is lifestyle change that can stand as an example; and (c) the physical structures that allow the psychopaths to run this culture are somehow immutable and cannot be physically dismantled.

Meanwhile, a very small group of half-starved, poverty-stricken people in Nigeria have brought the oil industry in that country to its knees. They remember what it is to love their land and their communities. Perhaps because they are not drowning in privilege, but in the toxic sludge of oil extraction. MEND has said to the oil industry: "It must be clear that the Nigerian government cannot protect your workers or assets. Leave our land while you can or die in it." And they are actualizing that.

Andrea Dworkin once said, "I found that it is always better to fight than not to fight, always no matter what."[110] This is the last moment to feel that passion, to defend whatever you love as a form of grace. Far too many people on the left claim that resistance never works. Some combination of cynicism, despair, ignorance, and cowardice has taken hold and even taken root. Some of the claimants have a solid radical analysis of capitalism, racism, patriarchy, civilization. They understand that the planet is being killed, that all we hold dear is under assault.

And yet. Resistance—its possibility, its activation—is unthinkable. These people obstruct any attempt to conceptualize how resistance to industrial civilization or, indeed, any form of oppression, could be organized. There are historical reasons for this: the obstructors do not act alone. Behind them are the Adamites, the Ranters, the Romantics, the Bohemians, and the *Wandervogel*, and some borrowings from Buddhism, building a cultural framework that channels despair, alienation, and even analysis away from direct action and toward individual life. That life may be built on quiet contemplation and good works, on the outraging of mores and boundaries, or on poetic suffering, but it's not built on confronting systems of power. Without a culture of resistance, alternative cultures and the antipolitical values they promote are all that the alienated and oppressed will find, and they aren't enough. Trees need rain; resistance needs a culture.

Across history, wherever there is oppression, there is resistance: let that be our first, drought-ending drop. We need to learn from those who have come before so we can decide where we need to build and what we need to abandon. Successful movements follow broad patterns, and one strong element in their success is the surrounding culture of resistance. Cultures of resistance mobilize existing cultures of survival, building on networks of community support and material exchange, the resilience that the oppressed must develop under the indignities of injustice, and the spiritual wellsprings that often occupy the center of cultures of survival. If we come from such a culture, we

can bring needed skills and experiences on which to model a specifically deep green culture of resistance. To date, most radical
environmentalists are white and globally privileged, which means that
the cultural norms they bring will be greatly lacking. But acknowledging that lack is the first step toward building something better,
something that this movement desperately needs if it's to win.

To make a successful cultural transition from survival to resistance
requires two related processes. One is an active, collective, and political embrace of direct confrontations with power. The other is a
psychological break with an identification with the oppressor. Malcolm
X was the eloquery of this experience for black Americans; Dee
Graham and her theory of Societal Stockholm Syndrome outlines a
similar process for the male identification structured into women's psychology.[111] This emotional remodeling often demarcates one generation
from the next, and can be a source of pain and conflict. When you have
survived by keeping your eyes down and your mouth shut, when the
consequences to speaking out and fighting back have been serious, it
feels uneasy at best when your cohorts start refusing to submit. But
that refusal is the foundation of resistance, and it has to happen.

Throughout the '60s, the left was split between the counterculture of
hedonism, drugs, and "mystical apoliticism,"[112] and, on the other hand,
a protest culture which had a critical analysis but failed for lack of long-
term strategy. Both sides of the split were predictable given their genesis
as youth cultures. That the hippie current would give rise to the New
Age navel gazers is no surprise, as its lineage from the *Wandervogel* and
Asconia is direct. But the protest culture was also a youth culture, creating "delusions of revolutionary grandeur" and "subsequent frustration
and disillusionment."[113] There was no long-term plan because the actionists didn't yet have the brains that could think long-term; while the
rejection of authority and everyone over thirty meant they allowed no
guidance from people who could have provided it. Left to their own
youthful devices, secular millennialism took hold, and poorly articulated
if intensely felt calls for militance were where the antiwar movement
and the left in general dead-ended. Those of us who try to propose a
thoughtful and strategic militant resistance—for instance, the targeting
of industrial infrastructure—are always arguing against the legacy of the

Weather Underground and the Black Panthers. The DGR strategy is *not* one of militant action to magically usher in generalized social chaos and revolt, nor is it a call to action because it feels better, nor is it militance to shore up masculinity. The DGR strategy is instead a recognition of the scope of what is at stake (the planet); an honest assessment of the potential for a mass movement (none); and the recognition that industrial civilization has an infrastructure that is, in fact, quite vulnerable. If you want to take issue with any of those three premises, well and good. But at least give us the respect of differentiation from other movements whose strategic goals we have clearly rejected.

Without long-term strategy—and let's be clear, a just and sustainable culture will take generations to achieve—some combination of millennialism and personal purity are what the disaffected will turn to, and they're both dead ends. In *From Slogans to Mantras*, Stephen A. Kent traces how '60s radicals adopted both. He writes, "The revolution would still come, but its arrival would be heralded by a personal transformation of purified individuals, and its appearance would (have to) be a divinely orchestrated event (since bitter experience had taught them that it could not be a *socially* orchestrated occurrence)."[114]

A bitter experience of perhaps five to ten years.

A culture of resistance must meld the idealism and courage that youth typically bring with the knowledge, experience, and long-term thinking of maturity. It also must believe in resistance if it's going to plan for it; beyond that, it must understand and embrace its other functions as a crucible of the resistance. In order to produce activists who last beyond youth, a culture of resistance must provide a range of emotional and material supports or people will give up and retreat to whatever personal solace they can find.

Central to that support is a framework that provides meaning. Humans are storytelling animals; we build narratives and then live inside them. It is no accident that the Irish independence struggle arose from the Gaelic Revival or that the civil rights movement followed the Harlem Renaissance. People need stories; people who resist need stories of resistance. But right now, calls for changing the culture are set *in opposition* to resistance. "Political change accomplishes nothing; it's the underlying culture that needs to change" is the assertion. But both are necessary to each other. Without a culture of resistance, actionists will give up in exhaustion, which will not be a personal failing but a collective one. Likewise, without resistance at the core, cultures stay locked into positions of mere (if Herculean) survival or are relegated to irrelevancy. A DGR strategy acknowledges the essential and symbiotic relationship between the culture of resistance and the actionists engaged in resistance. The authors of this book are *not* the people rejecting one for the other, as neither can exist without the other.

The tasks of a culture of resistance include holding and enforcing community norms of justice, equity, commitment, and solidarity; encouraging vibrant political discussion and debate; producing cultural products—poems, songs, art—that create a mythic matrix organized around the theme of resistance; and building individual character based on courage, resilience, and loyalty.

Specific material projects encompass everything from prisoner support to alternative schools to the creation of institutions capable of running civic society as the old system collapses. Along the way, from personal relationships to small groups to our larger institutions, a culture of resistance has got to embody justice and firmly reject domination. This means that white people have to own up to white privilege, ally with people of color, and commit to dismantling racism. It means that people from settler cultures have to acknowledge that the Americas are stolen land in an ongoing genocide, a genocide we must stop. It means men have got to cease in their sexual atrocities against women and girls, atrocities as quotidian on the left as on the right, and it means women have to stand in solidarity with each other. It means that men must ally themselves with women and against those who would abuse them.

We're up against a system that is not only unjust, but insane. A culture of resistance must collectively face the layers of horror embedded in history; the daily acts of sexual sadism that comprise slavery, conquest, and rape; the knowledge that these acts are not the mistakes of confused, tragic children. Forgive them or not: they know what they do. A culture of resistance believes in resistance because no amount of love or compassion or earnest education, no shining example of communal sustainability or individual self-respect has ever stopped the powerful.

Continent by continent, sustainable and egalitarian cultures have been wiped clean off the map, mere smudges of history that stood in the way of wealth; and the individuals who led them—brave, self-respecting, stalwart—took their place at the hanging end of a noose or beneath the heavy bodies of the hateful. "Power concedes nothing without a demand. It never did and it never will," wrote Frederick Douglass with the dense eloquence of those who know too well.

A culture of resistance is the mitosis of those demands, where the twin strands of pain and courage are triggered into life. As every living cell carries the message *life wants to live*, so, too, a culture of resistance is a determined miracle. However long the odds, life will live, and people will fight.

The odds are longer now than they've ever been, a shadow stretched with vanishing species and rising carbon. But there are warriors who might yet throw their bodies between the last of our future and its destroyers, if only they have a viable strategy and visible support. So the question is: Will the rest of us help them? Will we cast our lot with them, speak in their defense, shelter them in danger, sing songs of their stories, raise our children to take their place, prepare the way for their victory, claim them as our bravest and brightest?

Another 200 species went extinct today. Make your choice.

◙　◙　◙

Q: Is there a solidarity/support network in place to support someone who goes to prison for activism? Is there a support system in place to support someone's family if an activist goes to prison and is the breadwinner?

Derrick Jensen: For the former, there is. For example, Anarchist Black Cross and other organizations support political prisoners. But the truth is we need to build a much broader base than that. Prisoner support is actually pretty lacking. And it's pretty easy to do the basic stuff. My mother, every year, writes to many political prisoners on their birthdays and around winter solstice. Many of these people have been in prison for thirty and forty years, and her letters may be one of two or three that they receive throughout the year. So there are organizations in place, but those organizations have to be much more robust. As far as support for families, no, there isn't. But there should be. These are things that can and should be done by those who are entirely aboveground. We have emphasized throughout this book that not everyone needs to take actions that directly expose them to overt state repression. But we need a culture of resistance, and part of a culture of resistance is a robust prisoner support network for those who are on the front lines. We need a system where we support the troops, those who are actually fighting for the planet. That needs to be in place and so far it's not.

Chapter 5

Other Plans

by Lierre Keith

> We know that relying solely on argument we wandered for forty years
> politically in the wilderness. We know that arguments are not enough . . .
> and that political force is necessary.
>
> —Christabel Pankhurst, suffragist

The hour is late. It's too late for the creatures who went extinct today. Somewhere a tiny green frog sang the song of his species one final time. A small bird found no mate, and her last ovum is withering inside her. Another eighty-one million tons of carbon were added to the fragile blanket of our atmosphere, that long, ancient work of our good, green ancestors who made animal life possible.[1] A cascade of starvation strained the links of the food chain again, from plankton to salmon to grizzly bears; it's anyone's guess how long it will hold.

Our exploration of other plans for social transformation is informed not by a vague, protoutopian hope in a spiritual transformation passed from book club to bumper sticker, or a belief in the goodness that lies deep in every human heart, or, most especially, not in a *deus ex Akhashica*. In the simplest terms, a viable plan requires stopping the destruction that is civilization, actively repairing the damage done to biotic communities across the globe, and renewing and repairing human cultures that are truly sustainable—all within a framework of human rights.

None of this is technically difficult. Socially, politically, psychologically, even spiritually difficult? Sure. But what needs to happen to save this planet is not hard to understand.

Burning fossil fuels has to stop.
Its damage to the atmosphere is hell waiting to happen; its extraction, from drilling for oil to mining for coal, creates a swath of sludge and destruction that are essentially permanent on any scale that matters; and the easy energy it releases makes the rest of industrial civilization's horrors possible.

All activities that destroy living communities must cease, forever.

This includes clear-cutting forests, plowing up prairies, overgrazing grasslands, draining wetlands, damming rivers, vacuuming the oceans, and mining. It includes agriculture and it includes life in cities. All of those activities reside in one word: civilization. Instead, humans need to get sustenance as participants inside intact biotic communities, and not as destroyers of them.

Human consumption has got to be scaled back.

And drastically so. Since it's the rich countries doing most of the consuming, the rich's ability to steal from the poor is what must be confronted and stopped. Their resource transfer is currently organized into a system called capitalism and institutionalized into systems of law around the globe. Law is, of course, backed by the armed power of the state. Comprehending this is not intellectually difficult. With capitalism as one of the dominant religions of the planet, it is certainly psychologically and politically challenging. But this is a challenge to which we must rise if our planet is to have any hope of survival.

Human population must be reduced.

If we don't do it voluntarily, the world will reduce it for us. Even at Stone Age, solar-fueled levels of consumption, there are billions more people than the planet can support.

There is hope for our worn and weary planet, but to qualify as hope it must apprehend the facts. Without reality, hope is only a story for grown-up children. The people—animal, vegetable, and mineral—being consumed don't have the luxury of fairy tales. The privileged doing the consumption seem content to accept a happily ever after of wind, solar, and recycled tote bags. And the powerful are pleased that no one is threatening their conversion of the last of the living biomes into their own private wealth. Hope—real hope—is for the brave, because hope's only true action is to be that threat.

Without the brave and the willing, and without real engagement with the depth and scale of the problem, we're left with proposed solutions that will not save our planet. These alternate proposals break down into three basic categories.

1. *Tilters*, so named because they're tilting at windmills. These technofixers would leave industrialization and corporate capitalism in place, replacing fossil fuels with wind, solar, geothermal, and other so-called renewables. Lester Brown and Al Gore are two prime examples. They see that institutional change is necessary, which is true, but that change is identified as industrial culture switching to renewables as it continues to devour the earth.

2. *Descenders*. In his book *The Long Descent*, John Michael Greer argues that the oil economy will slow to a halt over a few generations. For Descenders, there is nothing much to fear and certainly nothing to be done beyond personal and local community preparation for energy descent. Cataclysmic climate change and ecosystem collapse are eerily absent from the future, and fighting back, of course, is never mentioned.

3. *Lifers.* They acknowledge resource depletion, energy descent, the destructive nature of industrial civilization, and the looming catastrophe of global warming, yet institutional change is foreclosed, and fighting back is discouraged if it's even considered. They urge personal lifestyle change and the concept of "lifeboats" as the only possible solution.

TILTERS

The problem with the Tilters is that they leave industrialization, capitalism, and, ultimately, civilization in place. All of these are disasters for the planet and for human rights. The Tilters urge us to accept that we are *all* equally responsible for the destruction of the planet.

Civilization is the destruction of the living world, and industrialization is an acceleration of that process. By harnessing the energy available in fossil fuels, both the speed and the scale of the devastation are dramatically increased.[2] It took the inhabitants of Easter Island a few hundred years to destroy their 63 square miles of forest using stone axes. A chainsaw can do that in a few weeks. Capitalism adds another accelerant: wealth. To be clear, when we say capitalism we aren't talking about all market economies. We're talking about the specific economies organized for *the accumulation of private wealth*. It may surprise readers to learn that this idea is quite new in the history of human affairs. As Ted Trainer points out,

> In almost all previous societies economic activity was determined mostly, and usually entirely, by social rules and procedures, not by the market. What a person produced, what he or she was paid, the price of the object, hours of work and who the work was done for, were all decided mainly by custom and tradition. . . . Most production and distribution were not determined by what would be most profitable, but rather by rules set by tradition, the church or organisations such as guilds. These nonmarket procedures set "fair prices." Labor, land, and capital were not sold. They were exchanged but the arrangements made were determined by social roles and not by bargaining in the market for the highest bidder. "Moral" considerations governed production and distribution. . . . It is therefore quite mistaken to assume that humans have always been motivated primarily by profit or have always had a market economy, or that such an economy is the natural, or the only way to organize economic affairs. In fact, virtually all previous societies . . . had exactly the opposite economic philosophy to ours.[3]

Chapter 3, "Liberals and Radicals," gave a brief history of how the merchant-barons arranged the US Constitution to support the accumulation of wealth, especially by the enforcement of contracts. The Constitution empowered the rising merchant class against both traditional constraints on accumulation and community protection of the commons, like forests and rivers. Those commons have been dismantled systematically and turned into private wealth. In the county where I live, the last run of salmon—a species that has been feeding forests for forty million years—is a whisper on its way to being a memory: only 500 returned last year, not enough for genetic diversity, and only a fading promise of nourishment for the 350-foot-tall redwoods. The timber companies are allowed to destroy salmon runs because the law declares the trees their property instead of our collective community.

Capitalism is an economic system based on extracting and accumulating wealth, not based on the provision of human needs. Writes Fritz Capra,

> Before the seventeenth century money lending for interest was immoral, it was expected that prices would be "just," personal gain and hoarding were discouraged, work was for the "well-being of the soul" and to produce things for direct local use was not to make profit. Throughout most of history, food, clothing and shelter and other basic resources were produced for use value and were distributed within tribes and groups on a reciprocal basis. The motive of individual gain from economic activities was generally absent; the very idea of profit, let alone interest, was either inconceivable or banned.[4]

There is no lack of critiques of capitalism or ideas about economic systems that would provide for human needs and human rights. But in brief, here are the main problems with capitalism.

Capitalism is based on endless growth. In our economic system, those who have capital invest it to make a profit. The problem, as Trainer points out, "is that as they make profits their capital grows and it is not possible for them to invest all profitably unless an increase in the value of producing and consuming takes place."[5] The economy must grow, or the system crashes. But our planet is finite. We cannot consume more of everything—trees, fish, soil—each year and have anything left.

Capitalist investment does not provide for human needs like food, housing, or health care; it goes where the investors might make a profit. What the rich want is what will be produced; what the poor need, well, the poor had better die and decrease the surplus population, as English literature's most famous capitalist said. Globally, one-fifth of the world's people get the lion's share of the resources, including fossil fuels, food, and even land.

Capitalism destroys democracy and human rights. Any arrangement where a tiny fraction of the population consumes most of the resources will require violence. People are not willingly separated from their sustenance. That violence is woven into every Walmart T-shirt, from the rivers drained dead for cotton to the farmers driven to suicide as corporations destroy their livelihoods to the farmer's children with no options but emigration to the nearest slum where a sweatshop is the *best* option for survival. A friend of mine is a professor whose students work at clothing stores like Old Navy. The students regularly find notes hidden in the jeans made in Asia: "Please help us."

In rich countries as well as poor, the power of concentrated wealth will distort and destroy democratic processes. Wealth can buy the laws, the courts, the government that it wants: the rest of us have essentially no access. With the commons privatized and local economies destroyed, people have no choice but to "bargain" with those corporations for their livelihoods. Only a free market fundamentalist could believe the results from such unequal bargainers could be fair.

This is what the Tilters fail to apprehend: leaving capitalism in place will never produce a just and sustainable world. A growth-based economic system will continue to turn living beings into dead consumer goods, local self-sufficient economies into corporate colonies of serfs, and democracies into commodities. Why would we want this system to continue? Yet the Tilters do.

Ted Trainer is worth quoting at length. He writes as an "apology to green people,"

> A sustainable and just society cannot be a consumer society, it cannot be driven by market forces, it must have relatively little international trade and no economic growth at all, it must be made up mostly of small local economies, and its driving values cannot be competition and acquisitiveness. Whether or not we're likely to achieve such a transition is not crucial here . . . The point is that when our "limits to growth" situation is understood, a sustainable and just society cannot be conceived in any other terms. Discussion of these themes is of the utmost importance, but few if any green agencies ever even mention them.
>
> The "tech-fix optimists," who are to be found in plague proportions in the renewable energy field, are open to the same criticism. . . . Despite the indisputably desirable technologies

all these people are developing, they are working for the devil. If it is the case that a sustainable and just world cannot be achieved without transition from a consumer society to a Simpler Way of some kind, then this transition is being thwarted by those who reinforce the faith that technical advances will eliminate any need to even think about such a transition.[6]

Lester Brown has a plan, currently updated to version 4.0.[7] He clearly recognizes that the planet is in severe distress. He is equally clear that overconsumption is driving the destruction. He is also frank in confronting overpopulation, which is often a very contentious issue for progressives. But he is attempting to save the thing which must be stopped: civilization itself.

Brown's *Plan B: Mobilizing to Save Civilization* involves four components: reducing carbon emissions by 80 percent by 2020; stabilizing the human population at eight billion; eliminating poverty; and repairing the planet's natural communities, "including its soils, aquifers, forests, grasslands, and fisheries."[8]

What is salutary in his plan is his understanding that the problems we face are systematic and interrelated. He writes, "We are not . . . likely to stabilize population unless we can also eradicate poverty. Conversely, we cannot restore the earth's natural systems without stabilizing population and climate, and we're not likely to stabilize climate unless we also stabilize population. Nor can we eradicate poverty without restoring the earth's natural systems."[9]

The problem with *Plan B* is that it leaves the overlapping accelerants of capitalism, industrialization, and civilization in place. This is the core fallacy of the Tilters, even when they acknowledge that something might be wrong with market forces. Writes Brown, "We rely heavily on the market because it is in some ways such an incredible institution. It allocates resources with an efficiency that no central planning body can match, and it easily balances supply and demand."[10] Allocates resources to whom? To the people who can buy them. And what capitalism calls "resources" other people consider their communities and, indeed, their lives. As Brown admits, "The market does not respect the carrying capacity of natural systems. For example, if the fishery is being continuously overfished, the catch eventually will begin to shrink and prices will rise, encouraging even more investment in fishing trawlers. The inevitable result is a precipitous decline in the catch and the collapse of the fishery."[11]

But the solution of the Tilters is not to dismantle the power schemes of capitalism, industrialization, or civilization. Their solution involves substituting renewables for fossil fuels, using incentives and penalties to try to make the market shift toward renewables. A carbon tax and cap and trade proposals are the favorites. In cap and trade, a regulatory body sets a limit on the allowable amount of a specific activity and then permits are auctioned off to the highest bidder. In general terms the problem is the usual capitalist pyramid: the people with the most money will get to buy the permits. If governments agree that only a set amount of carbon can be released each year, why do the rich get to use that carbon? Why isn't that set amount of carbon distributed equally to every human being?

In real life, cap and trade programs have proved unworkable at best, and damaging at worst. According to environmental lawyers Laurie Williams and Allan Zabel, the problems are legion.[12] They state bluntly, "We do not think a reliably accurate system can be put in place for enough sources of emissions and offsets within the necessary time frame." In Europe, fraudulent underreporting has helped render the Kyoto treaty ineffective. Indeed, cap and trade creates another source of wealth—carbon—that has, as usual, been shifted upward. The European carbon market has only "enriched polluting industries and their consultants, while producing minimal decreases in their emissions."[13]

The idea behind both cap and trade and a carbon tax is to make the market respond away from fossil fuels and toward (presumably cheaper) renewables, but it *leaves capitalism in place*. It assumes that both industrial civilization and capitalism are good arrangements of human affairs, that these arrangements can be redeemed by a simple switch to renewables, and that such a substitution is possible. We have already discussed why industrialization and capitalism are based on domination and destruction. But the issue of renewables is worth a critical look.

Renewables are the Promised Land for progressives, where

megawatts will flow like milk and honey. They can power everything—the consumer goods, the suburbs, the agriculture—and leave us with a viable atmosphere. All we have to do is direct the market to invest in some combination of solar, wind, geothermal, and biomass energy, and we can have our planet and eat it, too.

But reality is a harsh corrective, especially when people have staked both their emotional well-being and their future on hopes with no substance. The Tilters are scientific millenarianists: a new day will dawn on our solar panels as long as we purify our personal lives and all believe.

But no amount of belief will change the math and physics. Wind energy, for instance, is the "centerpiece" of Brown's *Plan B*, version 4.0. He asserts that "harnessing one fifth of the earth's available wind energy would provide seven times as much electricity as the world currently used."[14] He proposes building 3,000 gigawatts worth of windfarms, which would provide roughly 40 percent of world demand. These are the sorts of statements that lull the alarmed back to sleep. There are serious problems with wind energy that Brown and other Tilters ignore, and they ignore them at the peril of the planet.

As Ted Trainer makes clear, "Even in good wind areas, wind will not be able to provide more than a rather small fraction of electricity demand."[15] The first major problem is variability. Yes, the potential harvest might be larger than demand in some places. But the vast majority of that potential is useless because the wind is an intermittent force. Wind farms produce a "spiky" output, meaning an all or nothing generation pattern. If the winds are up and the turbines are spinning at capacity, the grid will be overloaded. Most of the power will need to be kept out of the grid to keep the system from frying. Why not turn off the fossil fuel plants and use the wind input? Because it takes twelve to twenty-four hours to get those powered up or down. If the excess electricity generation could be stored, the problem would be mitigated, but electricity is essentially impossible to store. The inherent variability problem means that in order to produce the amount of energy that industrial societies are used to, fossil fuel generating capacity has to be almost equal to the wind capacity. Without that backup, using only wind, "peak [wind] capacity would have to be something like twenty times average demand. The number of windmills needed would be impossibly large."[16]

This has been borne out by experiences in Europe, where variability means that a threshold as low as 5 percent from wind power causes "significant integration difficulties into the grid."[17] Denmark is often referenced as a positive example, as problems aren't reached until wind output reaches 18 percent. This happy number falls apart on examination, as most of the output is exported. The rest is dumped. In fact, only 4 percent can be taken in by the Danish grid. The two reasons Denmark can export the excess are because its neighbors aren't using wind power, and hence have a little more allowance, and because Denmark is small compared to those neighbors and so generates a small amount of power overall.

Most damning, one researcher believes that wind power would result in more fossil fuel usage than if windmills hadn't been built. Explains Trainer, "This is because the most efficient gas plants (combined cycle gas turbines) must be run at a constant output but the plants capable of varying their output to follow wind changes quickly are much less efficient. In addition frequent variation reduces the life of gas turbines."[18]

Some Tilters have a crush on hydrogen as a storage method. But even under the best circumstances, only about 25 percent of the energy going into the fuel cell comes out. It's a very poor storage mechanism. Take the twenty times average demand quoted above, and multiply that by three to get the number of windmills needed. It's not possible. With all apologies to Bob Dylan, the answer is not blowin' in the wind.

Solar energy fares little better. Brown again proselytizes for techno-millenarianism with statements like, "There are enough solar thermal resources in the US Southwest to satisfy current US electricity needs nearly four times over."[19] We need not fear for the future, only place our faith in the technological priesthood. But upon deeper investigation, the miracles promised by solar power fall apart like parlor tricks. As with wind, storage and integration make solar generation more useful as a backup energy source then as a prime source. Solar thermal energy costs more than 7.5 times as much as a coal-fired plant.[20] Solar photovoltaic (PV) panels could cost thirteen times as much.[21] Winter presents an insurmountable problem, leading some, including Brown, to suggest North Africa as the best site for *Europe's* electricity. One pithy bumper sticker asks, "How did our oil get under their sand?" The renewables version might ask, "How did our sunlight fall on their land?" This only works morally if, like Brown, you still advocate neoliberal globalization: poor countries should attract capital from the rich, and integrate themselves into global markets by selling whatever "resources" they can. The new world of renewables will look exactly like the old in terms of exploitation.

But even putting aside the basic issue of justice, the physics renders the scheme unworkable, as it involves enormously long transmission lines, which would include a stretch under the Mediterranean Sea. For Europe, with its more northern location, the presence of both clouds and winter mean that solar power cannot begin to replace fossil fuel levels of energy consumption. In the US, the situation is similar in that the best sites are in the less populated Southwest. To get that power to the population centers would require storage and long lines, with their "parasitic losses, energy costs, transmission losses and the cost of a backup system."[22] Solar thermal has the advantage of energy storage (oil, molten sand, or crushed rock). Right now, the storage can last up to twelve hours. But data shows that cloud cover can last for days even at the best sites, requiring backup capacity. PV systems have the same variability and storage problems as wind. They are also costly. Trainer's figures show that PV systems, including both household and industrial generators, can take anywhere from 150 to 294 years to pay back costs. He runs through the numbers on a household system and concludes that "if the electricity generated was sold at the same price as coal-fired electricity it would take 452 years to pay these costs."[23] As he states, "These long dollar payback periods indicate the magnitude of the increases in electricity price that would have to be accepted in an economy based solely on renewals."[24] Trainer examines a PV solar option for a 1,000 megawatt PV plant meeting twenty-four-hour demands. Figured for a good location, and assuming hydrogen storage, it could cost as much as thirty-four times that of coal.[25] At a certain point, the cost of energy would lead to the collapse of the industrial economy. The authors of this book have no problem with that outcome.

But the Tilters rallying behind renewables are trying their hardest to hold off that exact possibility.

These costs, of course, are all based on the still-cheap energy provided by fossil fuels. Windmills, PV panels, the grid itself are all manufactured using that cheap energy. When fossil fuel costs begin to rise, such highly manufactured items will simply cease to be feasible: *sic transit gloria* renewables. The elements used in some key technologies—gallium, indium, and tellurium—simply don't exist in the quantities that would be necessary for PVs to supply any meaningful amount of world electricity consumption. The basic ingredients for renewables are the same materials that are ubiquitous in industrial products, like cement and aluminum. No one is going to make cement in any quantity without the easy energy of fossil fuels: cement is so energy intensive that each pound of it releases a pound of carbon into the atmosphere.

And aluminum? The mining itself is a destructive and toxic nightmare from which riparian communities will not awaken in anything but geologic time. And like cement, production of aluminum and steel is saturated in embodied energy. These are not ingredients with which we can build a sustainable way of life. Their extraction leaves broken rivers behind them; their refining demands the heat of hell; and their intended usage is for more of the same, the continued consumption of the planet.

That I have to address biofuels at all tells me that mainstream environmentalists are dwellers in the land of fantasy, a fantasy built on entitlement to 3,000 pounds of personal steel.

Corn ethanol may not, in fact, provide any net energy. If it does, it's a tiny amount. More important, every acre of corn used for ethanol requires that a corresponding acre somewhere else must be cleared to make up for the food lost. The only "somewhere else" left is the tropics. A team at Princeton University did the math: biofuels based on land clearing in the tropics "dramatically" increased greenhouse gas emissions.[26] A study in *Science* put the number of the "biofuel carbon debt" at thirty-seven. Understand: converting both grasslands and rain forest to corn, soy, or palm oil for biofuels results in carbon emissions *thirty-seven times greater* than the reduction in greenhouse gases afforded by switching from fossil fuels to biofuels.[27] As Trainer puts it, "The limits to liquid fuel production have not primarily to do with the energy return ratio for producing fuels from biomass. They have to do with quantity, i. e., the areas of land available and the associated yields."[28] There is no more land, and, frankly, if there was, the plants and animals who lived there would vastly prefer their lives and communities to a monocrop of switchgrass. Perennials fare no better in the end than annuals: the continual harvesting of all cellulotic material—assumed in biofuel calculations—will degrade the land very quickly.

Again, there is that basic ignorance of how life actually works, knowledge that is endemic to agricultural societies. Soil is alive, profoundly so. It is not an inert material for humans to use or manipulate, and treating it as such has brought us to the end of the world. Because soil is alive, it needs to eat. Continuously removing the plant material means the soil starves. With starved soil, the plants in turn will have nothing to eat. Providing nitrogen from fossil fuels will temporarily let plants grow, but both the mineral content of the soil and the body of the soil itself will still be degrading.

Those fertilizers are, of course, part of the fossil fuel drawdown as well as the greenhouse buildup. A Nobel Prize winner, Paul Crutzen, found that the nitrous oxide emissions from the petrochemical fertilizers necessary for corn and rapeseed nixes any carbon savings.[29] Biofuels are just another agricultural assault against the planet.

The road to hell is paved with fossil fuels. And there is no energy source that can provide for the continuation of industrial culture. The Tilters have got to face the truth. Sun, rivers, wind, and trees can provide us with a home. They cannot provide for a personal empire of energy, not beyond a few generations, and the last generation is already here.

The Tilters have more proposals, but these are vampire ideas that turn to moral dust under daylight. We can keep our cars, they promise, as long as they are electric. It turns out that such cars can take up to *five times* as much energy to produce as a regular car. For a Prius, the figure to note is 142 percent more energy over the life of the car. In fact, hybrids consume more energy than an SUV. According to Richard Newman, the energy cost per mile over the lifetime of a Prius is 1.4

times the energy cost of the average car in the US.[30] Lester Brown suggests energy-saving appliances that can "talk" to the grid. And how much extra energy is embodied in the complicated circuitry? Meanwhile, we know how much energy is embodied in a cold cellar or springhouse, in food produced daily and locally: none. But no one dares suggest a different way of life, one that lays down the weapons of civilization, the sword and the plowshare both.

The other major failure of the Tilters is their assessment of overpopulation. On the positive side, most of the Tilters are at least willing to engage with the issue and to tell some difficult truths. Population is not an easy topic for people who care about human rights. Historically, some very nasty elements have used population as an excuse for "population control" policies constructed around a simmering racist metanarrative: the problem is really that brown people are too stupid and/or too sexual to control themselves. Those of us who come to the population discussion from the perspective of resource depletion, human rights, or feminism have to distinguish ourselves from the racist history entwined in the issue. When we say "overpopulation" we need to define what we mean and why it matters.

What I personally mean is that the earth is a bound sphere. The planet is finite. There are absolute limits to the numbers of individuals that any species can attain. That is what carrying capacity means: how many members of a species the environment can support indefinitely. Too many members and that species is drawing down resources, degrading the landbase for itself and for other species, and will most likely end in extinction. That is physical reality. For most of human history, we were very aware of the limits of our surrounding community. Hunter-gatherers know the ratio of productive adults to dependents that must be maintained to stave off hunger and ultimately degradation of the biotic community. Everything from abstinence to herbal abortifacients are mobilized, with infanticide as the fall-back plan. An Inuit woman whose husband died was expected to kill any children she had under the age of three.[31] The Arctic is a harsh climate and too many dependents means the whole community will suffer. That's actually true for all human societies, but in a more demanding environment the ill effects (hunger) of a skewed dependent-to-producer ratio will be felt immediately.

What broke the cultural knowledge of those relationships was agriculture. By drawing down entire ecosystems, humans were able to dramatically increase their numbers. Remember that agriculture is the replacement of biotic communities with monocrops for humans. Agriculture has let vast amounts of resources accumulate into more and more humans—sunlight, rain, rivers, soil. With the soil used up, the monocrops are now fertilized by fossil fuels. If you're eating grain, you're eating oil on a stalk. With the rivers drained and the water tables falling, the crops are now irrigated by fossil acquifers. The water is so inaccessible that oil drilling equipment is necessary to reach it. Huge swaths of our planet, once lush with forest, are nothing but scrub and salt. That profound drawdown is what is supporting our current numbers.

And here's a problem in the discourse about the dilemma. Many sustainability writers take the current level of resource extraction as an unquestioned baseline. They assume the amount of grain now being produced can simply go on indefinitely. It can't. It's based on drawdown and long-term destruction of entire continents, a destruction that is about to hit bottom.

I appreciate how Brown understands that historically, soil destruction and salinization have brought down previous civilizations, and the grim possibility that this future awaits us. The United Nation projects a world population of 9.2 billion by 2050. Brown addresses this head on:

> I do not think world population will ever reach 9.2 billion . . .
> The land and resource base is deteriorating and hunger is spreading. Simply put, many support systems . . . are already in decline, and some are collapsing. The question is not whether population growth will come to a halt before reaching 9.2 billion but whether it will do so because the world shifts quickly to smaller families or because it fails to do so—and population growth is checked by rising mortality.[32]

He states with the clarity of emergency, "If we cannot . . . stabilize population and climate, there is not an ecosystem on Earth that we can save."[33]

Brown is equally clear that raising the status of women and eliminating poverty are key to lowering the birthrate. He writes, "If the goal is to eradicate hunger and illiteracy, we have little choice" but to lower our numbers. He includes debt relief, universal health care, primary education especially for girls, and access to family planning as basic steps toward reaching that goal. All that is worth fighting for. All of it would help move us toward a truly sustainable future. None of it is enough.

Capitalism, especially the corporate version, has got to be dismantled, to be replaced by democratically controlled economies. If that includes a market economy, the markets must be nestled inside subsistence economies. Tilters like Brown can identify poverty as a factor in population overshoot, but *they don't identify capitalism or civilization as the leading cause of poverty.* Brown's solution is that, along with their sunshine, "low-income countries" are supposed to sell off whatever is left of their so-called resources. Brown urges funding to allow the third world to "develop their unrealized potential for expanding food production, enabling them to export more grain."[34] Brown is worried that peak oil may interrupt "international flows of raw materials." Such flows of raw materials are the model that has condemned the majority of the world to poverty and the earth to destruction. Those "raw materials" need to remain what they are: living forests, grasslands, rivers—soil-building communities that are the matrix of life. And the human members of those communities need relief from the relentless assaults of the globally powerful.

But what the planet needs most is relief from the relentless assault of agriculture. Like almost everyone alive today, the Tilters don't realize that agriculture is biotic cleansing, drawing down species, ecosystems, and soil to temporarily increase the planet's carrying capacity for humans. This is also the blind spot endemic to claims that shifting grain from animals to humans would solve world hunger: that grain is only temporary.

Brown proposes increasing food supplies by raising land productivity through fertilizers, irrigation, and higher-yield varieties. The disconnect in this thinking makes my head hurt. He knows that humans are destroying the climate with fossil fuels, yet his solutions depend on more of the same. The fertilizers are all derived from gas and oil; and their day is done. Irrigation results in soil death by salinization and has brought down a great number of preindustrial civilizations. It also results in river death by dewatering: a fish out of water is a dead fish. Eighty percent of China's rivers, for instance, now support no life. Irrigation also brings devastation to the surrounding wetlands, which should be the most species-dense habitats on the planet and are now historic oddities. Water tables have dropped so far that half of India's hand wells are dry, forcing people into desperate urban slums.[35] Agriculture provides its final insult to the land when water tables drop below the reach of tree roots. Trees are the backbone of their biotic communities: without them, the world is emptied to a monoculture of dust. Oil drilling equipment, which requires the cheap power provided by fossil fuels, is then necessary to get the water.

Brown also suggests no-till agriculture. Somewhere in all of this he recognizes that plowing is destructive. Some forms of no-till agriculture require specialized equipment that can drill through plant residues, equipment that is both industrial and costly. Other no-till methods also use herbicides instead of mechanical means to kill the invading plants, plants that are nature's desperate attempt to repair the world. Setting aside that such schemes will keep poor people dependent on industrial infrastructure, a cash economy, and the hierarchies behind both: do I really have to explain that coating the world in poison is a bad idea? There is no future for humans, for soil, for the winged and gilled in these proposals, or no future worth enduring. If we are going to face the truth about population overshoot, we need to actually face it.

No solutions that rely on agriculture will be real solutions. The soil will continue to collapse into sterile dust. Irrigation, the final tears of rivers, the last sigh of exhausted aquifers, will leave the land strangled with salt. The animals, from the awesome grace of the megafauna down to the tiny miracle of copepods—almost too small to see but aggregating into the largest animal biomass on the planet—have nowhere to go except the abyss of extinction, and not two by two but 200 at a time. This is what agriculture is: a funneling of biomes, once verdant with life and the resilient promise of more, into a monocrop of humans. The process is now nearly complete, its swan song a catastrophic failure of this once-living planet.

Despite the declarations of an inexplicably popular book, the world was not created for us. No marginally rational person would believe such insanity. As apex predators, we are utterly dependent on the work of millions of other creatures who took a cold rock and turned it into a home. "Go forth and multiply" is the clarion call of entitlement. We don't have a right to more than our share. We will not save this planet as long as agriculture—its religion, its psychology, its entitlement—continues.

Eight billion people are dreaming, except that such a dream would be more like a nightmare. So how many people could this planet support sustainably? In 1800, the beginning of the fossil fuel age, there were one billion people. Many resource depletion writers choose one billion as a benchmark. Such people have faced some hard truths, but they have not gone all the way to the bottom. In 1800, vast swaths of the planet had already been destroyed by agriculture and overshot by humans. A truly sustainable number would be somewhere between 300 and 600 million. It may sound impossible; it may be impossible, given the time we have left. On the positive side, the same social and political processes need to be set in motion whether the goal is eight billion, one billion, or 300 million. If we can do it at all, we might as well do it right.

One positive fact about being alive is that we're all going to die. If we can start reproducing at below replacement numbers, the problem would take care of itself. And it won't even take that long. At just over two children per woman, population continues to rise because of the number of existing children who have yet to enter their reproductive years. But at one child per woman, the global population would decline to about a billion around 2110. That billion could be reached in fifty years if adult mortality triples—and the human race has faced far worse mortality rates over its history. Some places are already undergoing increased mortality, and though personal suffering is obviously involved, the societies as a whole are not collapsing into lawless disorder as the population contracts (see our discussion of Russia, below).

Currently there are seventy-five countries with populations reproducing at replacement levels (2.1 children per woman). Thirty-three countries, in fact, have negative population growth. In some cases, that's due to high standards of living and civil rights for women, both

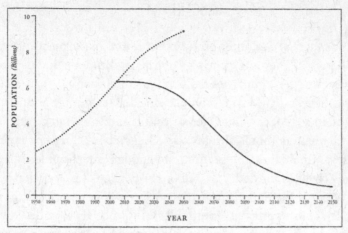

Projections, Population of the World

---------- UN scenario: fertility declining from 2.6 children per
woman in 2004 to slightly over 2 children per woman in 2050. Source:
Population Division of the Department of Economic and Social Affairs of the
United Nations Secretariat (2005).
——————— Scenario assuming that all fertile women are henceforth limited to 1
child. Source: Dr. Sergei Scherbov, research group leader, Vienna Institute of
Demography, Austrian Academy of Sciences.

CHART BY JONATHAN BENNETT

of which have a huge impact on population levels. In other countries,
though, the reasons are grim: disease, especially AIDS, the social
strains of war and poverty, and ecological collapse. In very stark relief,
these facts show us the possible futures—we voluntarily and peacefully
address the social forces behind population overshoot, or nature will
do it for us.

The questions we must face are: Will we—planetwide, species
"we"—recognize the problem? Will governments and other major institutions set the necessary policies in motion? Or will the Catholic
Church continue to condemn condoms? And will our planet be able to
withstand the power shifts and strains until our numbers begin to
decrease?

DESCENDERS

The Descenders are another group often found in permaculture and peak oil groups. They cite examples of past civilizations that collapsed due to resource depletion, especially soil, but their point is that collapse is not cataclysmic. Mayan civilization, for example, didn't end overnight; in fact Mayan cities "took a century and a half to go under."[36] This is the main point of John Michael Greer's book *The Long Descent*. He writes, "Gradual disintegration, not sudden catastrophic collapse, is the way civilizations end."[37] Based on the decline of past civilizations, he predicts that the end of industrial society will be a series of miniature crises and respites as energy decline proceeds in a downward stairstep. The crises will not be fun, but they will also not be apocalyptic freefall.

The problem with this basic thesis is twofold. Industrial society is industrial; it's based on vast quantities of fossil fuels. This condition does not match anything that has come before. The ancient Greeks, Mayans, and Chinese did not depend on fossil fuels for basic sustenance. Previous civilizations were at least human in scale, even if they were based on drawdown. Human-scale civilizations could end in human-scale collapse. But now, entire continents, and indeed six billion people, are dependent on fossil fuels for basic foodstuffs. What they don't import—at distances only made possible by fossil fuels—they grow using fossil fuel as fertilizer. This is what the Green Revolution has wrought: a quadrupling of a human population that was already overshot. When oil production starts its inevitable slide down the dark side of Hubert's curve, six billion people will have nothing to eat.

In previous collapses, there were intact biotic communities into which the civilized could fade. There were living forests, grasslands, rivers, and coastal areas inside of which people were able to subsist as they always had. That is over, over on a scale that no one seems willing to acknowledge, the emptiness as profound as the numbers are complete: fish, 90 percent gone; forests, 98 percent gone; prairies, 99 percent gone. No past civilization could even dream of this level of conquest, limited as they were by the distances that supply lines made of pack animals could travel. That protective boundary was broken by the steamship and the internal combustion engine. There is no verdant cushion of forest, no estuary dense with nourishing fish and fowl. We are living on oil which at a point not too distant will take more energy to extract than the energy it contains. This is a cliff, not a soft stair of descent.

The other chasm between the Descenders and reality is the collapse of ecosystems and basic life-support functions across the planet. Greer's book talks about "collapse," but his collapse only refers to human societies. Meanwhile, life is fraying at the seams from the surge of carbon, the clear-cut of species. Greer urges us away from the concept of apocalypse, claiming this is just a favored narrative of Abrahmic religions. But the murder of my planet is not a story: it's an ongoing outrage that demands committed action, and now. As the temperatures and sea levels rise, as coastlines and countries begin to drown, as the soil bakes past what bacteria and fungi can endure, at what point am I allowed to say "apocalypse"? There may well come a day in my lifetime when the last polar bear, still in her ancestral white against a world melted to brown, dies. I would call that *possibility* an apocalypse. Its reality I will call *hell*. What words Greer and the Descenders might prefer, I can't guess.

I am not attempting to create panic or survivalism. Neither will help. I am attempting to create a resistance movement with a strategy that can address the scale of the problem. The Descenders, like the Tilters, are attempting to create a way out of the horrible facts before us, but their way out is not to face and then attempt to alter those facts. The Descender's way out is essentially emotional, a lulling story that it will all be okay: it's happened before, and the world didn't come to an end.

Except this time, the world is coming to an end. That last polar bear may be here already, a cub enfolded in the evolutionary warmth of her mother's fur. Or perhaps she's still an ovum inside a yearning intelligence of dividing cells, protected by an ancient, mammalian sea. When she emerges, the world that she finds will not be the one that her mother or her mother before her found. There will not be enough ice for her to stand on. There will not be enough seals to make fat, or fur, or babies. There will be hunger and cold, until she dies of one or the other, in a sea stripped to desert by a culture that no one was willing to stop.

This is the reality of mass extinction. It is happening not just species by species, but one creature at a time: bear by bear, bird by bird, the exhaustion of too many miles and no ice, no river, no shelter in sight. The desperate nestlings, all open mouths and the future entire. They are hungry, thirsty, cold, and they are dying. This is the slow hemorrhage of life from our planet: those creatures are dying one at a time. Reducing physical reality to a narrative is, of course, one of the core components of liberalism. To suggest switching narratives as a political plan is a dead end of insane proportions. The murder of my planet is not a bad movie I can turn off. It's not a book I can take back to the library. It's not a story. Those creatures—each one a miracle of cells coordinating feathers and flight, patience and roots, joy and pain—those creatures are dying. They are real. And they need real defenders in the real world.

The Narrators have gained a fair amount of purchase amongst the environmentally concerned. They claim that human domination is simply a story rooted in Genesis, and humans are not powerful enough to destroy the earth. But for domination to be a mere story means that its victims are only characters. I disagree, as do black terns and Arctic foxes. This reduction of reality to a narrative breeds an odd passivity in its adherents. Nature will take charge: we are wayward children not responsible for our actions, and, indeed, we will be stopped before real damage is done. And the concept of "saving the earth" is, they claim, simply the Western individualist hero in all his masculine glory. The Narrators ignore, of course, the narrative of masculine entitlement, where women, animals, and the earth are consumables barely noticed even as "resources," on which the male ego is built. Maybe we could abandon *that* narrative instead?

Or maybe we could just abandon the narrative that the world is made of narratives. It's not. It's made of living creatures entwined in a vast complexity of giving and taking, a consanguinity of sunlight and carbon, a Great Communion. There is a prayer of participation in every animal breath, in every fragile, reaching radicle, every dividing cell. But our thanksgiving is collapsing to a plainsong, 200 species at a time.

So if we need a narrative, it's a simple one: resistance is possible. If you want to add some suspense, try: and we're out of time. Beyond that, can we stop telling stories and get to work?

The Tilters usually believe in political engagement. From Al Gore to Lester Brown to Bill McKibben, they encourage civic participation to force institutional change. There is often a fierceness to their urging that matches the seriousness of the situation. Even better is the underlying recognition that institutional change is primary, that personal change will never begin to address the situation. The problem with the Tilters is that they're attempting to save industrial civilization. Reduced consumption levels are part of their plan, but capitalism and its perpetual growth is an unquestioned—indeed, unquestionable—part of the future. And as already shown, there is no combination of solar, wind, or biofuel energy that will equal the dense, easy energy of fossil fuels, and no Patronus of technological breakthroughs to save the day. This way of life is over and they are not facing that.

The Descenders, on the other hand, have an assessment of energy—and the low-energy society of the future—that is reality-based. Writes Greer, "As fossil fuel stops being cheap and abundant, standards of living throughout the industrial world will shrink toward the level of the nonindustrial world."[38] Absent from most of the Descenders is any awareness of the biotic emergencies the planet is facing or any clarion call to action. (Am I allowed to say they are caught in the "Resistance Is Futile" narrative?) Indeed, political action is actively discouraged and dismissed. Ted Trainer, for instance, insists that "there is no other possible way" to a sustainable future besides personal lifestyle choices.[39] The claim is that our political institutions will never respond, and all we can do is prepare ourselves as individuals and maybe as local communities as the system collapses.

If our political institutions aren't working, then we need new ones. But the actions the Descenders suggest are the usual personal-scale adjustments: get used to less energy, plant a garden, learn a nonindustrial trade. The only larger-scale solution Greer encourages is on the community level: "Since governments have by and large dropped the ball completely, it's up to individuals, families, groups, and local communities to get ready for the future ahead of us."[40]

This is the other main drawback of the Descenders. As critical as they are of survivalism—the ultimate individualism—they are equally as dismissive of political activism. On the occasions that political resistance comes up, it is firmly erased as an option. I don't know if there has ever before been a movement that understands the problem is political yet unilaterally rejects political solutions, and I don't understand why this rejection has taken hold of so many smart, engaged minds.

Daniel Quinn urges "walking away."[41] To where? And more importantly, why? Richard Heinberg writes that "efforts to try to bring industrialization to ruin prematurely seem to be pointless and wrongheaded: ruin will come soon enough on its own. Better to invest time and effort in personal and community preparedness."[42] Contrast these words with the courage of Henning von Tresckow, who said that even though the Nazi state was doomed, the efforts to bring down this evil regime must continue because it was daily murdering more innocent victims. The current victimization of both human and nonhuman creatures is an order of magnitude larger, which should imply that our moral responsibility is that much greater.

Pat Murphy, in *Plan C: Community Survival Strategies for Peak Oil and Climate Change*, a book that stands out for its political intelligence and keen moral outrage, writes, "In terms of corporate globalization, a good local action might be to avoid purchasing goods from international corporations as much as possible."[43] In a book that is willing to name technology worship as a religion, the automobile as a destructive parasite, and goes so far as to recommend withdrawal from the mass media, the best political action he offers is a generalized personal boycott that will have zero effect on power. He urges readers to "begin the personal process of changing our lifestyle. This is truly thinking globally: choosing a healthy planet and a sustainable lifestyle over the short-term pleasure of excessive consumption."[44] Indeed, he has a whole chapter called "Post Peak—Change Starts with Us."[45] He further urges us to "make the personal changes needed to live in a post peak world, providing authentic leadership for those who will follow."[46]

Don Fitz calls this "exhortationism," which he defines as "the belief

that environmental Jesuits must convert individuals to piously consume less, a belief which ignores the economic, political and social
realities which force us to consume more."[47] To flog a very dead liberal
horse: personal change will never equal political change. We do need
authentic leadership, but toward political actions that dismantle unjust,
destructive institutional power. It would take massive numbers of
people withdrawing from corporate-produced goods for this to have
any effect, and if you're going to organize people to do that, why not
direct them toward an action capable of knocking a brick out of corporate power and industrial civilization? As Fitz points out,
exhortationism is "a call to build a new society without building social
movements."[48] And without those movements, nothing will change.

Murphy writes, "Changing personal habits should come first, at least
go hand-in-hand with lobbying for government and institutional
change. For it will only be with the experience that comes from personal change that people will develop the wisdom to make the proper
societal changes."[49] With all due respect, this is not how a single liberation movement in history has worked, nor how a single human right
has been won. Education and consciousness-raising are necessary to
build the ranks of activists who will do the work, but the work they do
is on an institutional level. Because of institutional change, hearts and
minds change on a society-wide level. That is the progression.

It is our job as activists to supply the necessary force. That is always
the job of activists: to make demands and back them up. Once again,
that force can be totally nonviolent, but the strategic and tactical questions are secondary to the knowledge that power has to be confronted,
and that it will not give up willingly.

The Tilters and the Descenders are both offering liberal solutions.
Since the Tilters are not willing to name the hierarchical power structures of capitalism, industrialism, or, ultimately, civilization, their
proposals cannot address the real problem. The Descenders are clearer
on the problem, but their insistence on the efficacy of a switch in narratives is idealist. Again, idealism is the belief that reality is constituted
by ideas, not material conditions. And the Descenders' foreclosing of
political solutions in favor of personal lifestyle choices is unalloyed individualism, the other core tenet of liberalism. Liberalism will always fail

to produce radical change, and if there was ever a moment when that change was needed, it is now.

LIFERS

The Transition Town model comes closest to the culture of resistance component of a Deep Green Resistance movement. But there is a deep contradiction in the Transition Town movement: the program implicitly calls for institutional change, yet many of its writers insist on a personal "Lifeboat" concept. The Lifeboat model was originally proposed by Richard Heinberg in his book *Powerdown: Options and Actions for a Post-Carbon World.* The idea is to accumulate skills and knowledge for small-scale community survival as well as "preserving the cultural achievements of the past few centuries."[50]

The Transition Town concept was created by Robert Hopkins as a framework for organizing a community response to peak oil and global warming. It was one way to answer the question "What can I do?" with a concrete plan. The plan is an Energy Descent Action Plan (EDAP). There's much to be said for the twelve steps that *The Transition Town Handbook* lays out as the process to create the EDAP. Local groups are directed to break down into working groups to address whatever they feel is relevant to the process of "building community resilience and reducing their carbon footprints."[51] Along the way they're encouraged to network with other related groups in their area, work on projects that are visible and practical for the public (e.g., planting nut trees in the town center), offer "reskilling" of lost and soon-to-be-needed traditional subsistence skills, and build bridges to local governments. They also recommend that Transition Town groups include the UN Declaration of Human Rights in their statement of purpose.

The Transition Timeline by Shaun Chamberlin is the second generation Transition Town, laying out the possible futures we face. With no major derailment of the current course, environmental and economic collapse, with its attendant civic breakdown, is the default setting. The Timeline has the government directing investments in infrastructure like mass transit, outlawing factory farming, and decommissioning coal plants. Chamberlin further envisions "a binding and sufficient global agreement" that "atmospheric concentrations never break through 400 ppm CO_2," including rationing and various carrot-and-stick tax and market approaches.[52] Thus the foundational documents of the Transition Town movement recognize that the population at large does not willingly give up fossil fuels and industrial levels of consumption. They are forced to, by the reality of depletion and by the government instituting rationing.

The Transition Timeline implicitly accepts that there will not be a voluntary transformation. Unfortunately, far too many of the Transitioners perseverate with the usual liberalisms: personal change is political change or personal change is the only change. The Transition Town movement is a decentralized, loosely organized network and the people involved hold a wide range of opinions. It may be that the people who insist on personal change only form a small but vocal minority, and that there is a broad consensus building about the necessity of deep, institutional change—and the activism that will require. But right now, the numbers are on the side of the antipolitical OIMBYs (Only In My Backyard) despite the fact that some of the foundational writings are clear about the necessity of institutional change. This is the deep contradiction in the Transition Town movement.

I would like this to read as more of an observation than a criticism, and, ultimately, an invitation. The Transitioners are trying to create at least some of the local infrastructure with which cultures of resistance are tasked: food, education, methods of economic exchange. What's missing is the recognition that political resistance is necessary. Even if the Transitioners can't see their way clear to militance, they should acknowledge the truth in their own timeline: institutional change, not personal change, is necessary to force this culture away from ecocide and its attendant horrors. No amount of "new stories" will apply the requisite pressure. To revisit Maud Gonne and the Irish struggle, she did not just tell new stories by acting in plays. She fought to win massive land reform, smuggled supplies into prison, nearly died on a hunger strike, served as a judge, forced the British government to feed Irish schoolchildren, and raised a son who won a Nobel Peace Prize. We need the permaculture wing to be Sinn Fein. We need an aboveground group that will vociferously defend direct action and militance, plan for it, support it, work beside it. We need massive pressure aboveground to dismantle corporate personhood, capitalism, civilization, and patriarchy. This includes building alternative institutions to take their place and to structure our cultures on justice and sustainability.

We also need to recognize that aboveground efforts may not be enough, that we're running out of time, 200 species at a time, and a hundredth monkey will not be the answer. This means a realistic assessment, not cloying platitudes or the community confirmation bias of those who think seed swaps are the revolution. It means accepting that as of now we don't have the numbers for a peaceful regime change. It means a stalwart solidarity with the few cadres and combatants who are willing to attempt direct attacks on the infrastructure that is killing our planet. When the governments fail to stop the transformation of carbon into heat and biomes into corporate wealth—and around the world they're failing catastrophically—the OIMBYs will be faced with a choice, as their backyards are drained of amphibians and bled clean of trees that can no longer reach water, along with the rest of the planet. The choice is to fight or to stand with those who fight. Anything else means the world will be left to die.

The case of Cuba is referenced repeatedly by the OINBYs and Transitioners and is worth a serious look. Cuba went through a collapse of its economy in 1989 when the Soviet Union stopped exporting oil (along with financing and manufactured goods) to the dependent nation. Cuba's onshore oil reserves are limited, and it has no ability to drill offshore. The islands have been subjected to the same deforestation that civilization inflicts on every piece of land it touches. Castro believed in industrialization, and he directed resources toward large-scale mechanized farming with near-catastrophic results for the soil and waterways. Mining, cement, and metal industries have also caused their attendant damage.

The scale of the cutbacks was dramatic. Between 1989 and 1993, fuel imports dropped 76 percent and consumer goods, 82 percent. Malnutrition became apparent in children under five in just a few weeks.

The first point that the OIMBYs refuse to grasp is that Cubans did not voluntarily give up an oil economy. On that basis alone, using Cuba as an example for the Transition movement is utterly fallacious.

Did civic order hold in Cuba? Yes. The country did not dissolve into the failed state horrors of the Democratic Republic of the Congo. But Cuba is ruled by a dictator, a point that is conveniently overlooked by the Transitioners who take up Cuba as a positive example. Cubans have lived with food rationing *since 1962*. It was therefore not difficult to keep the social order more or less intact. Though the Transition Timeline includes government rationing as a necessity, this is not a reality that the Transition Town rank and file, with their fervent belief in voluntary simplicity, seemed to have grasped. It also helps that Cuba is small— the size of Virginia—with a population of only eleven million.

The grim facts are that food intake may have fallen as low as 1,863 daily calories. For children and the very old, calories may have dropped to 1,450. Protein intake dropped by 40 percent, to 15–20 g a day, dietary fat dropped by 64 percent, vitamin A by 67 percent.[53] This is a famine. And yet I've witnessed far too many praises of the "health benefits" of this dietary regime in print and in person. Besides the malnourished children evident after a few weeks—deprivation that may well have damaged them for life—there were other broad-scale epidemics that must put the lie to the supposed superiority of this enforced diet. Fifty thousand Cubans were affected by a mysterious outbreak of symptoms: some went blind, others deaf, some lost bladder control, and still others were unable to walk for months or years. A team of physicians from the Pan American Health Organization declared the cause neuropathy due to "spare diet with great physical exertion."[54] The severe vitamin B deficiency, especially thiamine, from lack of animal foods damaged people's nerves. One report states, "The weight of evidence seems to point to a decline in health standards as a consequence of the severe deterioration in food intake." I should not have to cite medical reports to tell people that starvation has negative health consequences. Yet that is the position I find myself in.

For instance, Pat Murphy writes,

> Cubans learned to eat more fruits and vegetables . . . Cubans have been large consumers of meat, but meat required fossil fuel inputs to which they no longer had access. The amount of meat was reduced significantly, and their focus turned to

growing basic nutritious foods. The result has been a much
healthier diet (which reduced rates of heart disease and diabetes) . . . Healthy low energy foods typically imply more fresh
vegetables and fruits while giving up high-fat and sweetened
manufactured foods.[55]

Cuba's domestic food supply was based on an industrial model,
including factory-farmed animals with the attendant grain-feeding
and ethical horrors. Neither factory farming nor grain-feeding are
intrinsic to meat: indeed, for our first four million years, humans
were not in competition with animals for food. We worked in tandem
as participants in soil-building communities. Cuba did not turn to
"growing basic nutritious foods." It turned to growing as many bulk
calories as could be squeezed out of the land. That provided enough
basic energy to keep mass starvation at bay. Cheap carbohydrates will
do that, and nothing more. Cubans' rations contain rice, beans,
sugar, potatoes, and twelve eggs a month. Every fifteen days there is
half a pound of beef mixed with soy or one pound of chicken. Children get some milk. I should not have to argue that these are
starvation rations. The suffering inherent in that list should be
obvious. Since it's not, perhaps this will bring it into stronger relief:
domestic cats disappeared from Cuba's streets, and animals were
stolen from the Havana Zoo. Or maybe this: the direct maternal mortality rate increased 60 percent, and the total maternal mortality rate
increased 43 percent.[56] Rates of tuberculosis, hepatitis, and chicken
pox spiked, and old people died.[57]

The claim that Cuban health improved under starvation rations
traces back to a single study published in the *American Journal of Epi-
demiology*, which claimed that the weight loss reduction due to "reduced
energy intake and increased physical activity" caused a drop in mortality from diabetes and cardiovascular disease. Meanwhile, a
commentator in the *Canadian Medical Association Journal* states, "It
is . . . uncertain whether the all-cause mortality rate . . . and the rates of
death from diabetes mellitus and cardiovascular disease cited . . . have
in fact declined as much as they claimed in parallel with the popula-
tionwide weight loss."[58] There was a 20 percent increase in elder

mortality, which is to say the frailest people died, leaving the survivors as a whole healthier.

How can we know for certain how many people died? We can't. Criticizing the government, including its health care, is a crime in Cuba, which has more journalists in jail than any country except China. Once upon a time, health statistics from China and the former USSR were also quoted favorably and uncritically, while people starved and ate the things that starving people eat: grass, bark, corpses, and children. We should have learned this lesson by now. It's repugnant that anyone could put their emotional needs for an energy descent with a happy ending above the unassailable facts of human suffering.

And for those still clinging to the notion of voluntary transformation, consider: Cuba has contracts with numerous companies in Russia, China, India, Norway, and Brazil to explore oil reserves in the Straits of Florida.[59] Cuba has few fossil fuel reserves on its lands. It does, however, have oil potential offshore. In the US, this drilling would be illegal in such fragile ecosystems. Yet Cuba currently has three offshore production sites, and the explorations continue, proving once and for all that both "voluntary" and "transformation" are rather inapplicable to Cuba.

Russia is another example that's referenced by Transitioners and Descenders. Dmitry Orlov, in his book *Reinventing Collapse*, writes as an "eyewitness," having watched the economic collapse of the Soviet Union. He grew up in Russia and emigrated to the US, and during his periodic visits back he was able to watch the disintegration of the Soviet economy. His description in many ways mirrors John Michael Greer's template of collapse as a series of declines rather than one catastrophic event. The Soviet Union certainly endured economic ruin but its history brings little usable insight to the biotic and climactic collapse that is the subject of this current book.

Cuba's fossil fuel supply was cut off overnight: the Soviet Union's was not. The former Soviet Union's own reserves may have peaked, but the oil is still flowing. The collapse of its economy was due to economic and political policy, where Cuba's hardships were caused by the sudden, drastic lack of fossil fuels and cheap manufactured goods, including food. In the Soviet Union, both production and distribution faltered but never ground to a halt. Indeed, Orlov points out that "Russia was able to bounce back economically because it too remains fairly rich in oil and natural gas, and will probably continue in relative prosperity for at least a few more decades."[60]

Still, the experience of economic and social disintegration is useful to study, because we will certainly be facing those as biological emergencies build into collapse. The first fallout of economic collapse is, in fact, profoundly hopeful. Writes Orlov of his trip in 1990, "I . . . found a place I did not quite recognize. First of all, it smelled different: the smog was gone. The factories had largely shut down, there was very little traffic and the fresh air smelled wonderful."[61] Without the continuous assault of industrialism, the atmosphere and landbase were starting to heal. He continues, "There were very few gas stations open and the ones that were had lines that stretched for many blocks. There was a ten-liter limit on gasoline purchases." As in Cuba, and as in *The Transition Timeline,* government rationing is what forced change while keeping some semblance of civic order, not a sudden outbreak of voluntary goodness.

Orlov is instructive in his description of the black market and barter economies the Russians developed. Vodka was rationed, and a half liter was worth ten liters of gas, "giving vodka far greater effective energy density than rocket fuel."[62] He reaffirms one of the central impulses of the Transitioners: "When faced with a collapsing economy, one should stop thinking of wealth in terms of money. Access to actual physical resources and assets, as well as intangibles such as connections and relationships, quickly becomes much more valuable than mere cash."[63]

He also describes the human misery of old women selling grandchildren's toys to get money for food, of once-professional people digging through public trash bins, and of workers continuing to go to jobs which no longer paid salaries but had a cafeteria and hence a free lunch.

Russia is a country with a negative population growth caused by "a collapse of the birth rate and a catastrophic surge in the death rate."[64] The country has a 0.6 percent population decrease, which means it will lose 22 percent of the population by 2050. That adds up to thirty million fewer people.[65]

One reason for the decline is that Russia has an extremely high involuntary infertility rate. Somewhere between 13 and 20 percent of married couples are infertile, and that number may be rising.[66] For women, one of the main causes was a society-wide reliance on abortion as a form of birth control, abortions often done under substandard medical conditions. The literal scars of such procedures have left many women unable to conceive or carry to term. Sexually transmitted diseases are also a culprit—rates of syphilis are literally hundreds of times higher in Russia than in other European countries.[67] Marriage rates have dropped and divorce rates risen, and 30 percent of Russia's babies are being delivered to single mothers—this in a country too poor to offer public benefits. Women can't afford to have more children.

Add to that a mortality rate that is "utterly breathtaking."[68] Tuberculosis, AIDS, alcoholism, and the disappearance of socialized medicine have pulled the numbers up. The main two causes of death, though, are cardiovascular disease (CVD), which in thirty-five years increased 25 percent for women and an astounding 65 percent for men, and injury. The increases in CVD is traceable to smoking, poor diet, sedentarism, and severe social stress. The injury category includes "murder, suicide, traffic, poisoning and other violent causes."[69] The violence is so bad that the death rate for injury and poisoning for Russian men is twelve times higher than for British men. And both CVD and the violence are helped along by vodka, which Russians drink at an extraordinary rate, equivalent to 125 cc "for everyone, every day."[70]

Population in Russia is dropping dramatically without a cataclysmic event or a Pol Pot–styled genocide, which the authors of this book are often accused of suggesting. Though each individual death is its own world of tragedy, the deaths have not collectively brought daily life—or even the government—to a halt.

Russia may best illustrate the kind of slow decline of which Greer writes; and Russia's disintegration is not even based on energy descent, as oil and gas are still abundant. The former USSR may give us good insights into people's responses to economic decline, and how best to survive it, but as an example it does not address the conditions of biotic collapse that are our fundamental concern.

Except in one instance: Chernobyl. Ninety thousand square miles were contaminated with radiation; 350,000 people were displaced; and there is a permanent "exclusionary zone" encompassing a nineteen-mile radius and the ghosts of seventy-six towns.

But other ghosts have come back from the dead. Because despite the cesium-137 that's deadly for 600 years and the strontium-90 that mammal bones mistake for calcium, Chernobyl has become a miracle of megafauna: the European bison have returned, as well as, somehow, the Przewalski's horse. There are packs—that's plural—of wolves. There are beavers coaxing back the lost wetlands. There are wild boar. There are European lynx. There are endangered birds like the black stork and the white-tailed eagle, glorious in their eight-foot wingspans. All this even though ten years after the accident, geneticists found small rodents with "an extraordinary amount of genetic damage." They had a mutation rate "probably thousands of times greater than normal."[71] Yet twenty years after the accident, and with multiple excursions into the contaminated area, the same researcher, Dr. Robert Baker, said flat-out, "The benefit of excluding humans from this highly contaminated ecosystem appears to outweigh significantly any negative cost associated with Chernobyl radiation."[72] Witnessing the return of bison and wolves, who could say otherwise? Even a nuclear disaster is better for living creatures than civilization. And the real, if fledgling, hope: this planet, made not by some Lord God but instead by the work of all those creatures great and small, could repair herself if we would just stop destroying.

There are better ways to reduce our numbers than through alcoholism, syphilis, and nuclear accidents. We don't need to wring our hands in helpless horror, stuck in a wrenching ethical dilemma between human rights and ecological drawdown. In fact, the most efficacious way to address the twin problems of population and resource depletion is by *supporting* human rights.

One of the great success stories of recent years is Iran. People's desire for children turns out to be very malleable. Even in a context of religious fundamentalism, Iran was able to reduce its birthrate dramatically. In 1979, Ayatollah Khamenei dissolved Iran's family planning efforts because he wanted soldiers for Islam to fight Iraq (and n.b. to those who still think they can be peace activists without being feminist). The population surged in response, reaching a 4.2 percent growth rate, which is the upper limit of what is biologically possible for humans. Iran went from 34 million people in 1979 to 63 million by 1998.[73] Let's be very clear about what this means for women. Girls as young as nine were legally handed over to adult men for sexual abuse: for me, the word "marriage" does not work as a euphemism for the raping of children.

The population surge proved to be a huge social burden immediately, and Iran's leaders "realized that overcrowding, environmental degradation, and unemployment were undermining Iran's future."[74] Health advocates, religious leaders, and community organizers held a summit to strategize.

They knew that free birth control was essential, but it wouldn't be enough. All the major institutions of society had to get involved. Family planning policies were reinstituted and a broad public education effort was launched. Government ministries and the television company were brought into the project: soap operas took up the subject. Fifteen thousand rural clinics were founded and eighty mobile health care clinics brought birth control to remote areas. Thirty-five thousand family planning volunteers were trained to teach people in their neighborhoods about birth control options, and there were also workplace education campaigns. The government got religious leaders to proclaim that Allah wasn't opposed to vasectomies; after that, vasectomies increased dramatically. In order to get a marriage license both halves of the couple had to attend a class on contraception. And new laws withdrew food subsidies and health care coverage after a couple's third child, applying the stick as a backup to the carrots.

The biggest social initiative was to raise the status of women. Female literacy went from 25 percent in 1970 to over 70 percent in 2000. Ninety percent of girls now attend school.[75]

In seven years, Iran's birthrate was sliced in half from seven children per woman to under three. So it can be done, and quickly, by doing the things we should be doing anyway. As Richard Stearns writes, "The single most significant thing that can be done to cure extreme poverty is this: protect, educate, and nurture girls and women and provide them with equal rights and opportunities—educationally, economically, and socially. . . . This one thing can do more to address extreme poverty than food, shelter, health care, economic development, or increased foreign assistance."[76]

There is no reason for people who care about human rights to fear taking on this issue. Two things work to stop overpopulation: ending poverty and ending patriarchy. People are poor because the rich are stealing from them. And most women have no control over how men use our bodies. If the major institutions around the globe would put their efforts behind initiatives like Iran's, there is still every hope that the world could turn toward both justice and sustainability.

Vaxjo, Sweden, is another case that Tilters and Transitioners like to reference. In one primary way, it is a better example than Cuba or Russia: all of Vaxjo's initiatives have been voluntary. And as we shall see, it also serves as an example of how the best renewable options are useless when the goal is industrial civilization.

Sweden is one of the truest democracies on the planet. Comparing their constitution to the current US Constitution is instructive if you know what you are looking for. Article 1 states:

> (1) All public power in Sweden proceeds from the people.
> (2) Swedish democracy is founded on freedom of opinion and on universal and equal suffrage. It shall be realized through a representative and parliamentary polity and through local self-government.

As discussed in "Liberals and Radicals," the vibrant ferment of democracy in the British colonies in 1776 was displaced by the merchant-barons in their quest to privatize and gut this continent. They won, and the continent, and indeed the planet, has been turned into wealth for a very few. Power organized at a decentralized and local level was purposefully written out of the US Constitution: "local self-government" is not a phrase that appears anywhere.

In the US, so many people have given up in apathetic despair; this is understandable and, indeed, predictable. The US Constitution was not set up to empower the vast majority of us. But the Swedish government was, and people in Sweden have a reason to try. Witness the town of Vaxjo. Their emissions of CO_2 decreased a "fantastic" 32 percent from 1993 to 2007.[77] Vaxjo's per capita contributions to global warming stand at three tons of CO_2 per citizen, below the global average of four tons and well below the European average of eight tons. These advances were in large measure due to a switch from oil to biomass for heating; Sweden's forests now supply 90 percent of Vaxjo's heating needs. They have a centralized town heating system that burns sawdust and wood chips, waste products from paper mills and sawmills in the area.[78] Fifty-one percent of Vaxjo's energy comes from "biomass, renewable electricity, and solar."[79]

That sentence sounds so good, so hopeful, but the happy feeling doesn't hold. First, the biggest proportion of Vaxjo's energy comes from outside the city, and is, in fact, generated by hydropower and nuclear power in a 50/50 split. Are hydropower and nuclear power "renewable?" I'm happy to grant the point (only for the sake of discussion), but I will add that just because something is renewable doesn't mean it's good for life on the planet.

Imagine someone inserting a piece of concrete into one of your arteries, stopping the flow of blood. That is what a dam is to a river, except you are only one creature, one life: a river is a multitude of lives and their lives are at serious risk. Ten percent of all animals, including more than 35 percent of all vertebrates, live in freshwater ecosystems, and they are currently going extinct four to six times faster than other animals.[80]

And I don't know what image to use for nuclear power nor why anyone would need an image to understand plutonium-239, a substance so toxic that one-millionth of a gram is carcinogenic. A nuclear power plant will produce 200 kg of it annually, and those 200 kg will last 500,000 years.[81]

And as for "biomass": Sweden is 66.9 percent forest, with 17.2 percent being primary forest, which is the most biodiverse form. Arable land only totals 6.54 percent—Sweden is a cold place. Only 2 percent of the GDP is agriculture.[82] The basis of the economy is "timber, hydropower, and iron ore," and its economy is "heavily oriented toward foreign trade." Other "natural resources" include copper, zinc, lead, uranium, and, to round out the fun, arsenic.

Reforestation is certainly better than the opposite, but let us be very clear about what these "forests" contain:

> These areas have been reforested with single-storied middle-aged and old pine stands . . . during the 20th century. Fire suppression and changes in land use from subsistence-to-industrial forestry facilitated Norway spruce regeneration as undergrowth in open Scots pine stands after logging. This natural regeneration has, to a large extent, been cut down and replaced by pine afforestation. During the second half of the 20th century, the standing timber volume has steadily increased, while the mean age of the forest has decreased. Today's young dense forests will result in higher timber values in the coming decades, but the forest has lost a range of ecological niches.

Or, as the authors state bluntly, "The forest has been transformed into a production unit."[83]

So here's what is really happening in Sweden. There are some important cornerstones in place: real local democracy with citizens who participate, an extensive social safety net, a birthrate of only 1.66 percent, and income redistribution toward an equitable, stable society. Swedes rejected the euro over concerns about the possible destruction of their democracy. Sweden is also the originator of the Swedish Model, which recognizes prostitution as a human rights abuse and has made remarkable strides toward shutting down the commercial sex industry and ending sexual slavery.[84] In many regards this is a society with the right values.

But no one is telling the truth in Sweden, not anymore than in the rest of the world. The Swedish economy is an industrial economy, based on mining, manufacturing precision parts for industrial machines, and wood products. Mining is near *sui generis* in its combination of devastation and extraction. Industrial manufacturing to enable more industrial manufacturing is adding an accelerant to the fire already consuming the planet. And turning the forest into a monocrop tree farm leaves the trees to stand alone, a once-living community reduced to a production unit.

Adding some solar power and a high-tech boiler does not change the nature of the Swedish economy, which is an industrial economy based on globalization. They don't make what they themselves need; instead, they make what other industries need, and then buy what they need with the earnings. Sweden provides for very little of its own food. The Swedish newspaper *Svenska Dagbladet* calculated the distance traveled for fifty common food products. Combined, they went further than the distance between Earth and the moon. All of this—the mining, the manufacturing, the food imports—is only possible because of fossil fuels. A centralized boiler in one small town is nice, but the entire economy is floating on oil. And paper mill refuse cannot possibly scale up to heat the houses of eleven million people during a Nordic winter. Of course, the people who live near the mill should use the waste, but this is not a solution. To suggest it is one is like suggesting dumpster diving as a solution to world hunger.

As for the hydropower, on the entire European continent only four free-flowing rivers are left. They are all in the far north, with distance and cold as their only protection. The rest, civilization has left in shreds. "Fragmentation" (dams) and "regulation" (flow control) are the "manager's" terms. "Starvation" and "asphyxiation" might be the river's. One set of researchers calls dams "the biggest threat" to reindeer and the riparian forest communities on which they depend. The dams on the Lule River, built in the 1960s, "have changed the ecosystems completely."[85] The Lule River has been dammed by fifteen power plants which produce 16 percent of Sweden's electric needs. The Lule is "a massively reengineered" river now. One of the dam's "managers" says, "All of Sweden's lamps are powered by the river."[86] What they are powered by is a dead river.

And as is also true the world over, this destruction has had a terrible impact on the indigenous people. In Sweden these are the Sámi, who are among the last indigenous of Europe. The Sámi are the original inhabitants of much of northern Europe, from the Atlantic side of Sweden to the White Sea in Russia. The archaeological evidence, the beautiful utilities of bone and antler, is 10,000 years old.[87] Their society consists of extended family groups, *siida*, led by an elder, often the oldest person, and both men and women are eligible. Their religious leaders, *noaidi*, can also be both men and women.

The lives of the Sámi depend on reindeer, moose, beaver, and, once upon a time, salmon. Everything we need to remember about being indigenous is here:

> All the Sámi dialects contain a rich vocabulary related to the natural environment. They have numerous, very precise words to describe land, water and snow. There is also a rich, varied vocabulary for reindeer and reindeer breeding. For example, the appearance of a reindeer can be described using a large number of words. Its fur, antlers, sex and age can be conveyed in such detail that in a herd of several thousand animals, only one reindeer fits that particular description.[88]

Nation-states and Christianity have tried to break these people for centuries, but they "did not change the lives of the Sámi to any significant degree." What did, of course, was "when industrialization took off in Sweden and the country needed Sámi's natural resources: metal ores, hydroelectric power and timber."[89]

Land, animals, and people in a tender and sturdy entwinement that lasted 10,000 years, once again destroyed by the relentless assault of civilization and its endless hungers. I can keep asking why, but there's never an answer. What insanity would drive someone to kill a river? What entitlement could justify the scoured wound of a bauxite mine? Sweden is trying to keep these activities viable, and the bright green hope of the globally privileged along with them. But neither the activities nor the hope have a future. A way of life based on drawdown—of soil, species, of life itself—cannot last.

The authors of this book are repeatedly asked, "How do you want people to live?" The question is often thrown like a challenge. The assumption is that civilization is the only way and once pinned to the wall we will be forced to admit that. But while progressives and environmentalists propose solutions that are really just grasping at industrial straws, there are people living sustainably in Sweden, and doing it so intimately they can describe one reindeer out of a thousand. The civilized and the industrialized are still trying to destroy them—the people, reindeer, and rivers—to turn a lacework of interdependence into production units and consumer goods. Still, the Sámi persist. If the civilized could learn by example, surely of all people the Swedes would. But it is not the lack of examples of sustainable, egalitarian, and peaceful cultures that is the problem and it never has been. The problem is power, and the bottomless well of psychopathology that is eating the planet alive.

Our final and in some ways most hopeful example is the small, snowy state of Vermont. Beyond the purchase of a pint Ben and Jerry's (now owned by Unilever), Vermont isn't on many people's minds. It's remote and sparsely populated, with 625,000 citizens. It's also breeding a progressive populism organizing into a serious secession movement. Welcome to the Second Vermont Republic.

Vermont should be getting a lot more attention amongst the Transition Town movement and the left in general. As progressives seem content to swap seeds or wring their hands over the collapse of hope and change, the Second Vermont Republic is building a viable movement to withdraw from the United States and create "a moral, sovereign, and sustainable commonwealth of Vermont towns."[90]

They're quite clear as to the reasons. "The United States leadership is no longer amenable to change through representational democracy. It is bent to the task of preservation of a doomed idea. Our elites are committed to full spectrum dominance on the world stage, to a zero-sum game they're determined to 'win' at any cost, a cost of millions of lives and trillions of dollars, in order to preserve for themselves a moribund 'American way of life.'"[91] Both their founding documents and their newspaper, *Vermont Commons*, lay out the problem in plain Yankee talk. The US is too big, ruled by corporations, and bent on global domination through imperialist wars, while peak oil looms and the planet strains under the demands of a growth economy. Frank Bryan calls the time of the industrial revolution "the two most vicious centuries the world has ever known, ending with the hierarchical, totalitarian *industrial* horrors of Hitler and Stalin."[92] Thomas Naylor, the originator of the movement, writes of "technofascism" and its "affluenza, technomania, e-mania, megalomania, robotism, globalization, and imperialism."[93]

They are equally clear that a sovereign state of Vermont could be won. They point repeatedly to the breakup of the Soviet Union, which was almost entirely nonviolent. The balance sheet is on the side of the benefits, with democracy, sustainability, and human rights at stake. All that is required is the belief in the possibility: at last poll, 13 percent of Vermonters supported secession.

They have their own currency, a silver token stamped with the face of Scott Nearing, he of the Good Life. They have a foreign minister, already establishing relationships abroad. They have a solid statement of principles, ranging from Human Scale to Entrusting the Commons to Food Sovereignty.[94] They currently have a slate of candidates running for governor, lieutenant governor ("It's all about profit and getting the last drops of oil on earth and trampling people's rights"), and seven state senators on a secession platform. They have an A to B plan, and they intend to win, one town meeting at a time.

Town meetings were the original direct democracy by which New Englanders governed themselves. This political form has its origins in the Puritan movement. The Puritans practiced a system of church governance in which each congregation was sovereign and hence governed itself. This was in stark relief to other forms of hierarchical Christianity that were governed by Episcopal or Presbyterian polities. Towns across New England were founded by Puritans and their practice of direct democracy carried over into all local decision-making. Like Sweden and Switzerland, New England's foundations in direct democracy have helped form a regional culture that's tolerant and civic-minded. Unlike Sweden and Switzerland, though, New England town-level decision-making has no structural power and indeed is not even mentioned in the federal system as it was created in 1787.

Town meetings are what Frank Bryant and John McClaughry call *human-scale democracy*. Past a certain number of participants, the process breaks down. Kirkpatrick Sale, the original defender of the human scale, and a stalwart critic of technology, posits that somewhere between 5,000 and 15,000 works as a district size for direct democratic voting. Larger than that, the political process must revert to representative democracy. Representatives of fifty to one hundred districts are workable for nations, which means an upper limit of one million citizens.[95] Vermont's 625,000 people is perfect.

Town meetings are an annual event where residents gather to make decisions about operating budgets, elect municipal officers, and legislate policy. In Vermont, Town Meeting Day, the first Tuesday in March, is a state holiday—that's what people do when they take their democracy seriously. (And election day in the US isn't a federal holiday. Why?) For the skeptical, 90 percent of Swiss municipalities are run by town meetings. Switzerland is serving as a model for the Second Vermont Republic, and it's a model the aboveground wing of DGR—the Transitioners and the permaculturists—could also emulate.

Switzerland is essentially a federation of small towns, with unique supports for direct democracy. Switzerland is not a homogenous society; indeed, the Swiss have four national languages, all with distinct cultures. Yet they have managed to meld themselves into peaceable coexistence and a single political entity. The last bloodshed was in 1847, when civil war broke out. The conflict was over in a month with fewer than one hundred dead, and most of those through friendly fire. This level of conflict, when compared to the vast bloodletting that would continue to soak Europe, seems almost quaint, like hobbits killing hobbits. But their civil war had a huge impact on Swiss psychology and culture. It brought all parties to the table and resulted in a federal constitution that empowered local self-government. The constitution included the provision that the entire document could be scrapped and rewritten if it wasn't working, which the Swiss have done twice. Not everything the Swiss have done is perfect (though their chocolate comes close)—placating the Nazis with financing, for instance, falls rather short of anyone's moral mark—but they do provide a living, breathing model of a peaceful, multicultural society. Such models are not, in fact, in short supply. All that we are missing is people willing to believe in the possibilities and to fight for them.

Besides its living tradition of direct democracy, Vermont has a few other historical currents on its side. Robert Putnam ranks Vermont number one on his scale of "tolerance for gender, racial, and civil liberties."[96] Vermont also ranks first on measures of a civil society. Frank Bryan calls this combination "a living nexus between liberty and community."[97] The Vermont Constitution was the first to outlaw slavery and to remove property ownership as a barrier to voting. Vermont held onto its human-scale democracy in large part because of its location and climate. As the rest of the country embraced urban industrialism, Vermont was "left behind," as Bryan puts it. "This turned out to be a blessing." He explains, "The concentration of socio-economic life, which was necessary to sustain the urban-industrial era, relied on hierarchy—the classic 20th-century pyramid of roles and duties arranged to control organizational activity from the top down. Hierarchy requires authority, which promotes symmetry, which causes rigidity. The result is awkward, reactionary and (most important) insensitive—and thus inhumane."[98]

Young people left Vermont in record numbers for jobs in industrial areas, leaving it the most rural state in America by 1950. Then Helen and Scott Nearing called up a movement to reject mass society, with its militarism and materialism, and embrace self- and local sufficiency, mutual aid, and radical anti-imperialism. Their 1954 book, *Living the Good Life*, was the foundation of the back-to-the-land movement. Between 1967 and 1973, as many as 100,000 people heeded the call and headed for Vermont. The cold, rocky soil would never grow much, but the leftist embrace of the rural found fertile ground in Vermont. Like so many cultures of resistance, this one took time to put down roots and begin to branch, but fifty years later the tree is bearing fruit. Its challenge to empire, to corporate capitalism, and even to industrialization is a serious one.

Like everywhere else, Vermont has been gutted.

> Vermont resembles an economic colony more than a sovereign state. Our major minerals are owned by foreign corporations (Omya), our ground water is exported by out-of-state bottling companies (Coca-Cola and Nestle), our hydropower resources are owned by TransCanada, and 88% of surface-water withdrawals in Vermont are used by Vermont Yankee [nuclear power plant] for cooling water at no charge. The federal government, meanwhile, has given away 98 percent of our "public airwaves" for free, and allows private banks to create 93 percent bid of the currency with interest attached. Citizens and businesses are subject to taxation of earned income, which

impact job creation and economic productivity, while resource owners collect massive amounts of unearned income.[99]

And that's not including the Connecticut River, which should be 400 miles solid with Atlantic salmon, absent since 1798 because of the dams of industrialization.

But there are people in Vermont—citizens first and foremost of the Green Mountains they named themselves after, and second of a community of neighbors—who are not standing by bewildered or hopeless. This is no anemic walking away into psychological withdrawal. This is a gauntlet of withdrawal, thrown down by a tiny David of patriots—true patriots, defending their land and their community from a Goliath of power that will not stop and cannot be reformed. The Second Vermont Republic "rationalizes our instincts, electrifies our commitment, and sustains our courage."[100]

It also stands as a challenge to the permaculture wing and the Transitioners who want to do something to save the world, but have yet to understand the nature of power. Don't just swap seeds: swap the US Constitution for local direct democracies confederated across your bioregion. Swap capitalism and its sociopathic corporate personhood for local economies based on human needs and human morality. Swap the rapacious drawdown of civilization for a culture nestled inside a repaired community of forests and grasses, filling once more with species with whom we must share this home.

This will require a resistance movement, which is always greater than the sum of personal actions, no matter how noble or restorative those actions. The Second Vermont Republic takes its place in a long line of movements for justice that were willing to face the nature of power and then to face it down. This planet does not simply need more great gardens: it needs resistance against the forces that have been plundering our collective garden for 10,000 years.

Perhaps one day the people of Vermont will speak in a dialect that can identify one sugar maple out of a thousand, one hesitant salmon restored to a river bereft of her kin for 200 years, one decision well-made on a snowy Tuesday in March.

Chapter 6

A Taxonomy of Action

by Aric McBay

And here yet another temptation asserts itself. Why not wait until our cause becomes vivid and urgent enough, and our side numerous enough, to vote our opponents out of office? Why not be patient? My own answer is that while we are being patient, more mountains, forests, and streams, more people's homes and lives, will be destroyed in the Appalachian coal fields. Are 400,000 acres of devastated land, and 1,200 miles of obliterated streams not enough? This needs to be stopped. It does not need to be "regulated." As both federal and state governments have amply shown, you cannot regulate an abomination. You have got to stop it.

—Wendell Berry, author and farmer

We got further smashing windows than we ever got letting them smash our heads.

—Christabel Pankhurst, suffragist

What is at stake? Whippoorwills, the female so loyal to her young she won't leave her nest unless stepped on, the male piping his mating song of pure liturgy. They are 97 percent gone from their eastern range.

What is at stake? Mycorrhizal fungi, feeding their chosen plant companions and helping to create soil, with miles of filament in a teaspoon of earth. Bluefin tuna, warm-blooded and shimmering with speed. The eldritch beauty of amanita mushrooms. The mission blue butterfly, a fairy creature if there ever was one. A hundred miles of river turned silver with fish. A thousand autumn wings urging home. A million tiny radicles anchoring into earth, each with a dream of leaves, a lace of miracles, each thread both fierce and fragile, holding the others in place.

If you love this planet, it's time to put away the distractions that have no potential to stop this destruction: lifestyle adjustments, consumer choices, moral purity. And it's time to put away the diversion of hope, the last, useless weapon of the desperate.

We have better weapons. If you love this planet, it's time to put them all on the table and make some decisions.

What do we want? We want global warming to stop. We want to end the globalized exploitation of the poor. We want to stop the planet from being devoured alive. And we want the planet to recover and rejuvenate.

We want, in no uncertain terms, to bring down civilization.

As Derrick succinctly wrote in *Endgame*, "Bringing down civilization means depriving the rich of their ability to steal from the poor, and it means depriving the powerful of their ability to destroy the planet." It means thoroughly destroying the political, social, physical, and technological infrastructure that not only permits the rich to steal and the powerful to destroy, but rewards them for doing so.

◙ ◙ ◙

The strategies and tactics we choose must be part of a grander strategy. This is not the same as movement-building; taking down civilization does not require a majority or a single coherent movement. A grand strategy is necessarily diverse and decentralized, and will include many kinds of actionists. If those in power seek Full-Spectrum Dominance, then we need *Full-Spectrum Resistance*.[1]

Effective action often requires a high degree of risk or personal sacrifice, so the absence of a plausible grand strategy discourages many genuinely radical people from acting. Why should I take risks with my own safety for symbolic or useless acts? One purpose of this book is to identify plausible strategies for winning.

If we want to win, we must learn the lessons of history. Let's take a closer look at what has made past resistance movements effective. Are there general criteria to judge effectiveness? Can we tell whether tactics or strategies from historical examples will work for us? Is there a general model—a kind of catalog or taxonomy of action—from which resistance groups can pick and choose?

The answer to each of these questions is yes.

To learn from historical groups we need four specific types of information: their goals, strategies, tactics, and organization.

Goals can tell us what a certain movement aimed to accomplish and whether it was ultimately successful on its own terms. Did they do what they said they wanted to?

Strategies and tactics are two different things. *Strategies* are long-term, large-scale plans to reach goals. Historian Liddell Hart called military strategy "the art of distributing and applying military means to fulfill the ends of policy."[2] The Allied bombing of German infrastructure during WWII is an example of one successful strategy. Others include the civil rights boycotts of prosegregation businesses and suffragist strategies of petitioning and pressuring political candidates directly and indirectly through acts that included property destruction and arson.

Tactics, on the other hand, are short-term, smaller-scale actions; they are particular acts which put strategies into effect. If the strategy is systematic bombing, the tactic might be an Allied bombing flight to target a particular factory. The civil rights boycott strategy employed tactics such as pickets and protests at particular stores. The suffragists met their strategic goal by planning small-scale arson attacks on particular buildings. Successful tactics are tailored to fit particular situations, and they match the people and resources available.

Organization is the way in which a group composes itself to carry out acts of resistance. Resistance movements can vary in size from atomized individuals to large, centrally run bureaucracies, and how a group organizes itself determines what strategies and tactics it is capable of undertaking. Is the group centralized or decentralized? Does it have rank and hierarchy or is it explicitly anarchist in nature? Is the group heavily organized with codes of conduct and policies or is it an improvisational "ad hocracy?" Who is a member, and how are members recruited? And so on.

A TAXONOMY OF ACTION

We've all seen biological taxonomies, which categorize living organisms by kingdom and phylum down to genus and species. Though there are tens of millions of living species of vastly different shapes, sizes, and habitats, we can use a taxonomy to quickly zero in on a tiny group.

When we seek effective strategies and tactics, we have to sort through millions of past and potential actions, most of which are either historical failures or dead ends. We can save ourselves a lot of time and a lot of anguish with a quick and dirty resistance taxonomy. By looking over whole branches of action at once we can quickly judge which tactics are actually appropriate and effective for saving the planet (and for many specific kinds of social and ecological justice activism). A taxonomy of action can also suggest tactics we might otherwise overlook.

Broadly speaking, we can divide all of our tactics and projects either into acts of omission or acts of commission (Figure 6-1). Of course, sometimes these categories overlap. A protest can be a means to lobby a government, a way of raising public awareness, a targeted tactic of economic disruption, or all three, depending on the intent and organization. And sometimes one tactic can support another; an act of omission like a labor strike is much more likely to be effective when combined with propagandizing and protest.

In a moment we'll do a quick tour of our taxonomic options for resistance. But first, a warning. Learning the lessons of history will offer us many gifts, but these gifts aren't free. They come with a burden. Yes, the stories of those who fight back are full of courage, brilliance, and drama. And yes, we can find insight and inspiration in both their triumphs and their tragedies. But the burden of history is this: *there is no easy way out.*

In *Star Trek*, every problem can be solved in the final scene by reversing the polarity of the deflector array. But that isn't reality, and that isn't our future. Every resistance victory has been won by blood and tears, with anguish and sacrifice. Our burden is the knowledge that there are only so many ways to resist, that these ways have already been invented, and they all involve profound and dangerous struggle. When resisters win, it is because they fight harder than they thought possible.

And this is the second part of our burden. Once we learn the stories of those who fight back—once we *really* learn them, once we cry over them, once we inscribe them in our hearts, once we carry them in our bodies like a war veteran carries aching shrapnel—we have no choice but to fight back ourselves. Only by doing that can we hope to live up to their example. People have fought back under the most adverse and awful conditions imaginable; those people are our kin in the struggle for justice and for a livable future. And we find those people—our

A TAXONOMY OF ACTION
(political, social, and economic noncooperation)

Figure 6-1

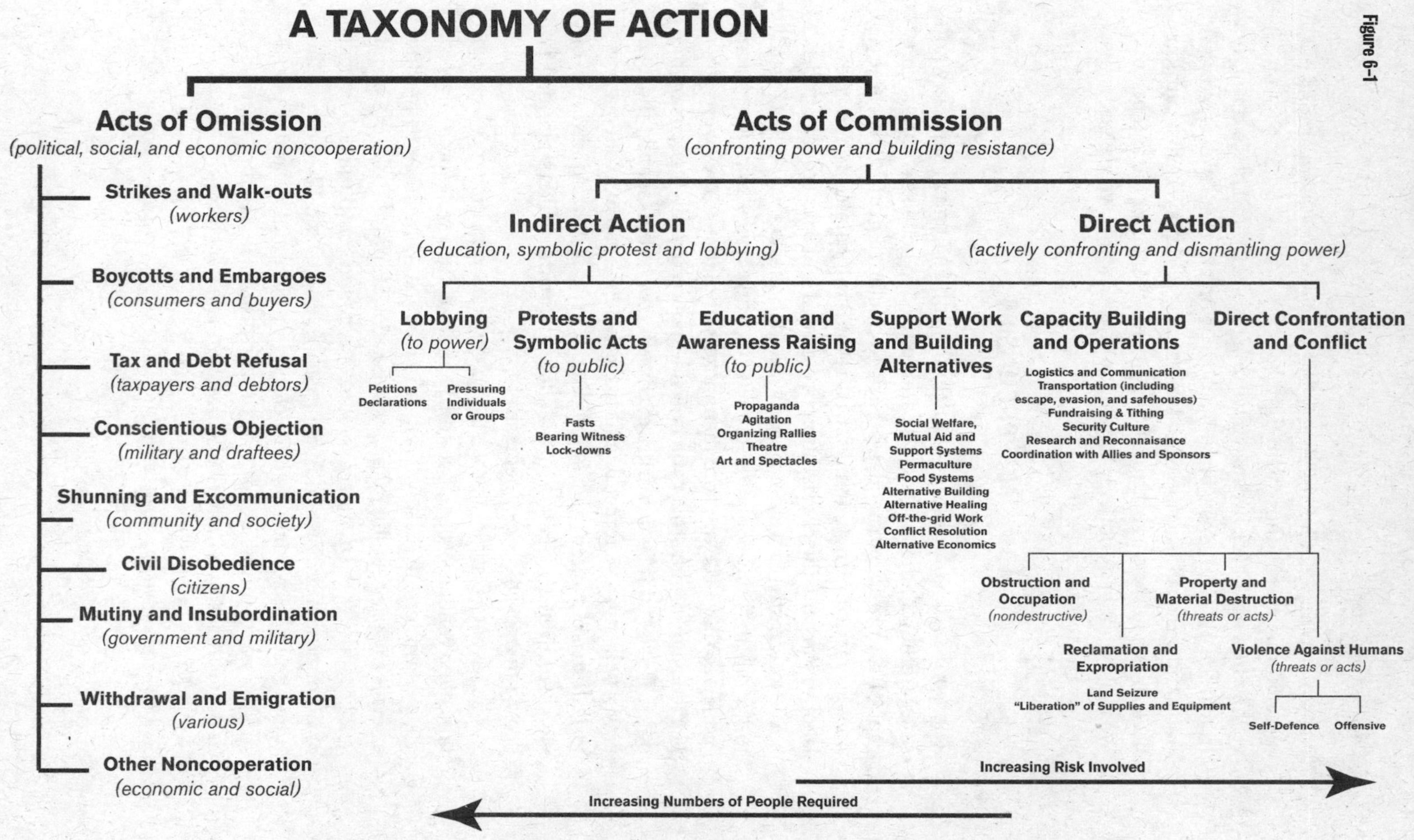

Acts of Omission
(political, social, and economic noncooperation)

- **Strikes and Walk-outs**
  (workers)
- **Boycotts and Embargoes**
  (consumers and buyers)
- **Tax and Debt Refusal**
  (taxpayers and debtors)
- **Conscientious Objection**
  (military and draftees)
- **Shunning and Excommunication**
  (community and society)
- **Civil Disobedience**
  (citizens)
- **Mutiny and Insubordination**
  (government and military)
- **Withdrawal and Emigration**
  (various)
- **Other Noncooperation**
  (economic and social)

Acts of Commission
(confronting power and building resistance)

Indirect Action
(education, symbolic protest and lobbying)

- **Lobbying**
  (to power)
  - Petitions
  - Declarations
  - Pressuring Individuals or Groups
- **Protests and Symbolic Acts**
  (to public)
  - Fasts
  - Bearing Witness
  - Lock-downs
- **Education and Awareness Raising**
  (to public)
  - Propaganda
  - Agitation
  - Organizing Rallies
  - Theatre
  - Art and Spectacles

Direct Action
(actively confronting and dismantling power)

- **Support Work and Building Alternatives**
  - Social Welfare, Mutual Aid and Support Systems
  - Permaculture
  - Food Systems
  - Alternative Building
  - Alternative Healing
  - Off-the-grid Work
  - Conflict Resolution
  - Alternative Economics
- **Capacity Building and Operations**
  - Logistics and Communication
  - Transportation (including escape, evasion, and safehouses)
  - Fundraising & Tithing
  - Security Culture
  - Research and Reconnaissance
  - Coordination with Allies and Sponsors
- **Direct Confrontation and Conflict**
  - Obstruction and Occupation
    (nondestructive)
  - Reclamation and Expropriation
    - Land Seizure
    - "Liberation" of Supplies and Equipment
  - Property and Material Destruction
    (threats or acts)
  - Violence Against Humans
    (threats or acts)
    - Self-Defence
    - Offensive

Increasing Risk Involved

Increasing Numbers of People Required

courageous kin—not just in history, but now. We find them among not just humans, but all those who fight back.

We must fight back because if we don't we will die. This is certainly true in the physical sense, but it is also true on another level. Once you *really* know the self-sacrifice and tirelessness and bravery that our kin have shown in the darkest times, you must either act or die as a person. We must fight back not only to win, but to show that we are both alive and worthy of that life.

ACTS OF OMISSION

The word *strike* comes from eighteenth-century English sailors, who *struck* (removed) their ship's sails and refused to go to sea, but the concept of a workers' strike dates back to ancient Egypt.[3] It became a popular tactic during the industrial revolution, parallel to the rise of labor unions and the proliferation of crowded and dangerous factories.

Historical strikes were not solely acts of omission. Capitalists went to great lengths to violently prevent or end strikes that cost them money, so they became more than pickets or marches; they were often pitched battles, with strikers on one side, police and hired goons on the other. This should be no surprise; any effective action against those in power will trigger a forceful, and likely violent, response. Hence, historical strikers often had a pragmatic attitude toward the use of violence. Even if opposed to violence, historical strikers planned to defend themselves out of necessity.

The May 1968 student protests and general strike in France—which rallied ten million people, two-thirds of the French workforce—forced the government to dissolve and call elections, (as well as triggering extensive police brutality). The 1980 Gdańsk Shipyard strike in Poland sparked a series of strikes across the country and contributed to the fall of Communism in Eastern Europe; strike leader Lech Wałęsa won the Nobel Peace Prize and was later elected president of Poland. General strikes were common in Spain in the early twentieth century, especially in the years leading up to the civil war and anarchist revolution.

Boycotts and embargoes have been crucial in many struggles: from boycotts of slave-produced goods in the US, to civil rights struggles and the Montgomery bus boycott in the name of civil rights, to the antiapartheid boycotts; to company-specific boycotts of Nestlé, Ford, or Philip Morris.

The practice of boycotting predates the name it itself. Captain Charles Boycott was the agent of an absentee landlord in Ireland in 1880. Captain Boycott evicted tenants who had demanded rent reductions, so the community fought back by socially and economically isolating him. People refused to work for him, sell things to him, or trade with him—the postman even refused to deliver his mail. The British government was forced to bring in fifty outside workers to undertake the harvest, and protected the workers with *one thousand* police. This show of force meant that it cost over £10,000 to harvest £350 of potatoes.[4] Boycott fled to England, and his name entered the lexicon.

As we have discussed, consumer spending is a small lever for resistance movements, since most spending is done by corporations, governments, and other institutions. If we ignore the obligatory food, housing, and health care, Americans spend around $2.7 trillion dollars per year on their clothing, insurance, transportation, and other expenses.[5] Government spending might be $4.4 trillion, with corporations spending $1 trillion on marketing alone.[6] Discretionary consumer spending is small, and even if a boycott were effective against a corporation, the state would bail out that corporation with tax money, as they've made clear.

But there's no question that boycotts can be very effective in specific situations. The original example of Captain Boycott shows some conditions that lead to successful action: the participation of an entire community, the use of additional force beyond economic measures, and the context of a geographically limited social and economic realm. Such actions helped lead to what Irish labor agitator and politician Michael Davitt called "the fall of feudalism in Ireland."[7]

Of course there are exceptional circumstances. When the winter's load of chicken feed arrived on the farm today, the mayor was driving the delivery truck, nosing carefully through a herd of curious cattle. But most people don't take deliveries from their elected officials, and—with apologies to Mayor Jim—the mayors of tiny islands don't wield much power on a global scale.

Indeed, corporate globalization has wrought a much different situation than the old rural arrangement. There is no single community that can be unified to offer a solid front of resistance. When corporations encounter trouble from labor or simply want to pay lower wages, they move their operations elsewhere. And those in power are so segregated from the rest of us socially, economically, culturally, and physically that enforcing social shaming or shunning is almost impossible.

Even if we want to be optimistic and say that a large number of people could decide to engage in a boycott of the biggest ten corporations, it's completely reasonable to expect that if a boycott seriously threatened the interests of those in power, they would simply make the boycott illegal.

In fact, the United States already has several antiboycott laws on the books, dating from the 1970s. The US Bureau of Industry and Security's Office of Antiboycott Compliance explains that these laws were meant "to encourage, and in specified cases, require US firms to refuse to participate in foreign boycotts that the United States does not sanction." The laws prohibit businesses from participating in boycotts, and from sharing information which can aid boycotters. In addition, inquiries must be reported to the government. For example, the *Kansas City Star* reports that a company based in Kansas City was fined $6,000 for answering a customer's question about whether their product contained materials made in Israel (which it did not) and for failing to report that inquiry to the Bureau of Industry and Security.[8] American law allows the bureau to fine businesses "up to $50,000, or five times the value" of the products in question. The laws don't just apply to corporations, but are intended "to counteract the participation of US citizens" in boycotts and embargoes "which run counter to US policy."[9]

Certainly, large numbers of committed people can use boycotts to exert major pressure on governments or corporations that can result in policy changes. But boycotts alone are unlikely to result in major structural overhauls to capitalism or civilization at large, and will certainly not result in their overthrow.

Like the strike and the boycott, *tax refusal* has a long history. Rebellions have erupted and wars have been waged over taxes; from the British colonial "hut taxes" to the Boston Tea Party.[10] Even if taxation is not the cause of a war, tax refusal is likely to play a part, either as a way of resisting unjust wars (as the Quakers have historically done) or as part of a revolutionary struggle (as in a German revolution in which Karl Marx proclaimed, "Refusal to pay taxes is the primary duty of the citizen!").[11]

The success of tax refusal is usually low, partly because people already try to avoid taxes for nonpolitical reasons. In the US, 41 percent of adults do not pay federal income tax to begin with, so it's reasonable to conclude that the government could absorb (or compensate for) even high levels of tax refusal.[12]

Even though tax refusal will not bring down civilization, there are times when it could be especially decisive. Regional or local governments on the verge of bankruptcy may be forced to close prisons or stop funding new infrastructure in order to save costs, and organized tax resistance could help drive such trends while diverting money to grassroots social or ecological programs.

Through *conscientious objection* people refuse to engage in military service, or, in some cases, accept only noncombatant roles in the military. Occasionally these are people who are already in the military who have had a change of heart.

Although conscientious objection has certainly saved people from having to kill, it doesn't always save people from dying or the risk of death, since the punishments or alternative jobs like mining or bomb disposal are also inherently dangerous. It's unlikely that conscientious objection has ever ended a war or even caused significant troop shortages. Governments short of troops usually enact or increase conscription to fill out the ranks. Where alternative service programs have existed, the conscientious objectors have usually done traditional masculine work, like farming and logging, thus freeing up other men to go to war. Conscientious objection alone is unlikely to be an effective form of resistance against war or governments.

For those already in the military, *mutiny and insubordination* are the chief available acts of omission. In theory, soldiers have the right, even obligation, to refuse illegal orders. In practice, individual soldiers rarely defy the coercion of their superiors and their units. And refusing an illegal order only works when an atrocity *is* illegal at the time; war criminals at Nuremberg argued that there were no laws against what they did.

Since individual insubordination may result in severe punishment, military personnel sometimes join together to mutiny. But large-scale refusal of orders is almost unheard of because of the culture, indoctrination, and threat of punishment in the military (there are notable exceptions, like the mutiny on the Russian battleship *Potemkin* or the mass mutinies of Russian soldiers during the February Revolution). Perhaps a greater cause for hope is the potential that military personnel, who often have very useful skills sets, will join more active resistance groups.

Shunning and shaming are sometimes used for severe social transgressions and wrongdoing, such as domestic or child abuse, or rape. These tactics are more likely to be effective in close-knit or low-density communities, which are not as common in the modern and urbanized world, although particular communities (such as enclaves of immigrants) may also be set apart for language or cultural reasons. The effect of shunning can be vastly increased in situations like that of Captain Boycott, in which social relations are also economic relations. However, since most economic transactions (either employment or consumption) are mediated by large, faceless corporations and alienated labor, this is rarely possible in the modern day.

Shunning requires a majority to be effective, so it's not a tool that can be used to bring down civilization, although it can still be used to discourage wrongdoing within communities, including activist communities.

Civil disobedience, the refusal to follow unjust laws and customs, is a fundamental act of omission. It has led to genuine successes, as in the civil rights campaign in Birmingham, Alabama. In the 1960s Birmingham was among the most racially segregated cites in the US, with segregation legally required and vigorously enforced.[13] The Commissioner of Public Safety was "arch-segregationist" Bull Connor, a vicious racist even by the standards of the time.[14] Persecution of black people

by the police and other institutions was especially bad. The local government went to great lengths to try to quash any change; for example,
when courts ruled segregation of city parks unlawful, the city closed
the parks. However, civil rights activists, including Martin Luther King
Jr., were able to conduct a successful antisegregation campaign and
turn this particularly nasty situation into a victory.

The Birmingham campaign used many different tactics, which gave
it flexibility and strength. It began with a series of economic boycotts
against businesses that promoted or tolerated segregation. Starting in
1962, these boycotts targeted downtown businesses and decreased
sales by as much as 40 percent.[15] Black organizers patrolled for people
breaking the boycott. When they found black people shopping in a
target store, they confronted them publically and shamed them into
participating in the boycott, even destroying purchased merchandise.
When several businesses took down their segregation signs, Commissioner Connor threatened to revoke their business licenses.[16]

The next step in the civil disobedience campaign was "Project C,"
the systematic violation of segregation laws. Organizers timed walking
distances between the campaign headquarters and various targets, and
conducted reconnaissance of segregated lunch counters, all-white
churches, stores, federal buildings, and so on.[17] The campaign participants then staged sit-ins at the various buildings, libraries, and lunch
counters (or, in the case of the white churches, kneel-ins). Businesses
mostly refused to serve the protesters, some of whom were spat on by
white customers, and hundreds of the protesters were arrested. Some
observers, black and white, considered Project C to be an extremist
approach, and criticized King and the protesters for not simply sticking
to negotiation. "Wasteful and worthless," proclaimed the city's black
newspaper.[18] A statement by eight white clergyman called the demonstrations "unwise and untimely," and wrote that such protests "incite to
hatred and violence" when black people should focus on "working
peacefully."[19] (Of course, they blamed the victim. Of course, they cautioned that an action like sitting down in a deli and ordering a sandwich
is only "technically peaceful" and warned against such "extreme measures." And, of course, it's never the right time, is it?)

The city promptly obtained an injunction against the protests and quadrupled the bail for arrestees to $1,200 per person (more than $8,000 in 2010 currency).[20] But the protests continued, and two days later fifty people were arrested, including Martin Luther King Jr. Instead of paying bail for King, the organizers allowed the police to keep him in prison to draw attention to the struggle. National attention meant the expansion of boycotts; national retail chains started to suffer, and their bosses put pressure on the White House to deal with the situation.

Despite the attention, the campaign began to run out of protesters willing to risk arrest. So they used a controversial plan called the "Children's Crusade," recruiting young students to join in the protests.[21] Organizers held workshops to show films of other protests and to help the young people deal with their fear of jail and police dogs. On May 2, 1963, more than a thousand students skipped school to join the protest, some scaling the walls around their school after a principal attempted to lock them in.[22] Six hundred of them, some as young as eight, were arrested.

Firehoses and police dogs were used against the marching students. The now-iconic images of this violence drew immense sympathy for the protesters and galvanized the black community in Birmingham. The situation came to a head on May 7, 1963, when thousands of protestors flooded the streets and all business ceased; the city was essentially defeated.[23] Business leaders were the first to support the protestors' demands, and soon the politicians (under pressure from President Kennedy) had no choice but to capitulate and agree to a compromise with King and the other organizers.

But no resistance comes without reprisals. Martin Luther King Jr.'s house was bombed. So was a hotel he was staying at. His brother's house was bombed. Protest leader Fred Shuttlesworth's house was bombed. The home of an NAACP attorney was bombed.[24] Some blamed the KKK, but no one was caught. A few months later the KKK bombed a Baptist church, killing four girls.[25]

And the compromise was controversial. Some felt that King had made a deal too soon, that the terms were less than even the moderate demands. In any case, the victorious campaign in Birmingham is widely regarded as a watershed for the civil rights movement, and a model for success.

Let's compare the goals of Birmingham with our goals in this book. The Birmingham success was achieved because the black protestors *wanted* to participate in economy and government. Indeed, that was the crux of the struggle, to be able to participate more actively and equally in the economy, in government, and in civil society. Because they were so numerous (they made up about one-third of the city's population) and because they were so driven, their threat of selective withdrawal from the economy was very powerful (I almost wrote "persuasive," but the point *is* that they stopped relying on persuasion alone).

But what if you don't want to participate in capitalism or in the US government? What if you don't even want those things to exist? Boycotts aren't very persuasive to business leaders if the boycotts are intended to be permanent. The Birmingham civil rights activists forced those in power to change the law by penalizing their behavior, by increasing the cost of business as usual to the point where it became easier and more economically viable for government to accede to their demands.

There's no doubt that we can *try* to apply the same approach in our situation. We can apply penalties to bad behavior, both on community and global scales. But the dominant culture functions by taking more than it gives back, by being unsustainable. In order to get people to change, we would have to apply a penalty proportionally massive. To try to persuade those in power to made serious change is folly; it's effectively impossible to make truly sustainable decisions within the framework of the dominant system. And persuasion can only work on *people*, whereas we are dealing with massive social machines like corporations, which are functionally sociopathic.

In any case, what we call civil disobedience perhaps is the prototypical act of omission, and a requirement for more than a few acts of commission. Refusing to follow an unjust law is one step on the way to working more actively against it.

The most generalized act of omission is a *withdrawal* from larger society or *emigration* to a different society. Both are common in history. These choices are often the result of desperation, of a sense of having run out of other options, of the status quo being simply intolerable. Of course, if the culture you are leaving is so terrible, good people leaving is unlikely to reform or improve it. Which doesn't mean people shouldn't emigrate or try to leave intolerable or dangerous social situations. It just means that leaving, in and of itself, isn't a political strategy likely to affect positive change.

Perhaps the biggest problem with withdrawal as a strategy now is that civilization is global. Where are you going to go? Where do you think you can escape climate change, for example? And what real effect will withdrawal have on the dominant culture? There is no shortage of labor, so huge numbers of people would have to withdraw in order to make a difference. Not buying things will not end the capitalist economy, and refusing to pay taxes will not bring down the government. If you *did* have enough people to do such things, you would become a threat, a dangerous example, and would be treated accordingly. As soon as enough people withdrew to become a bad example, civilization would go after them, thus ending their withdrawal and forcing them to engage with it, either by giving in or by fighting back.

History already tells us that withdrawing is not an option that the civilized will allow. First Nations people across Canada and the US, for instance, were not allowed to remain outside of the invading European civilization. Their children were taken by force to be abused—"enculturated"—and forced into settler culture.

It's a paradox. Withdrawal can only persist when it is ineffective, and so is useless as a resistance strategy.

Other acts of noncooperation can operate in smaller contexts such as individual religious temples, for example, or romantic relationships (as in the Lysistratan example). These smaller social structures may not have as great an impact on society at large, but smaller numbers of people are required to affect change. If nothing else, it's good practice.

◌ ◌ ◌

All acts of omission require very large numbers of people to be permanently effective on a large scale. There are plenty of examples of strikes shutting down factories temporarily, but what if you don't ever want that factory to run again? What if you work at a cruise missile factory or a factory that manufactures nuclear warheads? Is everyone working there willing to go on strike indefinitely? The large pool of unemployed or underpaid working poor means that there are always people willing to step in to work for a wage, even a relatively low one. Failing that, the company in question could just move the factory overseas, as so many have. All of this is especially true in a time when capitalism falters, and attempting to bring down civilization would definitely make capitalism falter.

The same problems apply to economic boycotts. You and I could stop buying anything produced by a given company. Or we could stop buying anything that had been sold through the global capitalist economy. We probably *will* see widespread acts of economic omission, but only when large numbers of people get too poor to buy mass-produced consumer luxuries. But because of globalization and automation, these acts of omission will be less effective than they were in the past.

Which isn't to say we shouldn't undertake such acts when appropriate. Acts of omission are commonly part of resistance movements; they may be implicit rather than explicit. Pre-Civil War abolitionists would not have owned slaves. But this was an implicit result of their morality and political philosophy rather than a means of change. Few abolitionists would have suggested that by refraining from personally owning slaves they were posing a serious or fundamental threat to the institution of slavery.

An effective resistance movement based on acts of omission might need 10 percent, or 50 percent, or 90 percent of the population to win. One in a thousand people withdrawing from the global economy would have negligible impact. Acts of commission are a different story. What if one out of a thousand people joined a campaign of direct action to bring down civilization? Seven million brave and smart people could ensure the survival of our planet.

$$\boxtimes \quad \boxtimes \quad \boxtimes$$

If we are going to talk about survival—or about courage, for that matter—we should talk about Sobibór. Sobibór was a Nazi concentration camp built in a remote part of Poland near the German border. Brought into operation in April 1943, Sobibór received regular train loads of prisoners, almost all Jewish. Like other Nazi concentration camps, Sobibór was also a work camp, both for prisoners skilled in certain trades and for unskilled labor, such as body removal. Sobibór was not the largest concentration camp, but it ran with murderous efficiency. Records show that by October 1944 a quarter of a million people had been murdered there, and some argue the casualties were significantly higher.[26]

Sobibór presented two distinct faces. Upon arrival to the camp, those selected to be killed received a polite welcoming speech from the Nazis (sometimes dressed in lab coats to project expertise and authority), and heard classical music played over loudspeakers. The door to the extermination "showers" was decorated with flowers and a Star of David. Touches like these encouraged them to go quietly and calmly to what some surely realized was their death. In contrast, those who were selected for work were shown a more overtly violent face, suffering arbitrary beatings and sometimes killed for even the smallest failure in cooperation. As at other concentration camps, if individual prisoners even attempted to escape, other prisoners would be killed as a reprisal. (At Auschwitz it was common practice for the SS to kill ten random prisoners for each escapee.)

Sobibór is a lesson for us because it became the site of the most successful—and also the most audacious—concentration camp uprising during the entire Holocaust. A small number of prisoners recognized that it was only a matter of time until they, too, were murdered, and decided that it was worth the risk to escape. However, they knew that those left behind would suffer the consequences of their act. So they hatched a bold plan to allow *everyone* in the camp to escape.

This was not an easy task. The camp was surrounded by multiple razor wire fences and a minefield, beyond which was forest. In addition to the SS, the camp had SS-trained guards of various Eastern European nationalities, guards who had themselves been brought in from POW camps. The perimeter of the camp had bright lighting systems and numerous machine gun towers.

A breakthrough came with the arrival of a group of Jewish-Russian POWs, with whom the long-time prisoners joined together and devised an escape plan. But to avoid being discovered, they had to keep the plan secret from all but a small group, meaning that the majority of the prisoners would be expected to escape at a moment's notice without preparation. A Russian POW leader, Alexander "Sasha" Aronowicz Pechersky, understood the benefits. "As a military man, I was aware that a surprise attack is worth a division of solders. If we can maintain secrecy until the last minute of the outbreak, the revolt is 80 percent accomplished. The biggest danger was deconspiration."[27] In preparation for the escape, the conspirators used their trade skills to make or steal knives and axes small enough to conceal in their clothes.

At four o'clock on the day of the escape, they sprang into action. Carefully but quickly, they began to lure SS guards into private locations one by one, under various false pretexts. Then, small groups of prepared prisoners would quickly and quietly kill the SS men by striking them on the head with an axe, or by covering their mouths and stabbing them to death. Within an hour they had killed eleven SS men, half of the SS guards present at the time, and concealed the bodies. At five o'clock they came together for evening roll call, but they arrived slightly early, before the remaining SS men had gathered. Their plan was to avoid the minefield by simply marching as a group to the front gate, as though they were on their way to a work detail. Upon reaching the gate, they hoped to shoot the two Ukrainian guards present and then rush out the front way.

Though they had been lucky so far, one of the bodies was discovered at the last moment, before they could make for the front gate. The Russian Sasha made a very brief "every man for himself" speech and encouraged everyone to escape immediately. The camp then burst into chaos, with some proceeding to the front gate, and others breaking their way through the fence and taking their chances with the mine field. All had to deal with machine gun fire from the guard towers.

Of the roughly 550 prisoners, 150 were unwilling or unable to escape. Some were separated in a different subcamp and were out of communication, and others simply refused to run. Anyone unable or unwilling to fight or run was shot by the SS. About eighty of those who did run were killed by the mines or by hostile fire. Still, more than 300 people (mostly with no preparation) managed to escape the camp into the surrounding woodlands.

Tragically, close to half of these people were captured and executed over the following weeks because of a German dragnet. But since they would have been killed by the SS regardless, the escape was still a remarkable success. Better yet, within days of the uprising, humiliated SS boss Heinrich Himmler ordered the camp shut down, dismantled, and replanted with trees. (See, they don't always rebuild.)[28] And a number of the escapees joined friendly partisan groups in the area and continued to fight the Nazis (including Sasha, who later returned to the Red Army and was sent to a gulag by Stalin for "allowing" himself to be captured in the first place).

The survivors would spend decades mulling over the escape. In many ways, they could hardly have hoped for better luck. If their actions had been discovered any earlier, it's very possible that everyone in the camp would have been executed. Furthermore, it's simply amazing that half of the group—very few of whom had any weapons, survival, or escape and evasion training—managed to avoid capture by the Nazis.

They certainly would have benefitted from further training or preparation, although in this case that was at odds with their priority of security. Another issue identified by survivors was that almost all of the firearms went to the Russian POWs, meaning that most escapees were defenseless. They also lacked prearranged cells or affinity groups, and many people who *did* know each other became separated during the escape. A further problem was the fact that the prisoners did not have contact with Allies or resistance groups who could have helped to arrange further escape or provide supplies or weapons. In the end, a large number of escaped prisoners ended up being killed by anti-Semitic Polish nationals, including some Polish partisans.

Despite these issues, we can learn a lot from this story. The prisoners made remarkable use of their limited resources to escape. The very fact that they attempted escape is inspiring, especially when literally millions of others went to their deaths without fighting back. Indeed, considering that so many of them lacked specific combat and evasion skills and equipment, it was solely the courage to fight back that saved many lives.

No withdrawal or refusal would help them—their lives were won only by audacious acts of commission.

ACTS OF COMMISSION: INDIRECT TO DIRECT

As we've made clear, acts of omission are not going to bring down civilization. Let's talk about action with more potential. We can split all acts of commission into six branches:

- lobbying;
- protests and symbolic acts;
- education and awareness raising;
- support work and building alternatives;
- capacity building and logistics;
- and direct confrontation and conflict.

The illustration (Figure 6-1, "Taxonomy of Action," page 243) groups them by directness. The most indirect tactics are on the left, and become progressively more direct when moving from left to right. More direct tactics involve more personal risk. (The main *collective* risk is failing to save the planet.) Direct acts require fewer people.

The first, *lobbying*, is attempting to influence or persuade those in power through letter writing, petitions, declarations or "speaking truth to power," protests, and so on. For the liberal, even atrocities are just big misunderstandings.[29] Lobbying informs those in power of their mistake (of course, since those in power are well-meaning, they will reform after being politely informed of their error).

Lobbying seems attractive because if you have enough resources (i.e., money), you can get government to do things for you, magnifying your actions. Success is possible when many people push for minor change, and unlikely when few people push for major change. But lobbying is too indirect—it requires us to try to convince someone to convince other people to make a decision or pass a law, which will then hopefully be enacted by other people, and enforced by yet a further group.

Lobbying via persuasion is a dead end, not just in terms of taking down civilization, but in virtually every radical endeavor. It assumes that those in power are essentially moral and can be convinced to change their behavior. But let's be blunt: if they wanted to do the right thing, we wouldn't be where we are now. Or to put it another way, their moral sense (if present) is so profoundly distorted they are almost all unreachable by persuasion.

And what if they could be persuaded? Capitalists employ vast armies of professional lobbyists to manipulate government. Our ability to lobby those in power (which includes heads of governments and corporations) is vastly outmatched by their ability to lobby *each other*. Convincing those in power to change would require huge numbers of people. If we had those people, those in power wouldn't be convinced—they would be replaced. Convincing them to mend their ways would be irrelevant, because we could undertake much more effective action.

Lobbying is simply not a priority in taking down civilization. This is not to diminish or insult lobbying victories like the Clean Water Act and the Wildlife Act, which have bought us valuable time. It is merely to point out that lobbying will not work to topple a system as vast as civilization.

Protests and symbolic acts are tactics used mostly to gain attention. If the intent of an action is to obstruct or disrupt business as usual in terms of transportation, the enforcement of laws, or other economic and political activities, then it's *direct confrontation*. If the protest is a rally for discussion or public education, it's *education and awareness raising* (see the next section).

When effective, demonstrations are part of a broader movement and go beyond the symbolic. There have been effective protests, such as the civil rights actions in Birmingham, but they were not symbolic; they were physical obstructions of business and politics. This disruption is usually illegal. Still, symbolic protests can get attention. Protests are most effective at "getting a message out" when they focus on one issue. Modern media coverage is so superficial and sensational that nuances get lost. But a critique of civilization can't be expressed in sound bytes, so protests can't publicize it. And civilization is so large and so ubiquitous that there is no one place to protest it. Some resistance movements have employed protests, to show strength and attract recruits, but the majority of people will never be on our side; our strategy needs to be based on effectiveness, not just numbers.

All resistance groups engage in some type of *education and awareness raising*, often public. In the most repressive regimes, education moves underground. Propaganda, agitation, rallies, theater, art, and spectacle are all actions that fall into these categories.

For public education to work, several conditions must be met. The resistance education and propaganda must be able to outcompete the mass media. The general public must be able and willing to unravel the prevailing falsehoods, even if doing that contravenes their own social, psychological, and economic self-interest. They must have accessible ways to change their actions, and they must choose morally preferable actions over convenient ones. Unfortunately, none of these conditions are in place right now.

Another drawback of education is its built-in delay; it may take years before a given person translates new information into action. But as we know, the planet is being murdered, and the window for effective action is small. For deep green resisters, skills training and agitation may be more effective than public education.

Education won't directly take down civilization, but it may help to radicalize and recruit people by providing a critical interpretation of their experiences. And as civilization continues to collapse, education may encourage people to question the underlying reasons for a declining economy, food crises, and so on.

Resistance movements need internal *support* structures to win. This may take the physical form of sustainable local food systems, alternative construction, alternative health care, and off-the-grid energy, transport, and communications. It may also include socially focused endeavors such as mutual aid, prisoner support, conflict resolution work, alternative economics, and intentional communities.

These support structures directly enable resistance. The Quakers' Society of Friends developed a sturdy ethic of support for the families of Quakers who were arrested under draconian conditions of religious persecution (see Chapter 5). People can take riskier (and more effective) action if they know that they and their families will be supported.

Building alternatives won't directly bring down civilization, but as industrial civilization unravels, alternatives have two special roles. First, they can bolster resistance in times of crisis; resisters are more able to fight if they aren't preoccupied with getting food, water, and shelter. Second, alternative communities can act as an escape hatch for regular people, so that their day-to-day work and efforts go to autonomous societies rather than authoritarian ones.

To serve either role, people building alternatives must be part of a culture of resistance—or better yet, part of a resistance movement. If the "alternative" people are aligned with civilization, their actions will prolong the destructiveness of the dominant culture. Let's not forget that Hitler's V2 rockets were powered by biofuel fermented from potatoes. The US military has built windmills at Guantanamo Bay, and is conducting research on hybrid and fuel-cell vehicles. Renewable energy is a necessity for a sustainable and equitable society, but not a guarantee of one. Militants and builders of alternatives are actually natural allies. As I wrote in *What We Leave Behind*, "If this monstrosity is not stopped, the carefully tended permaculture gardens and groves of lifeboat ecovillages will be nothing more than after-dinner snacks for civilization." Organized militants can help such communities from being consumed.

In addition, even the most carefully designed ecovillage will not be sustainable if neighboring communities are not sustainable. As neighbors deplete their landbases, they have to look further afield for more resources, and a nearby ecovillage will surely be at the top of their list of targets for expansion. An ecovillage either has to ensure that its neighbors are sustainable or be able to repel their future efforts at expansion.

In many cultures, what might be considered an "alternative" by some people today is simply a traditional way of life—perhaps *the* traditional way of life. Peoples struggling with displacement from their lands and dealing with attempts at assimilation and genocide may be mostly concerned with their own survival and the survival of their way of life. And for many indigenous groups, expressing their traditional lifestyle and culture may be in itself a direct confrontation with power. This is a very different situation from people whose lives and lifestyles are not under immediate threat.

Of course, even people primarily concerned with the perpetuation of their traditional cultures and lifestyles are living with the fact that civilization has to come down for any of us to survive. People born into civilization, and those who have benefitted from its privilege, have a much greater responsibility to bring it down. Despite this, indigenous peoples are mostly fighting much harder against civilization than those born inside of it.

Every successful historical resistance movement has rested upon a subsistence base of some kind. Establishing that base is a necessary step, but that alone is not sufficient to stop the world from being destroyed.

Capacity Building and Logistics

Capacity building and logistics are the backbone of any successful resistance movement. Although direct confrontation and conflict may get the glory, no sustained campaign of direct action is possible without a healthy logistical and operational core. That includes the following:

Resistance groups need ways of *recruiting* new members. The risk level of the group determines how open this process can be. Furthermore, new and existing members require training in tactics, strategy, logistics, and so on. Some or all of that *training* can take place in a lower-risk environment.

Resistance movements of all kinds must be able to screen recruits or volunteers to assess their suitability and to exclude infiltrators. Members of the group must share certain essential viewpoints and values (either assured through screening or teaching) in order to maintain the group's cohesion and focus.

Resisters need to be able to *communicate* securely and rapidly with one another to share information and coordinate plans. They may also need to communicate with a wider audience, for propaganda or agitation. Many resistance groups have been defeated because of inadequate communications or poor communications security.

Resistance requires *funding*, whether for offices and equipment, legal costs and bail, or underground activities. In aboveground resistance, *procurement* is mostly a subset of fund raising, since people can buy the items or materials they need. In underground resistance, procurement may mean getting specialized equipment without gathering attention or simply getting items the resistance otherwise would be unable to get.

Of course, fund raising isn't just a way to get materials, but also a way to support mutual aid and social welfare activities, support arrestees and casualties or their families, and allow core actionists to focus on resistance efforts rather than on "making a living."

People and equipment need access to *transportation* in order to reach other resisters and facilitate *distribution* of materials. Conventional means of transportation may be impaired by collapse, poverty, or social or political repression, but there are other ways. The Underground Railroad was a solid resistance transportation network. The Montgomery bus boycott was enabled by backup transportation systems (especially walking and carpooling) coordinated by civil rights organizers who scheduled carpools and even replaced worn-out shoes.

Security is necessary for any group big enough to make a splash and become a target for state intelligence gathering and repression. Infiltration is definitely a concern, but so is ubiquitous surveillance. This does not apply solely to people or groups considering illegal action. Nonviolent, law-abiding groups have been and are surveilled and disrupted by COINTELPRO-like entities. Many times it is the aboveground resisters who are more at risk as working aboveground means being identifiable.

Research and reconnaissance are equally important logistical tools. To be effective, any strategy requires critical information about potential targets. This is true whether a group is planning to boycott a corporation, blockade a factory, or take out a dam.

Imagine how foolish you'd feel if you organized a huge boycott against some military contractor, only to find that they'd recently converted to making school buses. Resistance researchers can help develop a strategy and identify potential targets and weaknesses, as well as tactics likely to be useful against them. Research is also needed to gain an understanding of the strategy and tactics of those in power.

There are certain *essential services and care* that keep a resistance movement running smoothly. These include services like the repair of equipment, clothing, and so on. Health care skills and equipment can be extremely valuable, and resistance groups should have at least basic health care capabilities, including first aid and rudimentary emergency medicine, wound care, and preventative medicine.

Coordination with allies and sponsors is often a logistical concern. Many historical guerilla and insurgent groups have been "sponsored" by other established revolutionary regimes or by states hoping to foment revolution and undermine unfriendly foreign governments. For example, in 1965 Che Guevera left postrevolutionary Cuba to help organize and train Congolese guerillas, and Cuba itself had the backing of Soviet Russia. Both Russia and the United States spent much of the Cold War "sponsoring" various resistance groups by training and arming them, partly as a method of trying to put "friendly" governments in power, and partly as a means of waging proxy wars against each other.

Resistance groups can also have sponsors and allies who are genuinely interested in supporting them, rather than attempting to manipulate them. Resistance in WWII Europe is a good example. State-sponsored armed partisan groups and other partisan and underground groups supported resistance fighters such as those in the Warsaw Ghetto.

Direct Conflict and Confrontation

Ultimately, success requires direct confrontation and conflict with power; you can't win on the defensive. But direct confrontation doesn't always mean *overt* confrontation. Disrupting and dismantling systems of power doesn't require advertising who you are, when and where you are planning to act, or what means you will use.

Back in the heyday of the summit-hopping "antiglobalization" movement, I enjoyed seeing the Black Bloc in action. But I was discomfited when I saw them smash the windows of a Gap storefront, a Starbucks, or even a military recruiting office during a protest. I was not opposed to seeing those windows smashed, just surprised that those in the Black Bloc had deliberately waited until the one day their targets were surrounded by thousands of heavily armed riot police, with countless additional cameras recording their every move and dozens of police buses idling on the corner waiting to take them to jail. It seemed to be the worst possible time and place to act if their objective was to smash windows and escape to smash another day.

Of course, their real aim wasn't to smash windows—if you wanted to destroy corporate property there are much more effective ways of doing it—but to fight. If they wanted to smash windows, they could have gone out in the middle of the night a few days before the protest and smashed every corporate franchise on the block without anyone stopping them. They wanted to fight power, and they wanted people to see them doing it. But we need to fight to win, and that means fighting smart. Sometimes that means being more covert or oblique, especially if effective resistance is going to trigger a punitive response.

That said, actions can be both effective *and* draw attention. Anarchist theorist and Russian revolutionary Mikhail Bakunin argued that "we must spread our principles, not with words but with deeds, for this is the most popular, the most potent, and the most irresistible form of propaganda."[30] The intent of the deed is not to commit a symbolic act to get attention, but to carry out a genuinely meaningful action that will serve as an example to others.

N N N

There are four basic ways to directly confront those in power. Three deal with land, property, or infrastructure, and one deals specifically with human beings. They include:

> Obstruction and occupation;
> Reclamation and expropriation;
> Property and material destruction (threats or acts); and
> Violence against humans (threats or acts)

In other words, in a physical confrontation, the resistance has three main options for any (nonhuman) target: block it, take it, or break it.

Let's start with *nondestructive obstruction or occupation*—block it. This includes the blockade of a highway, a tree sit, a lockdown, or the occupation of a building. These acts prevent those in power from using or physically destroying the places in question. Provided you have enough dedicated people, these actions can be very effective.

But there are challenges. Any prolonged obstruction or occupation requires the same larger support structure as any direct action. If the target is important to those in power, they will retaliate. The more important the site, the stronger the response. In order to maintain the occupation, activists must be willing to fight off that response or suffer the consequences.

An example worth studying for many reasons is the Oka crisis of 1990. Mohawk land, including a burial ground, was taken by the town of Oka, Quebec, for—get ready—a golf course. The only deeper insult would have been a garbage dump. After months of legal protests and negotiations, the Mohawk barricaded the roads to keep the land from being destroyed. This defense of their land ("We are the pines," one defender said) triggered a full-scale military response by the Canadian government. It also inspired acts of solidarity by other First Nations people, including a blockade of the Mercier Bridge. The bridge connects the Island of Montreal with the southern suburbs of the city—and it also runs through the Mohawk territory of Kahnawake. This was a fantastic use of a strategic resource. Enormous lines of traffic backed up, affecting the entire area for days.

At Kanehsatake, the Mohawk town near Oka, the standoff lasted a total of seventy-eight days. The police gave way to RCMP, who were then replaced by the army, complete with tanks, aircraft, and serious weapons. Every road into Oka was turned into a checkpoint. Within two weeks, there were food shortages.

Until your resistance group has participated in a siege or occupation, you may not appreciate that on top of strategy, training, and stalwart courage—a courage that the Mohawk have displayed for hundreds of years—you need basic supplies and a plan for getting more. If an army marches on its stomach, an occupation lasts as long as its stores. Getting food and supplies into Kanehsatake and then to the people behind the barricades was a constant struggle for the support workers, and gave the police and army plenty of opportunity to harass and humiliate resisters. With the whole world watching, the government couldn't starve the Mohawk outright, but few indigenous groups engaged in land struggles are lucky enough to garner that level of media interest. Food wasn't hard to collect: the Quebec Native Women's Association started a food depot and donations poured in. But the supplies had to be arduously hauled through the woods to circumvent the checkpoints. Trucks of food were kept waiting for hours only to be turned away.[31] Women were subjected to strip searches by male soldiers. At least one Mohawk man had a burning cigarette put out on his stomach, then dropped down the front of his pants.[32] Human rights observers were harassed by both the police and by angry white mobs.[33]

The overwhelming threat of force eventually got the blockade on the bridge removed. At Kanehsatake, the army pushed the defenders to one building. Inside, thirteen men, sixteen women, and six children tried to withstand the weight of the Canadian military. No amount of spiritual strength or committed courage could have prevailed.

The siege ended when the defenders decided to disengage. In their history of the crisis, *People of the Pines*, Geoffrey York and Loreen Pindera write, "Their negotiating prospects were bleak, they were isolated and powerless, and their living conditions were increasingly stressful . . . tempers were flaring and arguments were breaking out. The psychological warfare and the constant noise of military helicopters had worn down their resistance."[34] Without the presence of the media, they could have been raped, hacked to pieces, gunned down, or incinerated to ash, things that routinely happen to indigenous people who fight back. The film *Kanehsatake: 270 Years of Resistance* documents how viciously they were treated when the military found the retreating group on the road.

One reason small guerilla groups are so effective against larger and better-equipped armies is because they can use their secrecy and mobility to choose when, where, and under what circumstances they fight their enemy. They only engage in it when they reasonably expect to win, and avoid combat the rest of the time. But by engaging in the tactic of obstruction or occupation a resistance group gives up mobility, allowing the enemy to attack when it is favorable to them and giving up the very thing that makes small guerilla groups so effective.

The people at Kanehsatake had no choice but to give up that mobility. They had to defend their land which was under imminent threat. The end was written into the beginning; even 1,000 well-armed warriors could not have held off the Canadian armed forces. The Mohawk should not have been in a position where they had no choice, and the blame here belongs to the white people who claim to be their allies. Why does the defense of the land always fall to the indigenous people? Why do we, with our privileges and resources, leave the dirty and dangerous work of real resistance to the poor and embattled? Some white people did step up, from international observers to local church folks. But the support needs to be overwhelming and it needs to come before a doomed battle is the only option. A Mohawk burial ground should never have been threatened with a golf course. Enough white people standing behind the legal efforts would have stopped this before it escalated into razor wire and strip searches. Oka was ultimately a failure of systematic solidarity.

The second means of direct conflict is *reclamation and expropriation*— take it. Instead of blocking the use of land or property, the resistance takes it for their own use. For example, the Landless Workers Movement—centered in Brazil, a country renowned for unjust land distribution—occupies "underused" rural farmland (typically owned by wealthy absentee landlords) and sets up farming villages for landless or displaced people. Thanks to a land reform clause in the Brazilian constitution, the occupiers have been able to compel the government to expropriate the land and give them title. The movement has also engaged in direct action like blockades, and has set up its own education and literacy programs, as well as sustainable agriculture initiatives. The Landless Workers Movement is considered the largest social movement in Latin America, with an estimated 1.5 million members.[35]

Expropriation has been a common tactic in various stages of revolution. "Loot the looters!" proclaimed the Bolsheviks during Russia's October Revolution. Early on, the Bolsheviks staged bank robberies to acquire funds for their cause.[36] Successful revolutionaries, as well as mainstream leftists, have also engaged in more "legitimate" activities, but these are no less likely to trigger reprisals. When the democratically elected government of Iran nationalized an oil company in 1953, the CIA responded by staging a coup.[37] And, of course, guerilla movements commonly "liberate" equipment from occupiers in order to carry out their own activities.

The third means of direct conflict is *property and material destruction—* break it. This category includes sabotage. Some say the word *sabotage* comes from early Luddites tossing wooden shoes (*sabots*) into machinery, stopping the gears. But the term probably comes from a 1910 French railway strike, when workers destroyed the wooden shoes holding the rails—a good example of moving up the infrastructure. And sabotage can be more than just physical damage to machines; labor activism has long included work slowdowns and deliberate bungling.

Sabotage is an essential part of war and resistance to occupation. This is widely recognized by armed forces, and the US military has published a number of manuals and pamphlets on sabotage for use by occupied people. The *Simple Sabotage Field Manual* published by the Office of Strategic Services during World War II offers suggestions on how to deploy and motivate saboteurs, and specific means that can be used. "Simple sabotage is more than malicious mischief," it warns, "and it should always consist of acts whose results will be detrimental to the materials and manpower of the enemy."[38] It warns that a saboteur should never attack targets beyond his or her capacity, and should try to damage materials in use, or destined for use, by the enemy. "It will be safe for him to assume that almost any product of heavy industry is destined for enemy use, and that the most efficient fuels and lubricants also are destined for enemy use."[39] It encourages the saboteur to target transportation and communications systems and devices in particular, as well as other critical materials for the functioning of those systems and of the broader occupational apparatus. Its particular instructions range from burning enemy infrastructure to blocking toilets and jamming locks, from working slowly or inefficiently in factories to damaging work tools through deliberate negligence, from spreading false rumors or misleading information to the occupiers to engaging in long and inefficient workplace meetings.

Ever since the industrial revolution, targeting infrastructure has been a highly effective means of engaging in conflict. It may be surprising to some that the end of the American Civil War was brought about in large part by attacks on infrastructure. From its onset in 1861, the Civil War was extremely bloody, killing more American combatants than all other wars before or since, combined.[40] After several years of this, President Lincoln and his chief generals agreed to move from a "limited war" to a "total war" in an attempt to decisively end the war and bring about victory.[41]

Historian Bruce Catton described the 1864 shift, when Union general "[William Tecumseh] Sherman led his army deep into the Confederate heartland of Georgia and South Carolina, destroying their economic infrastructures."[42] Catton writes that "it was also the nineteenth-century equivalent of the modern bombing raid, a blow at the civilian underpinning of the military machine. Bridges, railroads, machine shops, warehouses—anything of this nature that lay in Sherman's path was burned or dismantled."[43] Telegraph lines were targeted as well, but so was the agricultural base. The Union Army selectively burned barns, mills, and cotton gins, and occasionally burned crops or captured livestock. This was partly an attack on agriculture-based slavery, and partly a way of provisioning the Union Army while undermining the Confederates. These attacks did take place with a specific code of conduct, and General Sherman ordered his men to distinguish "between the rich, who are usually hostile, and the poor or industrious, usually neutral or friendly."[44]

Catton argues that military engagements were "incidental" to the overall goal of striking the infrastructure, a goal which was successfully carried out.[45] As historian David J. Eicher wrote, "Sherman had accomplished an amazing task. He had defied military principles by operating deep within enemy territory and without lines of supply or communication. He destroyed much of the South's potential and psychology to wage war."[46] The strategy was crucial to the northern victory.

The fourth and final means of direct conflict is *violence against humans*. Here we're using violence specifically and explicitly to mean harm or injury to living creatures. Smashing a window, of course, is not violence; violence *does* include psychological harm or injury. The vast majority of resistance movements know the importance of violence in self-defense. Malcolm X was typically direct: "We are nonviolent with people who are nonviolent with us."[47]

In resistance movements, offensive violence is rare—virtually all violence used by historical resistance groups, from revolting slaves to escaping concentration camp prisoners to women shooting abusive partners, is a response to *greater* violence from power, and so is both justifiable and defensive. When prisoners in the Sobibór extermination camp quietly SS killed guards in the hours leading up to their planned escape, some might argue that they committed acts of offensive violence. But they were only responding to much more extensive violence already committed by the Nazis, and were acting to avert worse violence in the immediate future.

There have been groups which engaged in systematic offensive violence and attacks directed at people rather than infrastructure. The Red Army Faction (RAF) was a militant leftist group operating in West Germany, mostly in the 1970s and 1980s. They carried out a campaign of bombings and assassination attempts mostly aimed at police, soldiers, and high-ranking government or business officials. Another example would be the Palestinian group Hamas, which has carried out a large number of violent attacks on both civilians and military personnel in Israel. (It is also a political party and holds a legally elected majority in the Palestinian National Authority. It's often ignored that much of Hamas's popularity comes from its many social programs, which long predate its election to government. About 90 percent of Hamas's activities are these social programs, which include medical clinics, soup kitchens, schools and literacy programs, and orphanages.[48])

It's sometimes argued that the use of violence is never justifiable strategically, because the state will always have the larger ability to escalate beyond the resistance in a cycle of violence. In a narrow sense that's true, but in a wider sense it's misleading. Successful resistance groups almost never attempt to engage in overt armed conflict with those in power (except in late-stage revolutions, when the state has weakened and revolutionary forces are large and well-equipped). Guerilla groups focus on attacking where they are strongest, and those in power are weakest. The mobile, covert, hit-and-run nature of their strategy means that they can cause extensive disruption while (hopefully) avoiding government reprisals.

Furthermore, the state's violent response isn't just due to the use of violence by the resistance, it's a response to the *effectiveness* of the resistance. We've seen that again and again, even where acts of omission have been the primary tactics. Those in power will use force and violence to put down *any* major threat to their power, regardless of the particular tactics used. So trying to avoid a violent state response is hardly a universal argument against the use of defensive violence by a resistance group.

The purpose of violent resistance isn't simply to do violence or exact revenge, as some dogmatic critics of violence seem to believe. The purpose is to *reduce the capacity* of those in power to do further violence. The US guerilla warfare manual explicitly states that a "guerrilla's objective is to diminish the enemy's military potential."[49] (Remember what historian Bruce Catton wrote about the Union Army's engagements with Confederate soldiers being incidental to their attacks on infrastructure.) To attack those in power without a strategy, simply to inflict indiscriminant damage, would be foolish.

The RAF used offensive violence, but probably not in a way that decreased the capacity of those in power to do violence. Starting in 1971, they shot two police and killed one. They bombed a US barracks, killing one and wounding thirteen. They bombed a police station, wounding five officers. They bombed the car of a judge. They bombed a newspaper headquarters. They bombed an officers' club, killing three and injuring five. They attacked the West German embassy, killing two and losing two RAF members. They undertook a failed attack against an army base (which held nuclear weapons) and lost several RAF members. They assassinated the federal prosecutor general and the director of a bank in an attempted kidnapping. They hijacked an airliner, and three hijackers were killed. They kidnapped the chairman of a German industry organization (who was also a former SS officer), killing three police and a driver in the attack. When the government refused to give in to their demands to release imprisoned RAF members, they killed the chairman. They shot a policeman in a bar. They attempted to assassinate the head of NATO, blew up a car bomb in an air base parking lot, attempted to assassinate an army commander, attempted to bomb a NATO officer school, and blew up another car bomb in another air base parking lot. They separately assassinated a corporate manager and the head of an East German state trust agency. And as their final militant act, in 1993 they blew up the construction site of a new prison, causing more than one hundred million Deutsche Marks of damage. Throughout this period, they killed a number of secondary targets such as chauffeurs and bodyguards.

Setting aside for the time being the ethical questions of using offensive violence, and the strategic implications of giving up the moral high ground, how many of these acts seem like effective ways to reduce the state's capacity for violence? In an industrial civilization, most of those in government and business are essentially interchangeable functionaries, people who perform a certain task, who can easily be replaced by another. Sure, there are unique individuals who are especially important driving forces—people like Hitler—but even if you believe Carlyle's Great Man theory, you have to admit that most individual police, business managers, and so on will be quickly and easily replaced in their respective organizations.[50] How many police and corporate functionaries are there in one country? Conversely, how many primary oil pipelines and electrical transmission lines are there? Which are most heavily guarded and surveilled, bank directors or remote electrical lines? Which will be replaced sooner, bureaucratic functionaries or bus-sized electrical components? And which attack has the greatest "return on investment?" In other words, which offers the most leverage for impact in exchange for the risk undertaken?

As we've said many times, the incredible level of day-to-day violence inflicted by this culture on human beings and on the natural world means that to refrain from fighting back will not prevent violence. It simply means that those in power will direct their violence at different people and over a much longer period of time. The question, as ever, is which particular strategy—violent or not—will actually work.

◙ ◙ ◙

Q: You can't force people to change. What we really need is a paradigm shift.

Aric McBay: Proponents of a chiefly educational strategy often assert that persistent work at building public awareness will eventually result in a global "paradigm shift," which will dramatically change the actions and opinions of the majority. The term paradigm shift comes from Thomas Kuhn's 1962 book *The Structure of Scientific Revolutions*, but it's inapplicable to our situation for a number of reasons. Although the phrase gained usage in the 1990s as a marketing buzzword, Kuhn wrote explicitly that the idea only applied to those fields usually called the hard sciences (physics, biology, chemistry, and the like). A paradigm, he said, was a dominant system of explanation in one of these sciences, whereas "a student in the humanities has constantly before him a number of competing and incommensurable solutions to these problems, solutions that he must ultimately examine for himself."[51] Scientists trying to use equations to explain, say, orbital mechanics, can come to agreement on which theory is best because they are trying to develop the most accurate predictive equations.[52] Social sciences and other fields do not have this luxury, because there is no agreement on which problems are most important, how to evaluate their answers, what kinds of answers are the most important and how precise they should be, and what to do when answers are arrived at.

Because of these differences, Kuhn argued that the true *scientific* paradigm shifts always lead to better paradigms—paradigms that do a better job of explaining part of the world. But in society at large this is not true at all—dominant worldviews can be displaced by worldviews which are considerably worse at explaining the world or which are damaging to humans and the living world, a phenomenon which is distressingly common in history.

Furthermore, Kuhn argued that even when a much better paradigm is supported by strong evidence, the scientific community doesn't necessarily switch quickly. Scientists who have been practicing the obsolete paradigm for their entire careers may not change their minds even in the presence of overwhelming evidence. Kuhn quotes Nobel laureate Max Planck, who said that "a new scientific truth does not triumph by convincing its opponents and making them see the light, but rather because its opponents eventually die, and a new generation grows up that is familiar with it."[53]

Even worse for us, Kuhn and Planck are assuming the people in question are genuinely and deliberately trying to find the best possible paradigm. Doing this is literally a full-time job. Do we really believe that the majority of people are spending their free waking hours trying to gain a deeper understanding of the world, trying to sift through the huge amounts of available information, trying to grasp history and ecology and economics? The very idea of a paradigm shift assumes that the majority of people are actively trying to find large-scale solutions to our current predicament, instead of being willfully ignorant and deeply invested in a convenient economic and social system that rewards people for destroying the planet.

Indeed, part of the problem with "education" is that it's not only left-ists who do it, and it's rarely unbiased. Studies have shown that on the right wing, *more* educated people are *less* likely to admit the existence of global warming.[54] This is probably because they have more sophisticated rationales for their delusions.

But let's pause for a moment and take the most optimistic (if somewhat mangled) interpretation of Kuhn's concept and assume that a beneficial paradigm shift *is* going to happen, rather than a worsening shift in dominant politics and worldviews. That shift would require abundant evidence that the dominant culture—civilization—is inherently destructive and doomed to destroy itself along with the living world. Since we can't do multiple experimental run throughs of a global industrial civilization, for many people the only inescapable empirical demonstration of the dominant system's fundamental unsustainability would be the collapse of that system. Only at that point would the majority of people be seriously and personally invested in learning how to live without destroying the planet. And even then, those people would likely continue to insist on their outdated worldview, until, as Max Planck observed, they die, resulting in a further decades-long delay beyond collapse before a beneficial paradigm was dominant. This means that even in the most optimistic and reasonable assessment, a "global paradigm shift" would be decades too late.

Q: How can I do something to help bring down civilization and not just throw away my life in a useless act?

Derrick Jensen: There are three answers. The philosophical answer is that we can't know the future. We can never know whether some action will be useful. We can pick what we think are the most effective actions, but that still doesn't guarantee any given act will succeed. What we can know is that if this culture continues in the direction it's headed, it will get where it's headed, which is the murder of the planet. There are already casualties, and they're called the salmon. They're called the sharks. They're called the black terns. They're called migratory songbirds. They're called oceans, rivers. They're called indigenous people. They're called the poor. They're called subsistence farmers. They're called women.

The second, historical answer is about the way resistance movements work. You lose and you lose and you lose until you win. You get your head cracked, get your head cracked, get your head cracked, and then you win. You can't know when you start how many times you have to get your head cracked before you win. But the struggle builds on struggle. It has to start somewhere and it has to gain momentum. That happens through organizing, it happens through actions. And it happens through victories. One of the best recruiting tools is some sort of victory. And you can't have a victory unless you try.

And now the pragmatic: we are horribly outnumbered and we do not have the luxury to throw away our lives. How we can be most effective? We have to be smart. Choose targets carefully, both for strategic value and safety. And we have to organize. A lone person's chance of sparking a larger movement is much lower than that of a group of organized people.

Whatever actions a person takes (and this is true in all areas of life) need to count. Many of the actions being taken right now are essentially acts of vandalism, as opposed to acts of active sabotage that will slow the movement of the machine. So choose. How can you make your actions (and your life) have the most significance in terms of stopping the perpetration of atrocity?

All those who begin to act against the powers of any repressive state need to recognize that their lives will change. They need to take that decision very seriously. Some of the people captured under the Green Scare knew what they were getting into, and some of them made the decision more lightly. The latter were the people who turned very quickly when they were arrested. One person turned within five seconds of getting into the police car. That person probably didn't seriously consider the ramifications of his actions before he began. The Black Panthers knew when they started the struggle that they would either end up dead or in prison.

Finally, we have to always keep what we're fighting for in sight. We are fighting for life on the planet. And the truth is, the planet's life is worth more than you. It's worth more than me. It is the source of all life. That doesn't alter the fact that we should be smart. We need to be very strategic. We need to be tactical. And we need to act.

Did John Brown throw away his life? On one hand, you could say yes. His project ultimately failed. But, on the other hand, you could say that it set up much greater things. Did Nat Turner throw away his life? Did members of the revolt at Sobibór throw away their lives? On one hand, you could say yes. On the other hand, you could say that they did what was absolutely right and necessary. And something we must always remember is that those who participated in the Warsaw Ghetto uprising had a higher rate of survival than those who didn't. When the whole planet is being destroyed, your inaction will not save you. We must choose the larger life. We must choose to do what is right to protect the planet. It is our only home.

PART II: ORGANIZATION

Chapter 7

The Psychology of Resistance

by Aric McBay

I hear many condemn these men because they were so few. When were the good and the brave ever in a majority?

—Henry David Thoreau, "A Plea for Captain John Brown."

How can we expect righteousness to prevail when there is hardly anyone willing to give himself up individually to a righteous cause. Such a fine, sunny day, and I have to go. But what does my death matter, if through us thousands of people are awakened and stirred to action?

—Sophie Scholl, The White Rose Society, her last words.

Our premise is that the majority of people will not engage in resistance. Some reasons are obvious: ingrained obedience, ignorance, and the benefits of participation in the dominant culture. But there are also specific psychological barriers to resistance, at least four of which have been explored in psychological research.

In the 1950s, psychologist Solomon Asch conducted a series of experiments into social effects on perception. Asch set out to prove that when faced with a crystal-clear, objective question, a person's judgment should not be affected by others.

Experimental subjects were brought into a room one at a time with people *posing* as participants: the experimenter's confederates. They were shown a set of lines: a "reference" line, and several comparison lines of varying length, one of which matched the reference line. The experimenter asked the participants to call out which line matched. They did this twelve times with twelve different figures. The trick was that the fake subjects—the experimenter's confederates—lied. They were instructed ahead of time to choose a line which was very clearly too long or too short.

After five false participants had stated their choice, the genuine participants would state their choice. The results of the experiment were completely the opposite of what Asch had expected. In more than half of the trials the subjects went along with the consensus, even though the correct answer was obvious. Some 25 percent of the participants *refused* to conform in every trial, but 75 percent of the participants gave the consensus answer at least once.[1] Interviewing the participants afterward, Asch found that most people *saw* the lines correctly, but felt that since the rest of the group was in consensus, they themselves must be wrong. Some knew that the group was wrong but went along with it anyway to avoid standing out. And some insisted, after the experiment had completed, that they actually *saw* the lines the same way as the rest of the group.

Later research by other psychologists found certain commonalities among those most likely to conform.[2] Such people, they observed, tended to have high levels of anxiety, low status, a high need for approval, and authoritarian personalities. That last part is particularly interesting—the people who are likely to boss others around are themselves psychologically pliable.

It's not just the prevailing opinion that affects whether we will conform or not. Authority plays a very important role. Yale psychologist Stanley Milgram famously began a series of experiments in 1961, shortly after the beginning of Nazi Adolf Eichmann's war crimes trial. He wanted to understand the degree to which those responsible for the Holocaust were "just following orders." In Milgram's experiment, the subject was instructed by an authority (the experimenter in a lab coat) to give increasingly powerful shocks to another person, an actor who sometimes claimed to have a heart condition. The actor was not actually shocked, but pretended to be, eventually screaming in pain, banging on walls, and then falling silent as the shocks passed a presumably lethal threshold.

Prior to the experiment, Milgram polled his students and colleagues, all of whom believed that only a tiny percentage of subjects would administer the maximum 450-volt shock. Of course, when the experiment took place, 65 percent of people administered successive shocks all the way up to the maximum voltage.[3] Of those subjects who refused to administer the maximum shock, no one demanded that the experiment itself should be stopped; no one questioned its existence. In later experiments, Milgram examined what would happen if more trappings of authority were added. He found that the more respectable the locale of the experiment was (say, a courthouse instead of a back-alley office), the higher the obedience rate. (Suspecting that the subjects may have realized the victim was faking, two other psychologists later conducted the same experiment using real shocks and a live puppy. They found an even higher obedience rate than in the original experiment.[4])

When a confederate performed the actual shocks, and the subject only had to assist them with other aspects of the experiment, virtually all subjects completed the full experiment. The good news is that when two confederates were introduced into the mix to defy the authority, almost all of the subjects refused to continue the experiment.

Milgram's experiment is one of the most oft-cited studies when trying to understand why people listen to those in power even when they are obviously doing wrong. And, of course, like Asch's experiment, real-world people face a worse situation than the subjects of the experiment. Milgram's lab-coated experimenter could use only verbal pressure to encourage obedience. The subject did not risk censure from their family or social group. They did not risk losing their jobs. They did not risk public ridicule. The experimenter could not use the legal system against them, or threaten them, or use physical violence to ensure their compliance. In the real world, all of these things are used against people who contemplate resistance.

Learned helplessness offers another insight into the psychology of resistance and nonresistance. The term comes from a series of experiments conducted by Martin Seligman in the late 1960s. In this experiment, several groups of dogs were put into restraining harnesses. One group, a control group, was soon released from the harnesses unharmed. The second group was given series of electric shocks, but had a lever that could be pressed to stop them. A third group was given shocks that appeared to start and end at random, with no way of controlling them. The first two groups soon recovered from the experiment, but dogs in the third group began to show symptoms similar to clinical depression.[5]

In the second half of the experiment, the dogs were put unrestrained into a "shock box" that they could easily jump out of. The dogs from the first two groups jumped out when the shocks began. Most of the dogs in the third group, however, simply lay down and whined, even though they could have easily escaped. They had *learned* to be helpless, the experimenters concluded. The good news is that about one-third of the dogs in the latter group did *not* become helpless, but managed to escape the box despite their previous traumatic experience.

When extrapolating the experience of these more resilient dogs to the experience of humans, Seligman and other psychologists found that their behavior correlated highly with optimism.[6] It was not, they cautioned, a naïve or Pollyanna-ish approach to optimism. This was no "cheermongering." Instead, overcoming learned helplessness is all about understanding and explaining the source of the trauma. People who believed their problems were pervasive, permanent ("things have always been this way, and they always will be"), and personal ("it's all my fault") were much more likely to suffer from learned helplessness and depression.

This, too, can be extrapolated to our own situation. Those in power encourage us to believe that the status quo is natural, inevitable, even the best possible society. If someone is dissatisfied with the way society works, they say, then it is that individual's personal emotional problem. Furthermore, the individual traumas perpetuated by those in power on individual people, on groups of people, and on the land, can seem random at first glance. But if we can trace them back to their common roots—in capitalism, in patriarchy, in civilization at large—then we can understand them as manifestations of power imbalance, and we can overcome the learned helplessness such horrors would otherwise create.

Further, those in power systematically try to get us to believe that environmental destruction is *our* fault (because we, too, use toilet paper) instead of being caused by the decisions and actions of those who run the economy. If those in power can convince us that "it's all our fault," they have pushed us one step closer to learned helplessness, depression, and, ultimately, a failure to resist.

The bystander effect, and the related diffusion of responsibility, is a final psychological effect at play in determining resistance or nonresistance. The concept is usually linked to the 1964 murder of a New York woman named Kitty Genovese. Genovese was stabbed to death, over a period of about half an hour, near her apartment building. A dozen people heard her screams for help and the sound of her struggle, and some actually saw portions of the attack in which she was stabbed. But no one intervened.[7] The bystander effect is surely something we've all seen at various times. I remember sitting in my apartment some years ago after dinner, reading a book, when the sound of a woman screaming came from the corridor outside. She called for help, banging on doors with her hands and feet as an assailant dragged her down the corridor by her hair. Of the ten or fifteen people living on the floor, I was the only one who left my apartment to stop the attack. No one even bothered to call the police.

After the murder of Kitty Genovese, psychologists John Darley and Bibb Latané carried out a series of studies to explore the diffusion of responsibility. They put college students in several different cubicles, speaking by intercom about an unrelated "decoy" topic. Early in the experiment, one of the participants—a confederate of the experimenters—mentioned that he sometimes had seizures. Then, later in the experiment, the confederate feigned a seizure over the intercom, begging for help, telling the others that he was having a seizure and thought he was going to die, and then falling silent. The chance that another participant would leave their own cubicle to go help the "seizure victim" directly correlated to the number of people involved in the intercom conversation. When only one participant was present, there was an 85 percent chance that this person would go to aid the victim. When two were present that dropped to 62 percent. When five were present, only 31 percent responded. The response time of the participants also increased significantly as the number of participants grew.[8] In other words, the more people present, the more their sense of responsibility became diffused. The experimenters found no difference between women and men.

Interestingly, Darley and Latané reported that the people who *did* act appeared *less* upset than those who did not. The people who left their cubicles appeared generally calm and "without panic," while those who remained in their cubicles often appeared visibly upset, sweating and trembling. It wasn't so much that those people had decided *not* to act, wrote the psychologists. Rather, they were unable to decide *to* act, to commit to action, worried that they would "make fools of themselves by overreacting."

In a second study, Darley and Latané decided to examine how the *attitudes* of bystanders affected how a person would respond. In this study, participants sat in a room filling out questionnaires. After a few minutes, the experimenters began to flood the room with smoke. Lone subjects left the room and reported the smoke 75 percent of the time. With three subjects present, the chance that a participant would report the smoke dropped to 38 percent. For the last part of the study, the experimenters put one subject in the room with two confederates who were instructed to notice and then deliberately ignore the smoke. In the final case, only 10 percent of people reported the smoke.[9]

John Darley wrote that in such situations a given bystander interprets the inaction of their comrades to mean that the situation isn't urgent or dangerous. "A kind of 'anti-panic mob' is formed in which individuals do not respond because they define the situation as *no emergency.*"[10]

We can again see the parallels for our situation. Those in power constantly promise—or more subtly, imply by their inaction—that everything is fine. That mass poverty is not a problem. That global warming is not an emergency. They claim that people who do warn about such problems are "fearmongers," and act as though acknowledging the serious global problems they cause would cause chaos and mass panic.

Even this patronizing attitude is not well-supported by history. In her book *Disaster: A Psychological Essay*, Martha Wolfenstein examined the attitude of WWII British government officials and consultants in the months before the bombing of England by the Germans began. When the bombing began, the officials expected, there would be mass panic, the masses would flee London in outright terror, and the number of psychological casualties would outnumber physical casualties three to one.[11] Of course, that did not happen. "There was no panic flight from London or any other city. Evacuation was orderly and fewer people than anticipated showed a wish to leave their homes for a safer location."[12] While this idea of mass panic is a common and vivid fantasy, Wolfenstein writes that instead, "Disaster-stricken populations . . . are apt to be quiet, stunned, and dazed."

Wolfenstein also examines the reasons that so many people, when faced with imminent danger or disaster, do nothing. Assurances by those in power—and to some degree the mere *existence* of those in power and their asserted expertise—help to keep people passive. Wolfenstein writes, "This confidence that the 'leaders' or the 'government' could and would do something was generally combined with a belief that there was nothing the private citizen could do. Such attitudes towards world affairs illustrate the trend of what has been called 'privatization.' The ordinary citizen tends to feel increasingly that he has neither the knowledge nor the means to take a hand in the great affairs which affect his destiny."[13]

Not only do they feel that they can do nothing, many people in this situation (like those in the psychologist's smoke experiment) appear to actually *feel* as though nothing is wrong: "The expectation that superior authorities will do something to ward off the threat, and the often combined belief that the individual himself can do nothing, are apt to be associated with absence of worry."[14]

Of course, not everyone falls for such cognitive falsehoods. Furthermore, some people—as the psychological research suggests—are not so prone to blindly follow authority, are not so vulnerable to the pressures of conformity. Instead, some people seem psychologically predisposed to resistance. This minority group includes those who are the first to fight against injustice, the first to join and organize resistance groups. Rather than "early adopters," such people are "early resisters."

Claude Bourdet (a leader in the *Combat* movement of the French Resistance during WWII) said that early resisters were people who had already "broken with their social and professional milieu."[15] Famed French *resistant* Emmanuel d'Astier de la Vigerie believed that "one could only be a resister if one was maladjusted."[16] However, in his history of the German occupation of France, Julian Jackson argued that most early resisters were "far from being outsiders," but they *were* people with strong moral convictions who may have been from traditional backgrounds or occupations. Jackson writes: "These were not maladjusted mavericks although clearly they were individuals of exceptional strong-mindedness, ready to break with family and friends."[17]

Although some postwar stories about France portray a broad base of resistance against the Nazis, in fact only a very small minority of the population participated. The French Resistance at most comprised perhaps 1 percent of the adult population, or about 200,000 people.[18] The postwar French government officially recognized 220,000 people[19] (though one historian estimates that the number of active resisters could have been as many as 400,000[20]). In addition to active resisters, there were perhaps another 300,000 with substantial involvement.[21] If you include all of those people who were willing to take the risk of reading the underground newspapers, the pool of sympathizers grows to about 10 percent of the adult population, or two million people.[22]

This is, of course, not unique to 1940s France. At the peak of Irish resistance to British rule, the Irish War of Independence (which built on 700 years of resistance culture), the IRA had about 100,000 members (or just over 2 percent of the population of 4.5 million), about 15,000 of whom participated in the guerrilla war, and 3,000 of whom were fighters at any one time. Among Jews in Nazi Germany, the number of people who actively fought back was often tragically outnumbered by the people who simply killed themselves. In Berlin, roughly 4 percent of Jews called up for "relocation" committed suicide, almost all of them upon the arrival of the notice (those who chose to kill themselves were mostly older and highly assimilated to German society).[23] Within Nazi Germany, resistance mostly consisted of small and isolated groups.

Even after the war, retroactive support for German resistance was limited. In 1952, after the Nuremberg Trials, and after information about the concentration camps, horrific medical experimentation, and other Nazi atrocities had become known, surveys of public opinion about resistance were made in West Germany. Members of the public were asked whether a person convinced that "injustices and crimes" were being committed by the Nazis would be justified in resisting them—whether any resistance of *any* sort was justifiable. Only 41 percent said it was. Worse, when asked whether resistance was defensible in wartime, only 20 percent of people said yes. Another 34 percent said that potential resisters should wait until the return of peace (which, under the Nazis, as under any empire, means never). The second-largest group of 31 percent was undecided about whether resistance against the Nazis could have been justified. They were not undecided about whether *they* would participate (we can safely assume they would not), they were undecided about whether resistance should have existed at all! And another 15 percent insisted that resistance was never justifiable, whether in peacetime or wartime.[24] I found all this sickening and deplorable. I deeply wish I could say I found it surprising.[25]

Those who are willing to undertake serious resistance are always a small minority regardless of circumstances, largely for the psychological and social reasons discussed above.[26] To put it bluntly: we have to get over the hope that resistance will ever be adopted by the majority and focus on doing what we can with who we have. Given all that, the purpose of a resistance organization is to enable as many of those people as possible to resist, and to organize those people in ways that makes maximum use of their limited numbers.

N N N

As we discussed a few chapters ago, we too often base our activism on the idea that we need to have a mass movement to overturn this wretched system. But Germany suggests the exact opposite: that overturning the system is the prerequisite to a mass movement. Even years after Germany's defeat, the great majority of Germans did not think resistance would have been justified. Only after the Nazis' authoritarian grip had been broken, and only after years or decades had passed, would the German people understand why resistance was not only acceptable, but needed.[27]

I can only believe that if there is ever a mass movement against those in power, it will happen after civilization collapses, and not before.

N N N

The effective resister has some important personality characteristics, with bravery, intelligence, and persistence among the most important. Intelligence alone is never enough. Though an intelligent person may be better able to see through propaganda and to understand the problem at hand, real courage is a requirement for action in the face of danger. The brilliant coward simply has a more sophisticated rationalization for inaction. And persistence is required to continue in the face of unfavorable odds against a powerful enemy in a struggle that is bound to be rife with setbacks and mistakes.

For those individuals who *are* psychologically predisposed and willing to resist, a number of factors influence whether or not they will actively engage in that resistance: the perceived benefits of resistance, the perceived chance of success, the perceived risk of participating, the perceived degree of personal responsibility for the problem (the bystander effect), the perceived legitimacy of the resistance organization or activities, and the availability of potential resistance comrades. You can probably think of more—just think about what would influence *your* decision.

In any case, a good resistance organization addresses all of these factors. It can propagandize about the problems with the status quo and the benefits that would come with its success. And the very existence of proper organization increases the chance of success. There is always some risk to resistance, but good organizing reduces that risk through a security culture and good tactics. Solid recruitment overcomes the bystander effect by addressing specific people and giving them specific means to act. A resistance organization can increase its own legitimacy through good decision-making practices, adherence to a moral code, endorsement by sympathetic authorities, and, most importantly, by its own longevity and effectiveness.

◙ ◙ ◙

Q: If we act effectively against those in power, won't those in power just come down on us harder?

Derrick Jensen: They will, but that's not a reason to submit. This is how authoritarian regimes and abusers work: they make their victims afraid to act. They reinforce the mentality, "If I try to leave him, my abusive husband, my pimp, may kill me." And that is a very good reason to not resist.

This question explicitly articulates what we all know to be true: the foundation of this culture is force. And the primary reason we don't resist is because we are afraid of that force. We know if we act decisively to protect the places and creatures we love or if we act decisively to stop corporate exploitation of the poor, that those in power will come down on us with the full power of the state. We can talk all we want about how we supposedly live in a democracy. And we can talk all we want about the consent of the governed. But what it really comes down to is if you effectively oppose the will of those in power, they will try to kill you. We need to make that explicit so we can face the situation that we're in. And the situation we're in is that those in power are killing the planet and they are exploiting the poor, they are murdering the poor, and we are not stopping them because we are afraid.

But there have to be some of us who are willing to act anyway. We should never underestimate the seriousness of attempting to stop those in power. And we also need to be very clear about the seriousness of what is happening to the world. If you're reading this book, you probably understand how desperate things are.

What is the legacy that we want to leave for those who come after? How do you want to be seen by the generations that follow? Do you want to be seen as someone who knew what the right thing was and didn't do it because you were afraid? Or do you want to be remembered as someone who was afraid and did the right things anyway? It's okay to be afraid. Almost everyone I know is afraid at some time or another. But there is tremendous joy and exhilaration that comes, too, from doing what is right. The fact that those in power will use their power against resisters is not a reason to give up the fight before we even begin. It is a reason to be really, really smart.

Chapter 8

Organizational Structure

by Aric McBay

> There is one thing you have got to learn about our movement. Three
> people are better than no people.
>
> —Fannie Lou Hamer, civil rights leader

Resistance organizations can be divided into aboveground (AG) and
underground (UG) groups. These groups have strongly divergent organizational and operational needs, even when they have the same goals.
Broadly speaking, aboveground groups do not carry out risky illegal
actions, and are organized in ways that maximize their ability to use
public institutions and communication structures. Underground
groups exist primarily to carry out illegal or repressed activities and are
organized in ways that maximize their own security and effectiveness.

Some aboveground groups do carry out illegal activities as part of a campaign of civil disobedience, or they break or bend lesser laws as a means
of causing disruption or confronting power (for example, through "illegal"
protests). These groups often occupy something of an awkward middle
ground, a subject we'll return to. As police become more draconian and
punishments more severe, such groups may split into underground and
aboveground factions, with some members refraining from illegal acts out
of fear of punishment, while others seek to escalate their actions.

There has to be a partition, a firewall, between aboveground and
underground activities. Some historical aboveground groups have tried
to sit on the fence and carry out illegal activities without full separation. Such groups worked in places or times with far less pervasive
surveillance than any modern society. Their attempts to combine aboveground and underground characteristics sometimes resulted in their
destruction, and severe consequences for their members.

In order to be as safe and effective as possible, every person in a
resistance movement must decide for her- or himself whether to be
aboveground or underground. It is essential that this decision be made;
to attempt to straddle the line is unsafe for everyone.

FUNDAMENTAL DIFFERENCES BETWEEN AG AND UG ORGANIZATIONS

CRITERION	ABOVEGROUND	UNDERGROUND
Membership	Membership is likely open, membership of any given member known by others in the organization.	Membership is closed or closely guarded. Members are not aware of the identity of members outside of their immediate area of the organization.
Public face and outward behavior	The group aims to attract attention and conducts public relations using "its own face." Members may strongly voice support for change and resistance.	The group aims to appear unremarkable or to deflect attention from itself (though probably not its action). Communication with the public happens through anonymous communiqués or press offices.
Decision making	May emphasize democratic, transparent, and participatory decision making. They tend to be more broadly participatory in nature.	Members are likely to appear outwardly apolitical or conservative. Decision making process is internally known but outwardly covert, many decisions based on internal rank and structure.
Internal communication and movement	Internal communication (with and between groups) may be open, frequent, and in the clear.	Communication between groups is likely to be limited, guarded, terse, and encoded.
Actions	Members may move between different groups routinely to share skills.	Movement between groups is very limited, but skill sharing is still important.
Goal with regard to general populace	Likely to announce in advance to maximize attention and media coverage. May target areas where enemy is strongest or most concentrated (i.e., demonstrations in financial districts). May hope to mobilize citizens or gain broader support.	No advance announcement, or perhaps disinformation about upcoming actions. Targets areas where enemy is weakest or most diffuse. Is not concerned with support of the majority, but may want to increase network of sympathizers. Hopes to avoid reprisals carried out on general population.

The differences between aboveground and underground organizing are expressed in every facet of a group's structure and practice. Some of these differences are summarized in the table to the left.

Regardless of whether they are aboveground or underground, any group which carries out effective resistance activity will be considered a threat by those in power, and those in power will try to disrupt or destroy it.

BASIC ORGANIZATIONAL STRUCTURES

Within both aboveground and underground activism there are several templates for basic organizational structures. These structures have been used by every resistance group in history, although not all groups have chosen the approach best suited for their situations and objectives. It is important to understand the pros, cons, and capabilities of the spectrum of different organizations that comprise effective resistance movements.

The simplest "unit" of resistance is the individual. Individuals are highly limited in their resistance activities. Aboveground individuals (Figure 8-1b) are usually limited to personal acts like alterations in diet, material consumption, or spirituality, which, as we've said, don't match the scope of our problems. It's true that individual aboveground activists can affect big changes at times, but they usually work by engaging other people or institutions. Underground individuals (Figure 8-1a) may have to worry about security less, in that they don't have anyone who can betray their secrets under interrogation; but nor do they have anyone to watch their back. Underground individuals are also limited in their actions, although they can engage in sabotage (and even assassination, as all by himself Georg Elser almost assassinated Hitler).

Individual actions may not qualify as resistance. Julian Jackson wrote on this subject in his important history of the German Occupation of France: "The Resistance was increasingly sustained by hostility of the mass of the population towards the Occupation, but not all acts of individual hostility can be characterized as resistance, although they are the necessary precondition of it. *A distinction needs to be drawn between dissidence and resistance.*" This distinction is a crucial one for us to make

Figure 8-1

Organizational Network Types

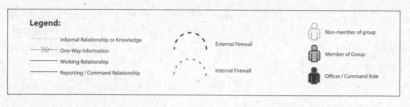

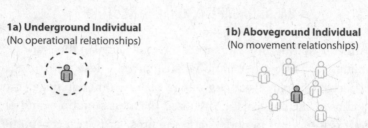

1a) Underground Individual
(No operational relationships)

1b) Aboveground Individual
(No movement relationships)

as well. Jackson continues, "Workers who evaded [compulsory labor], or Jews who escaped the round-ups, or peasants who withheld their produce from the Germans, were transgressing the law, and their actions were subversive of authority. But they were not resisters in the same way as those who organized the escape of [forced laborers] and Jews. Contesting or disobeying a law on an individual basis is not the same as challenging the authority that makes those laws."[1]

Of course, one's options for resistance are greatly expanded in a group.

The most basic organizational unit is the affinity group. A group of fewer than a dozen people is a good compromise between groups too large to be socially functional, and too small to carry out important tasks. The activist's affinity group has a mirror in the underground cell, and in the military squad. Groups this size are small enough for participatory decision making to take place, or in the case of a hierarchal group, for orders to be relayed quickly and easily.

The underground affinity group (Figure 8-2a, shown here with a distinct leader) has many benefits for the members. Members can specialize in different areas of expertise, pool their efforts, work together toward shared goals, and watch each others' backs. The group can also offer social and emotional support that is much needed for

Figure 8-2

2a) Underground Affinity Group
(Limited direct relationships
with broader movements)

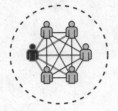

2b) Aboveground Affinity Group
(Socially embedded in
broader movements)

people working underground. Because they do not have direct relationships with other movements or underground groups, they can be
relatively secure. However, due to their close working relationships, if
one member of the group is compromised, the entire affinity group is
likely to be compromised. The more members are in the group, the
more risk involved (and the more different relationships to deal with).
Also because the affinity group is limited in size, it is limited in terms
of the size of objectives it can go after, and their geographic range.

Aboveground affinity groups (Figure 8-2b) share many of the same
clear benefits of a small-scale, deliberate community. However, they
may rely more on outside relationships, both for friends and fellow
activists. Members may also easily belong to more than one affinity
group to follow their own interests and passions. This is not the case
with underground groups—members must belong only to one affinity
group or they are putting all groups at risk.

The obvious benefit of multiple overlapping aboveground groups is
the formation of larger movements or "mesh" networks (Figure 8-3b).
These larger, diverse groups are better able to get a lot done, although
sometimes they can have coordination or unity problems if they grow
beyond a certain size. In naturally forming social networks, each
member of the group is likely to be only a few degrees of separation

Figure 8-3

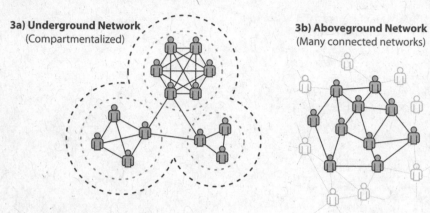

3a) Underground Network
(Compartmentalized)

3b) Aboveground Network
(Many connected networks)

from any other person. This can be fantastic for sharing information or finding new contacts. However, for a group concerned about security issues, this type of organization is a disaster. If any individual were compromised, that person could easily compromise large numbers of people. Even if some members of the network can't be compromised, the sheer number of connections between people makes it easy to just bypass the people who can't be compromised. The kind of decentralized network that makes social networks so robust is a security nightmare.

Underground groups that want to bring larger numbers of people into the organization must take a different approach. A security-conscious underground network will largely consist of a number of different cells with limited connections to other cells (Figure 8-3a). One person in a cell would know all of the members in that cell, as well as a single member in another cell or two. This allows coordination and shared information between cells. This network is "compartmentalized." Like all underground groups, it has a firewall between itself and the aboveground. But there are also different, internal firewalls between sections.

Such a network does have downsides. Having only a single link between cells is beneficial, in that if one cell is compromised, it is much more difficult to compromise other cells. However, the connection is also more brittle. If a "liaison" is removed from the network or loses

Figure 8-4

4a) Underground Hierarchy
(Compartmentalized)

4b) Aboveground Hierarchy
(Many internal and
external relationships)

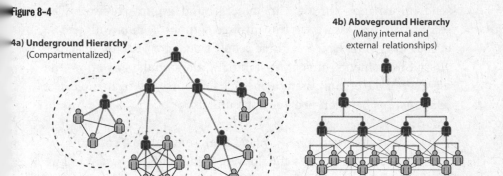

communication for whatever reason, then the network may be broken up. A backup plan for regaining communication can reduce the damage from this, but increase the level of risk. Also, the nonhierarchal nature of this network means that choosing actions can be more difficult. The more cells are involved, the larger the number of people who must have critical information in order to make decisions. That said, these groups can be very effective and functional. The famous Underground Railroad was a decentralized underground network.

Some of these problems are addressed in both aboveground and underground groups through the use of a hierarchy. In underground hierarchies (Figure 8-4a), large numbers of cells can be connected and coordinated through branching, pyramidal structures. These types of groups have vastly greater potential than smaller networks. Their numbers make for increased risk, yes, but that increased risk can be reduced by the use of specialized counterintelligence cells within the network and wide-ranging coordinated attacks.

Aboveground hierarchies (Figure 8-4b) are quite familiar and common, in part because they are highly effective ways of coordinating large numbers of people to accomplish a specific objective. As shown, aboveground hierarchies facilitate many relationships between people in different parts of the hierarchy. This lack of compartmentalization might be good in terms of productivity, but not in terms of security.

There are very specific situations in which it may be acceptable to send information through an underground group's firewall. The recruitment process necessarily involves communication with people outside the group. However, these people would not be active in aboveground movements, and, at least initially, they would only know one member of the organization in one cell. Of course, there are no direct relationships between people in the underground and aboveground groups.

In certain situations, one-way (and likely anonymous) communications may take place across the firewall. Informants who want to give information to the resistance network may pass on information to a member of an internal intelligence group. However, the intelligence group would not share information about identities or the network with those people. Information may also travel one-way in the opposite direction. The underground groups may want to send communiqués or other information to the media or press office. Of course, any communication across the firewall, even those thought anonymous, entails a certain small amount of risk. Therefore, the benefits must outweigh the risks.

All of the examples illustrated are simplified and generalized. Resistance groups in history have had a wide variety of internal structures based on these general templates. They often had to make a deliberate compromise between organizational security (which comes from loosely connected and decentralized cells) and organizational effectiveness (which comes from more densely connected and centralized cells).

◙ ◙ ◙

As well as belonging to different groups, members of a resistance movement can be divided into five general classes: *leaders*; *cadres* or *professional revolutionaries*; *combatants* or *frontline activists*; *auxiliaries*; and the *mass base*. Although the terminology stems from armed struggle, the basic division of roles can apply to any group that wants to confront and dismantle oppressive systems of power.

Leaders are those who work to organize and inspire the organization, either as administrators or ideologues, and serve important decision-making roles. In explicitly antiauthoritarian organizations, like those of the pre-WWII Spanish Anarchists, the leaders may be effectively integrated with the cadres. In the organizational illustration of Figure 8-5, the leaders (or officers) are colored black.

Cadres or *professional revolutionaries* form the backbone of a resistance organization. Though the term "cadre" has been prominently used in communist circles, it's also used in a more general organizational sense and especially in militaries. In the original military sense, cadres are "the key group of officers and enlisted personnel necessary to establish and train a new military unit," or, more generally, "a nucleus of trained personnel around which a larger organization can be built and trained."[2] Cadres (the term refers to both the group and individuals) have the skills needed to operate and perpetuate a resistance organization, and they take their job seriously. They carry out their resistance work as professionals, regardless of how they make their income. Most people who take on this role in community groups are called "organizers" or the like, but you can recognize them when you see them by their commitment, their experience, and their work ethic. As the organizational core group, they do what needs doing to move the group forward, including the recruitment and training of new members. Essentially anything in the taxonomy of action that falls under "capacity building and operations" is under the purview of cadres. Good cadres are distinguished by their psychological drive to succeed, their dedicated professionalism, their experience and history, and their concrete organizational work.[3] In Figure 8-5, cadres could be any permanent member of the resistance shown in dark grey, but would definitely include the affinity group on the left selecting new recruits.

Combatants or *frontline activists* are those who engage in direct confrontation and conflict with power. They are, in a word, warriors. This could be anyone who does that work in conjunction with resistance organizations, from people who do tree sits to people who confront and expose rapists. This kind of work can entail a very high level of risk, physical or otherwise. As we've already discussed, the people on the front lines are usually a small (but essential) percentage of those involved in resistance. This role can overlap with that of the cadres, but there are important differences. Work on the front lines may be more

Figure 8–5

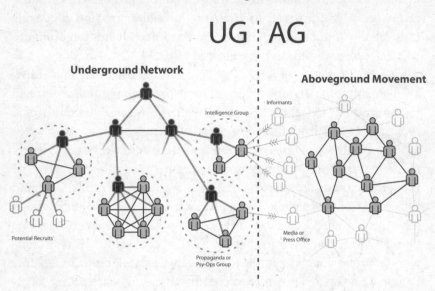

specialized than organizational cadre tasks, and it requires a narrower area of experience and responsibility. Despite this, the risk and stress involved means that not everyone who would make a good cadre would make a good combatant, and vice versa. Some people have families or children who need their support, and some people simply aren't psychologically suited to the roles on the front lines.

Know that the most effective combatants are those willing to give up their lives, whether through death or prison. Even aboveground activists engaged in confrontation activities may spend time in prison; in fact, they are more likely to be identified and arrested, although they may serve less time than underground activists who are caught. A man from the Mohawk Warriors Society once explained to me why the police were afraid of his group: "They aren't scared of us because we're willing to take up arms. They're scared of us because we're willing to die." Likewise, many Black Panthers knew that when they joined the Black Panther Party (BPP) they would either end up dead or in prison. The struggle was worth it to them.

Auxiliaries are sympathizers, people living otherwise normal lives who offer moral or material support to more active members of the resistance. Auxiliaries may or may not be considered a formal part of a resistance organization. They may provide funding, material support, shelter and safehouses, transportation, a pool of (and screening for) recruits, or health care and equipment maintenance. Auxiliaries may also pass information on to the resistance, including information they observe about occupier activities such as construction, troop movements, or personnel information. Auxiliaries can be candidates for recruitment to more serious roles. In Figure 8-5, the auxiliaries would include the light grey figures associated with the movement who act as informants or do press work without actually being part of a formal organization.

The *mass base* consists of the people who generally support or sympathize with the resistance, and follow its activities with interest, but who aren't organizationally involved and who don't offer direct material support. People who read the underground newspapers in occupied France might be a part of this group. The mass base usually supports the resistance in a more generalized or nebulous way through discussions with friends or neighbors, which increases its perceived popularity and legitimacy. They might also share literature or other materials of interest with each other. Aboveground organizations sometimes engage the mass base for fund raising. And the mass base in general can act as a pool for recruitment, either into the auxiliaries or for other roles.

The WWII Resistance in occupied France offers many lessons on how to create and organize resistance movements. Historian Julian Jackson notes that, prior to WWII, there were few organized resistance movements to emulate; liberation movements had been more nationalist or anticolonial in nature, were based in a specific distinct population, and often did not have an underground component. "What did 'resistance' mean to these people? One must cast aside romantic images . . . the hackneyed phrase 'he or she joined the Resistance,' is entirely inappropriate to 1940–1. Before it could be joined, resistance had to be invented."[4]

Jackson continues, "Structures did gradually emerge, and gave rise to two distinct types of resistance organizations: networks . . . and movements. The networks were set up with specific military objectives—the collection of information, sabotage, organizing escape routes," and so on. "Some networks developed from spontaneous local initiatives" while others "were set up from scratch by intelligence agents sent out from London. . . . For security reasons, networks had to be rigidly compartmentalized and hierarchically organized. They did not have newspapers because the overriding priority was secrecy. By contrast, newspapers were central to the existence of most movements. Although these also collected information and sought links outside France, their priority was to target the French population: to shake it out of its lethargy and eventually organize it for action."[5]

These networks and movements were, as you can surely see, roughly analogous to the underground and aboveground structures described above. Part of the difference was that the French population was clearly aware that it was under occupation and was sympathetic, so the movements were able to use structures that were hidden from the Germans. They were also able to organize in ways more structurally similar to aboveground groups. Indeed, in many areas (mostly in southern "Vichy" France, which was not directly occupied by the German army) Resistance movements were something of an open secret for the French people. There are stories of visitors walking into newspaper buildings and successfully asking directions to the "offices of the Resistance." Of course, these distinctions were not initially clear to a group of people trying to teach themselves how to organize against occupation.

"The differentiation between movements and networks crystallized gradually. The first resisters did whatever seemed possible. The Muse'e de l'homme group started by smuggling escaped prisoners to the Free Zone [southern France]; it then moved on to collecting information; then, finally, it founded a newspaper. In theory it had gone from being a network to a movement, but such distinctions did not yet exist. Once the networks became more professionalized and started receiving aid from London, *the rule was that their members could not also be in a movement.*"[6] That last part in particular is worth rereading. The French Resistance clearly recognized the need for a firewall between aboveground and underground activism.

This division was expressed in many aspects of their organization. As Jackson notes, "The distinction between movements and networks was fundamental. The networks were specialized, secretive, and usually small: effectiveness and security might be jeopardized by size. The movements, on the other hand, sought to increase their numbers. The networks had mysterious coded names—Ali-France, Jade-Fitzroy, Caviar, Brutus, Comète—while the names of the movements spoke for themselves: Libération, Défense de la France, Résistance."[7]

The networks, being smaller and distinct, had differing areas of focus and recruitment. "The networks' social composition also varied. Some specialized in infiltrating a particular institution, like the Ajax network which recruited among the police. Others, like Jade-Fitzroy, recruited eclectically: its members included railway workers, postal workers, garage owners, a prefect of police, hairdressers, restauranteurs, gendarmes, doctors, teachers, lawyers, priests, students, and engineers."[8]

Members of the Resistance recognized that the goals of the organization ultimately determined its structure and other characteristics. This was especially true a few years into the Occupation, by which time the Resistance had been able to shake out some of the initial bugs. "In 1942," notes Jackson, the Resistance leader Christian "Pineau distinguished between two kinds of Resistance: 'military resistance can only be performed by a real Secret Army . . . composed of men ready, outside their daily tasks, to undertake a specific mission . . . Political resistance, on the other hand, is performed by each Frenchman in the framework of his normal activities.' The former required 'a hierarchy, a discipline, a discretion incompatible with the idea of a mass movement'; the latter 'leaves a lot to individual initiative.'"[9] It's hard to put it more clearly than that.

Networks and movements also had very different ways of growing and recruiting. While the networks were secretive and highly selective, the movements could afford to recruit larger numbers of people with lower risk. They could also join forces with other movements without as much concern about infiltration. And that's what happened. As the Resistance grew, various independently formed movements gradually consolidated. In 1943, the three largest movements agreed to merge into a single organization. This was very beneficial for their main activities at the time, which were newspaper publishing and propaganda.

The movements were unsuitable for some forms of action, but they were still a vital part of the Resistance. And their greater numbers, and a relatively broad membership, meant that they could assign people to specialize in certain tasks. "Each movement had a section to manufacture false papers . . . ; a social service section to help the families of resisters who had been arrested; a section responsible for gathering intelligence; and embryonic paramilitary units."[10] It's also important to note that the two types of resistance were deliberately complementary. People in networks recognized the vital importance of movements, and movements recognized the vital importance of the networks. Pineau described two different types of organization working in parallel, not in opposition.

But we should also recognize that that wasn't always the case. Early in the development of the Resistance, the French people did not have a good idea of what would constitute real action against the occupation. As is often the case for people who lack effective organizations for resistance, many of them clung to individual or personal expressions of discontent. Early in the war, some observers even claimed that the "elegance" of Parisian women constituted a form of resistance against the Germans.[11] It seems laughable in retrospect, but people who aren't presented with real options for resistance will cling to whatever they can find. This same phenomenon expressed itself in a focus solely on "spiritual" resistance by some movements—that is, they believed that the French people should not actively resist the Germans, but instead focus on their own souls. Does this sound familiar?

In 1941, a year after the beginning of the Occupation, the majority of Resistance movements opposed violence and even sabotage. Over time, this changed. The Resistance grew, and so did the number and diversity of its attacks on the occupiers. In 1943, thousands of acts of sabotage took place, and assassinations became relatively common. Notes Jackson, "This radicalization of Resistance affected even a movement like Défense de la France which had originally privileged the idea of spiritual resistance. The Catholic convictions of its leaders made them suspicious of violence. But in November 1942 the movement's newspaper declared that everyone's duty was to bear arms; a year later, it approved . . . assassinations of individuals."[12] A prominent leader of the movement even wrote an article titled "The Duty to Kill," which at length encouraged people to kill Germans and collaborators. His advice on police who aided the Germans and particularly the members of the German-run paramilitary was to "exterminate them . . . strike them down like mad dogs . . . destroy them as you would vermin."[13]

This trajectory is one we see again and again in many resistance movements. They start with atomized dissidents and "leaders" who fear resistance and privilege personal change, then coalesce into dedicated affinity groups that carry out new or risky tactics, and finally escalate to large political movements and networks that can mobilize and strike with force.

Although the growth of the Resistance was initially slow, it eventually began to grow with greater and greater speed. This growth occurred in part because many of the French could *see* a path to victory. Jackson observes: "The expansion of the Resistance occurred at a time when it seemed increasingly likely Germany would lose the war. This does not mean that these comparatively late arrivals should be written off as opportunists. As the Germans became weaker, they became more dangerous: the growth of resistance was a function of opportunities more than opportunism."[14] As the Resistance grew, even the *movements* shifted from propaganda to military action and guerrilla action. And when the Allied forces landed at Normandy to begin to liberate France using conventional warfare, they were aided by Resistance members who served as guides as well as engaging in guerrilla strikes and widespread sabotage.

The Resistance grew relatively slowly until it looked as though Germany might lose. It's easy to draw parallels to our own situation. The cracks in the façade of industrial civilization are inspiring more resistance. As that system breaks down further, resistance will become more feasible, more effective, and more necessary.

⊠ ⊠ ⊠

Q: Will civilization just reassemble itself?

Derrick Jensen: I have several answers to that. The first is that, no, this is a one-time blowout. The easily accessible reserves of oil are gone.

There will never be another oil age. There will never be another natural gas age. There will never be another Iron Age or Bronze Age. Further, there will never be—or not for a very, very long time—an age of tall ships, for example, because the forests are gone. This culture has destroyed so much that there will not be the foundation upon which a similar civilization could be built. Topsoil is gone. No, there will never be another rise of a civilization like this. There might be—presuming humans survive—some small-scale civilizations, but there will never be another one like this.

Second, I don't really think that's the right question. It's like waking up in the middle of the night and hearing the screams of your family as they're tortured, and then you look up and you see an ax murderer standing over your bed. You turn to the person sleeping next to you and you say, "Darling, honeybunch, how can we make sure that ax murderers don't break into our home tomorrow?" Right now, we have a crisis and we need to deal with that crisis. I wish we had the luxury to worry about whether civilization will rise again in the future, but we don't have that luxury. Right now, we need to stop this culture from killing the planet and let the people who come after worry about whether it's going to rise again.

This question reminds me of another I was once asked: "How much time do you think we have left?" I gestured toward the person next to her. "Pretend she is being tortured in that room over there. We can hear her screaming. How much time do you think she has left before we need to act? How much time should we allow the torturers to continue before we stop them?" There are injustices happening right now. Two hundred species went extinct today. And how much time did they have? None. The question for them is not, will civilization rise again? The question is what can we do to protect them right now. If we see these injustices, we need to stop them.

Chapter 9

Decision Making

by Aric McBay

> Given the same amount of intelligence, timidity will do a thousand times
> more damage than audacity.
>
> —Carl von Clausewitz, strategist

Just as aboveground and underground groups have very divergent ways
of structuring themselves, they also have different ways of operating.
The way that a group makes decisions is crucial, and determines how
that group does almost everything else.

There is a continuum of decision-making methods, ranging from
the participatory consensus model to the militaristic hierarchy. The
more participatory methods are deeply ingrained for those of us
coming from progressive backgrounds. And for good reason; participatory methods can effectively include people of many different
backgrounds in a social movement, and can help to unseat power
imbalances like sexism. But, despite its appeal, the consensus model
(in which everyone must agree before a decision is accepted) is not
appropriate for every kind of resistance.

The more authoritarian methods of decision making—the hierarchies of businesses or the military—are common for a reason: they get
things done. Hierarchies may permit abuses of power, but they are very
effective at getting certain tasks accomplished. And if we want to be
effective as resisters, we have to decide what we want to get done, and
pick a decision-making process suited to that job.

Picture a group of people organizing to stop a new uranium mine
in their area. They need to share ideas about how to organize effectively,
they need to involve and mobilize many people from different backgrounds, and they need to develop a cohesive group so that they can
hold together in the face of opposition. Participatory models like consensus can be great for this kind of situation because they make sure
that everyone is involved, they draw on wisdom from the group, and
they build a project that everyone feels invested in.[1] But imagine a

group of French Maquis on a sabotage mission trying to use the same methods. If the SS show up in the middle of the job, can you imagine our freedom fighters sitting down for a go-around? Can you picture Pierre blocking the decision to return fire because Juliet left him off the speaker's list by mistake? Inclusive but rambling meetings are fine when the stakes are low, but prolonged discussion has no place in immediate life or death circumstances.

The key issues are information and timeliness. Underground resistance groups must keep secrets and make decisions quickly in emergencies. In the most authoritarian system, only one person need have all of the information to make a decision, and they needn't discuss the issue with anyone else. That way the information won't be spread around, and the decision can be made quickly. In the most participatory system, everyone in the group has access to all relevant information, and they need time to think about it and discuss it with each other so that everyone can agree on the specifics. This makes it hard to keep secrets, and well nigh impossible to make tough decisions quickly. And once the group gets beyond a certain size, collective discussion is impractical.

Fortunately, we don't have to use one approach for every circumstance. There is a spectrum of options available, all of which have been used by successful resistance movements in different circumstances.

A *permanent rank structure* is a basic military-style decision-making system. There is an organized hierarchy with orderly promotions and a recognized chain of command. Military and paramilitary organizations use this approach because it holds together even under extreme circumstances. In virtually every situation, there is a person clearly in charge and responsible for making decisions to ensure that a group can maintain effectiveness when there is no time for discussion. The downsides are obvious. Abuses of power, the reinforcement of existing hierarchies, and a smaller pool of thinkers are all potential failings that a hierarchy must work against.

A hierarchy can be scaled to any size, while ensuring that every member of the group is as close as possible to the command. This is not possible with models like consensus, which is not very scalable, nor is it functional in an emergency. The key lesson is that certain kinds of resistance—like armed resistance—only work when there is a hierarchy in place. This is a lesson demonstrated by groups from the African National Congress to the original Irish Republican Army. If someone can't make tough decisions fast in an emergency, then people get killed. It's an uncomfortable lesson for people who struggle against hierarchy and inequality, but the point is ultimately that people choose the resistance they undertake. If you can't tolerate a chain of command, choose a different group. Just remember that some avenues of resistance are only open to some types of groups.

A *dynamic rank structure* is a hierarchy with a difference—the hierarchy is not permanent. So when a group is actually carrying out an emergency action, one person might be in charge of giving orders. The rest of the time, another person might be in charge or the group might operate on a more participatory basis. This approach offers a compromise between the more rigid option above, and the participatory options below.

Some historical pirates followed a similar model; their "captains" were elected by the crew and were absolutely in charge during battle. Day to day operations were coordinated by the quartermaster. And if the crew was displeased with either person, they could call a vote and replace him—so long as the ship wasn't in the middle of a battle.

A dynamic rank structure can be scaled to pretty much any size, just like permanent rank structures. The compromise is that while this model is good for dealing with emergencies, it's not necessarily effective at building large networks for command and communication, because the "leaders" may change. It's also difficult to keep information on a "need-to-know" basis if the people who need to know keep changing.

A *majority-rules system* is a good way to make decisions "democratically" in groups that don't have time for extensive discussion, or that are too large or heterogeneous to use the consensus model. Pretty much everyone is familiar with this model, so it's easy to implement. The problem, of course, is that for a majority-rules system to work, everyone has to have enough knowledge and expertise about the matter at hand to make a good decision. This can be a solid approach for affinity groups, but is much less functional in underground networks. It's also too slow for emergencies.

Under the *consensus model*, every member of a group has to agree before a decision is made. (Some people may choose to stand aside, and there are variations, but that's the gist of the model.) This is an excellent way of making sure that everyone is included in a decision and in discussion, and a great way of entertaining all available perspectives. It also takes time—sometimes a very long time—to discuss all sides of an issue and arrive at a decision. The more people in a group, and the more varied their perspectives, the harder it is to build consensus. Further, consensus requires that everyone involved have access to all available information. These factors mean that consensus as a model is poorly suited to serious underground work; it simply doesn't function in emergencies.

None of these methods are good or bad; they're just suited to different situations. And sometimes the same group may use multiple methods at different times. Even an underground group could use consensus or voting to make certain general decisions about their goals and strategy. They might appoint one person to make tactical decisions in an emergency. All of these models have a place in resistance; the trick is to realize what that place is.

◊ ◊ ◊

Q: Why should I take large-scale direct action against the system when almost nobody else, especially in the first world, is?

Derrick Jensen: Because the world is being murdered. And because members of the so-called first world are the primary beneficiaries. It is not up to the poor to be on the front lines yet again. It is not up to the indigenous to be on the front lines. It is not up to the nonhumans to be on the front lines. It is our responsibility as beneficiaries of this system to bring a halt to the system.

MEND (Movement for the Emancipation of the Niger Delta) has been able to reduce oil industry output by up to 30 percent in Nigeria. They have done so because they love the land they live in and that land is being destroyed. We have much greater resources at our disposal. It's our responsibility to use those resources and to use the privilege that we have to stop this culture from killing the planet.

Chapter 10

Recruitment

by Aric McBay

When they asked for those to raise their hands who'd go down to the courthouse the next day, I raised mine. Had it high up as I could get it. I guess if I'd had any sense I'd've been a little scared, but what was the point of being scared? The only thing they could do to me was kill me and it seemed like they'd been trying to do that a little bit at a time ever since I could remember.

—Fannie Lóu Hamer, civil rights leader

Methods of outreach and recruitment vary depending on whether a group is aboveground or underground, how it is organized, and what role is being filled. There are really two kinds of recruitment, which you might call *organizational* and *mutual* recruitment. In organizational recruitment, an existing organization finds and inducts new members. In mutual recruitment, unorganized dissidents find each other, and forge a new resistance group. When resistance is well established, organizational recruitment can flourish. When resistance is rare or surveillance extensive, dissidents mostly have to find each other.

Recall that a movement can be divided into five parts based on roles: leaders, the cadres or professional revolutionaries who form the movement's backbone, combatants or other frontline activists, auxiliaries, and the mass base.

Leaders, if they are recruited at all, are likely to find each other early on or be recruited from within the organization (especially in the underground, for the obvious reasons that they are known, have experience, and can be trusted).

The *cadres* and *combatants* or frontline activists are recruited in person, screened, and given training. Recruiting such people may require the bulk of recruitment resources, but that commitment of resources is necessary; cadres form the backbone of the resistance as professionals who give their all to the organization, and combatants are, of course, on the front lines.

Auxiliaries may be easier to recruit because they require a lesser commitment to the group, and the screening process may be simpler because they do not need to be privy to the same information and organizational details as those inside the organization. However, there generally should be some kind of personal contact, at least to initiate the relationship.

The *mass base* does not require direct recruitment because they support the resistance because of their own circumstances or experience, combined with propaganda and outreach from the resistance. Outreach to the mass base can take place through inexpensive mass media like books and newspapers, so that they require minimal effort per person to "recruit," but they also offer little or no material support to the resistance. However, they may take some action on prompting from the resistance, and participate generally in acts of omission or noncooperation with those in power.

So how does one recruit? It depends. Aboveground groups have it pretty easy in terms of recruitment, because recruitment plays to their strengths. It's relatively easy for them to engage in outreach and to publicize their politics and actions. Of course, because of this they are more vulnerable to infiltration. Underground groups need a somewhat more involved recruitment procedure, largely for security reasons, and they have a much smaller pool of potential recruits. All of this brings us to one of the most important conundrums for modern-day militants, what you might call the paradox of militant radicalization.

Most people who want to change the world start with low-risk, accessible activities, things like signing petitions or writing letters. When those don't work, activists may escalate to protests, disruption, and civil disobedience. Maybe they are teargassed or beaten at a protest, and they become radicalized. If they care enough about their cause, they will continue to ratchet up their action until it works. Unless their issue is popular enough to be solved with legal action, activists eventually hit a wall at which further escalation is illegal or dangerous. At this point, some people choose to act underground. And here's the paradox: aboveground action is based on *getting attention*. The people who have been the most persistent and relentless and most successful at raising awareness—the very people with the dedication and drive needed to go underground—may be the people who are at the most risk in going underground.

People living in overtly oppressed groups do not have the privilege of ignorance, and are more likely to be radicalized younger and in greater numbers. But within a surveillance society that doesn't alter our fundamental problem: the process of militant radicalization is liable to draw counterproductive attention to the radical, simply because most people don't turn to militant action until they have personally exhausted the less drastic and lower-risk avenues. Many of the most serious and experienced members of aboveground resistance thus become cut off from further escalation.

There's no perfect solution; serious resistance entails risk, and all members have to decide for themselves what levels of risk they are willing to take on. Keeping a low profile is part of the answer. Someone who is considering serious underground resistance should avoid prominent, militant aboveground action; it's important not to draw unwanted attention in advance. That doesn't mean that people should stop being activists or stop being political, but militant aboveground action is a definite disqualifier for underground action.

This paradox must be addressed by individual communities of resistance having a culture of resistance. We must offer alternatives to the traditional routes of radicalization. Rather than simply following the default path, budding activists need to be told that there is a choice to be made between aboveground and underground action. Activists can privately discuss these options with trusted friends, but without planning specific actions (which would entail extra risk). This applies regardless of whether a movement is willing to use violence or not. As we have discussed, repression happens when a movement is effective, regardless of their tactics: witness Ken Saro-Wiwa.

Furthermore, it's our assumption that successful resistance will grow, gather attention, and progress toward more militant activity as needed. That growth will increasingly draw unwanted attention and infiltration from intelligence agencies. That means any resistance movement that plans to eventually succeed needs to incorporate excellent security measures from the very beginning. Because the situation has been worsened by the rapid development of electronic surveillance, we radicals have been a bit behind the curve on this. Recruitment is a crucial area to apply good security.

STAGES OF RECRUITMENT

There are three basic stages of recruitment. The first is outreach or "prospecting," in which a group tries to make contact with potential recruits (and make their pitch). The second is screening or selection, in which the available candidate pool is winnowed down and the best recruits are chosen. In the third and final phase, those recruits are offered training and integrated into the organization. These basic stages apply whether the group is a modern military, a business, an institution, or a resistance group.

Outreach

The outreach method depends on the number of people required and the skills and talents they need. For rallies or protests, a movement may simply need large numbers of people with no particular skills. In this case, "warm body" recruitment based on mass call outs, word of mouth, posters, etc., can work very well. However, if specific skills and attributes are needed, it is necessary to go out and find those people, often in more peripheral parts of the resistance movement (like the mass base or auxiliaries).

Aboveground recruiters have many ways to look for people. They watch to see who does good work, solicit volunteers, and seek out recommendations from comrades and colleagues. (Underground recruiters can also do so, albeit more subtly, as we will discuss.) Once a candidate has been found, it's the recruiter's job to make the pitch.

A good pitch has four distinct parts.

First, recruiters should hit their high points and explain the benefits of joining up. In personal terms, as already discussed, the recruitment may look beyond material benefits and focus on the social benefits (being part of a tight-knit group with similar beliefs and perspectives), esteem and accomplishment (actually getting things done, making a difference in the world, accomplishing goals they couldn't reach without the group), and self-actualization (putting their own special gifts and talents to use, actualizing their own potential as a human being and a member of the resistance, responding creatively to difficult and challenging situations, and so on). Recruitment may also focus on causes, anything from making a difference in a local community to saving a local landbase to building a more equitable society to stopping the destruction of the planet.

Speaking with a person who has experience in the organization can help convince the candidate. The US National Guard, for example, had increased recruitment rates when they used more people with actual overseas experience. As an alternative, or in addition to this, recruiters can use testimonials from other members, explaining the benefits of being involved. Most resistance groups don't have funding for wages, but offering special perks or incentives, if appropriate, can help bring a recruit into the movement.

Bard E. O'Neill lists seven methods by which insurgents and revolutionaries attempt to gain support. They apply to general public support and to recruitment in particular. These are:

a) Charismatic attraction, in which revolutionaries attract people through persuasion, example, and the force of personality of charismatic leaders.

b) Esoteric appeals put revolution in an ideological context, and are usually directed at more intellectual or educated groups.

c) Exoteric appeals focus on the concrete grievances of the people.

d) Attacks on those in power, which Bard calls "terrorism," to demonstrate weakness of the government.

e) Provocation of government repression, to alienate people from the government and show that the government is bad.

f) Demonstrations of potency by the revolutionaries, either through force or through administrative and social services, or both.

g) Coercion to force or threaten people into supporting them.

Although all of these are interesting on a broader strategic level, the first three are directly relevant to individual recruitment.

Second, the appeal needs to hit at a deep emotional level, not just an intellectual one. Many people are aware of problems and do nothing about them. A personal or emotional impact is required to spur people to action. Furthermore, current neuroscience research shows that we often make decisions on an unconscious level long before coming up with a conscious rationale for those decisions.[1] Recruiters are after the small minority of people who are predisposed to resist. They don't have to create new feelings; they just have to evoke or release strong feelings already present in the candidates.

These candidates *want* to resist, but often suppress that desire because they lack an effective outlet. By joining, aspiring resisters can meet their deep-seated desire to fight back and make a better world. For some, just the idea of being part of an effective resistance group— being trained, participating in actions, working with like-minded people—is exhilarating. For others, an emphasis on grander goals or narratives may resonate more.

Third, recruiters must address any concerns or anxieties. As volunteer recruiting experts McCurley and Vineyard put it, "Remove their reasons to say 'no.'"[2] Obviously, a recruiter should have good answers to common concerns in advance. Candidates may be concerned about the level of risk involved, the people they would be working with, their role in the group and how they would be trained, their existing relationships with friends and family, security issues, or the amount of time they'll be committing. Recruiters can start by asking only for an initial commitment, like a "tour of duty" for a specific campaign.

Some people may say they are just too busy, but I suspect most sympathizers wouldn't be too busy to help with something they felt would actually make a difference. Few people who care about the planet would turn down a chance to do something they felt was genuinely effective. But that's also the difference between auxiliaries and cadres. Auxiliaries engage in resistance around their normal schedule. Cadres *live* for resistance, and make the time and personal sacrifices needed to engage in serious resistance. That can mean not watching television. It can mean living minimalistically. It can mean not having children.

Lastly, the recruiter offers *next steps* to the candidate. The recruiter may want to make a follow-up appointment or have a particular follow-up process. It's the recruiter's job to take responsibility for the next step—although a recruiter should constantly be aware of how a candidate is feeling, that recruiter should also offer definite ways to proceed.

◙ ◙ ◙

A brief aside before we continue. Recruitment is only one side of the membership equation; the other side is activist retention. Many things can keep activists going, like success, camaraderie, and a sense of momentum. But there are just as many ways to lose members. One way is to fail to take care of people. Activists need emotional support and morale boosting, especially when things are not going well. Unless your group has a morale officer, that responsibility falls to everyone. Another way to lose people is to fail to appreciate them. Few activists get paid for what they do, and most campaigns are protracted. Good work and long-term commitment should be recognized and celebrated.

Some people are especially good at doing these things of their own initiative, and these people should be nurtured and encouraged for the good of the group. Of course, there are also people who are very *bad* at these things, who constantly criticize new members for doing things differently, who engage in self-righteous cliquishness, and who generally make people miserable by being poster children for horizontal hostility. More prevalent in groups without experienced and well-behaved role models, these cranky activists are poison for activist retention.

People like this should be politely told to cut it out. If they can't or won't stop, either kick them out of the group or start a different one. Any group where such people hold sway will stagnate or self-destruct in the sort of way that causes lasting animosity and bitterness.

You are much less likely to have these kinds of problems if you screen people in the first place.

Screening and Selection

All groups should engage in some screening of recruits (formally or informally), the underground being especially vigilant. Security concerns apply aboveground as well, but breaches in those groups are unlikely to be catastrophic. So here we give particular emphasis to techniques used by underground groups. There are many different screening methods (some superficial, some more rigorous), only some of which will be used by any given group. In roughly sequential order, these methods include:

Outreach prescreening / prospecting: Before approaching a potential recruit or beginning the larger screening process, the group may look for indicators that the candidate has promise, including the possession of preexisting skills, a history of voicing sentiments against those in power, a history of participating in actions against those in power, or a record of other reasons to dislike those in power (such as deaths of family members).

Physical checks: The group may physically check the candidate and their effects to look for listening devices, police union cards, and the like. Obviously, the candidate cannot be warned about this in advance.

Vouching or references: The resistance movement, or its auxiliaries, may already include people who have known the candidate for years, and can offer an opinion or vouch for the individual. However, vouching alone is not enough. (If it were, an infiltrator could easily bring in many other infiltrators. Further, vouchers may have a biased perspective on close friends or family, and especially romantic interests.)

Background checks: A member of the group may question the candidate about history, past actions, school or employment, residences, etc. The questioner will then check to make sure that the story is internally consistent and that it can be verified, to screen out informers who are fabricating or hiding parts of their history. This typically involves checking records as well as speaking to individual people in the candidate's background. Although government and online records may be convenient to check, they can be falsified in order to provide a cover for an informer, so they cannot be relied on alone. Checks in newspaper records and the like (as may be available in libraries) are less falsifiable, but high-profile actions in the past may make the candidate unsuitable for participation in an underground group. The background check may also serve to determine whether a candidate's past history indicates that the person is reliable.

Surveillance or tailing: Some groups have followed or otherwise engaged in surveillance of potential recruits. This surveillance can help verify their story, determine whether they are meeting with police or government agents, and gather more information. (Following a person is also a way of finding out whether someone *else* is also following them.)

Lifestyle or habit checks / warning signs: Some groups disqualify members on the grounds of drug addiction or other unacceptable habits or actions (such as abuse) that go against the group's code of conduct or that would put the group at risk.

Interview or political screening: Candidates may be asked questions about their politics, or they may be asked to study and agree with certain materials, points of unity, or conduct. Effective questions for candidates should be open-ended, and leading questions should be avoided, to get the most indicative responses. Interviews should take as much time as needed.

Intuition and trust: Though these methods of screening are essential, they are not infallible. The ultimate test of any candidate is the intuition—the gut feelings—of members of the group. If those in the group do not feel certain that they can trust the candidate, then it does not matter whether the individual is an informer or not—the recruit cannot join the group, because the existing members will not be able to work with that person. The group needs to be totally satisfied that the new group member can handle responsibilities.

Test task: Oftentimes a candidate may be given a test task. This may simply require the person to demonstrate potential and the ability to follow instructions. In other cases, they may be required to carry out a task that an infiltrator would not do. On a related theme, they may be asked to perform an illegal task in front of other members of the group. This inhibits them from potentially testifying against other members since those people could testify against *them* in court. Of course—this has happened before—infiltrators may be willing to go along with things in order to get closer to the group.

Induction and oath: If the candidate passes the preceding screening measures, the person may be provisionally inducted into the group. This may involve an oath of allegiance to the group or resistance movement, and a promise to maintain secrecy and good conduct. Implicit (or explicit) in this oath is the recruit's understanding of the consequences for breaking this oath. In armed groups, the consequence for collaboration has almost universally been death.[3] Such oaths have been so effective that the English government declared in the late eighteenth century that merely taking the Luddite oath of loyalty was itself punishable by death.[4]

Evaluation period: There may be a provisional or evaluation period after the recruit has joined the group. In this period, the new member may be required to undertake more missions, and identifying information about members of the group (or other sensitive information) may be withheld until the recruit has completed this period.

Be absolutely certain that a candidate is suitable and trustworthy before inviting the person to join. Underground groups cannot "disinvite" someone who knows who and where they are. Recruiters do not share this information freely. Recruiters may not reveal if they are already part of an underground group. Indeed, some recruitment may be done by auxiliaries with little dangerous information.

Recruits must have the psychological balance required to deal with stressful situations, and the social skills needed to work in a close cell or affinity group. They should be willing to accommodate new group norms, but have enough personal fiber to stand up to difficult situations. They must understand the consequences of capture. Members of an underground resistance should also be willing to go to jail if needed, whether that's for five years, for ten years, or longer. A person with dependents is often not a good match for underground work. A single parent with young children would be in a terrible bind if threatened with prison. At that point no decision could avoid bad consequences for the person's children, comrades, or both.

Be alert for warning signs in recruits. Be concerned if a candidate shows a lack of known history, or gaps in history—not just their stated history, but their verifiable history. Evasion or a failure to answer questions directly could indicate a problem. Recruiters should also be on the lookout for psychological or behavioral problems, especially abusive behavior. A history of impulsive or irresponsible behavior would be a danger to the group. Recruiters should be very concerned about a history of drug addiction, because underground groups are based on trust, and someone who is addicted to drugs cannot be trusted if captured. Candidates may also be turned down if they are already too high-profile as militant activists. Police are known to surveil such people looking for clues. Recruiters should also be wary of a history of collaboration or loyalty problems. Relatives with these problems, or relatives in the police, may also cause concern.

Resistance organizations have to decide what to do about "rejected" candidates. If there are too many good candidates to train with available resources, some candidates may be recruited fully at a later date. If the candidate is trustworthy but lacks skills or experience, the individual may be put into the auxiliaries or given further small tasks. If the candidate is a suspected infiltrator or informer, an underground organization may want to either sever communication or attempt to confirm their suspicions and pass on disinformation without letting the person into the group.

During screening candidates may also be assessed to identify how their skills and abilities best fit into the group, and what further training they need. Also, screening does not truly stop after the recruit has been inducted, but continues in a modified form on an ongoing basis. In her volunteer screening handbook, Linda L. Graff writes that "[s]imply put, it is nothing short of dangerous to assume that risks end when a candidate has been screened, even when the screening has been rigorous."[5] She continues by suggesting that organizations use "[m]echanisms such as buddy systems, on-site performance, close supervision, performance reviews, program evaluations . . . , unannounced spot checks, and discipline and dismissal policies" to ensure that candidates continue to be suitable for the organization.

In the 1980s the underground African National Congress used many of these different screening methods in recruitment. Steven Davis explains: "Propagating the underground has traditionally been considered extremely risky because of the danger of inadvertent recruitment of police informers. To minimize the danger, the Congress adopted rigorous intake screening while prescribing punishment for Blacks thought to be assisting the regime. A typical sequence of recruitment would normally begin with a clandestine meeting of the street cell to compile a list of potential enlistees who live on the block. The names may be those of residents who participated in a recent march or school boycott, thereby demonstrating to ANC observers a measure of political consciousness. Members initiate security checks on each candidate to determine his or her reliability and political opinions. One cell member is assigned the task of meeting secretly with each potential recruit. A test, such as acting as a marshal for a funeral protest rally, may be set for the candidate. If the person passes it, he or she may be provisionally invited to join the cell.

"Once the recruit accepts, an initiation process begins. The ANC places great emphasis on instructing its members in party history, philosophy, and strategy. 'We don't want someone who merely knows how to use a gun,' asserts . . . Thabo Mbeki, 'we need a political person, who understands what we stand for.'" (Apparently the ANC preferred to militantize radicals, rather than radicalize militants.)

The initiation process proceeds, continues Davis, and "[u]nder the tutelage of his contact, the new cadre is expected to study the Freedom Charter and accept standards of conduct outlined for all members, including the ban on targeting civilians and the need to maintain discipline. Should the recruit pass muster on these points, he or she is normally fully inducted into the ANC underground. The control agent assigns the enlistee a code name and provides training in methods of secret communication with the cell. In addition, the agent gives the new cadre rudimentary instruction in the use of firearms and explosives."

"The cell leader, perhaps in consultation with colleagues at higher levels, then assigns the enrollee one of a variety of missions."[6] This rigorous recruitment process worked very well for the ANC, and without it they would not have succeeded in abolishing apartheid.

Recruit Training and Enculturation

New recruits need two kinds of training. On one hand, they need cultural training, that is, they need to develop a shared culture with the other members of their group so that everyone can work together smoothly. On the other hand, they need training in the specific skills needed for their work. Some of the shared culture comes from a culture of resistance, and some is on a group-by-group basis. Many of the basic skills for resistance are also common across different groups.

This suggests the need for a sort of "basic training for activists," which would be generally available—and strongly encouraged—for people who want to be part of a culture of resistance. Some skills that belong on the list are already taught in many nonpolitical settings. And conversely, some political groups seem ignorant of key skills needed for successful resistance.

Skills that are legal and should be ubiquitous in a culture of resistance include the following:

- Antioppression analysis and training
- Group facilitation, decision making, conflict resolution, crisis intervention
- Basic history of resistance
- Basic grounding in resistance organizational styles and strategies
- Basic off-the-grid and survival skills
- First aid
- Reinforcement of culture of resistance norms and attributes
- Physical training and self-defense
- Communications, including secure communications

Some of these skills are technical, and so can be readily learned from many sources. Others are deeply political in nature, and need to be taught by people with a commitment to aboveground organizing— probably the people we'd call cadres.

Firearms training should be pretty much universal in a culture of resistance. The potential self-defense applications of this are one aspect; that's not the most important reason. It's difficult to make an informed choice on whether or not to own guns if you don't actually know the rudiments of how to use them. Handling guns is important in demystifying them, because anyone who comes up against power is going to encounter guns (or at least the implicit threat of their use) sooner or later. All this is also important for understanding the history of a culture that has spread and gained power through violence. You cannot truly understand the history of power in this culture—and the history of armed or even unarmed resistance to that power—without handling a firearm. Trying to develop resistance strategy without knowing how to fire a gun would be like trying to understand the impact of communication on human society without ever having spoken or written a word.

If these skills become commonplace in resistance cultures then very little "remedial" training will be required for new recruits. That will allow resistance cadres, especially underground cadres, to focus on training the particular skills needed for their strategy and tactics.

◘ ◘ ◘

Q: A resistance movement will be demonized and portrayed as ecoterrorists by the mainstream media. Is there an alternative media in place with a strategy to counter this?

Derrick Jensen: There is an alternative media in place, but will it counter this demonization? No. The alternative media is tepid and full of horizontal hostility. The larger question is, "Is there a media forum that is supporting serious resistance against this culture's murder of the planet?" And the answer, sadly, is no. Even so-called nature magazines have tremendous resistance to promoting anything other than composting or riding bicycles. Or rather, I should say, a lot of the readers do. One purpose of this book is to help create that literature of resistance—an absolutely necessary literature of resistance—that will help to put in place a larger media of resistance. It takes all forms, from comics to films to books to graffiti to people having conversations on their back porches. We need to be discussing this and we need to be discussing it openly. One of the absolutely necessary precursors to a resistance is to talk about it. This has been true of every resistance movement in the past and it will be true as long as there are resistance movements. We must put all the options on the table and discuss them openly, honestly, earnestly.

Chapter 11

Security

by Aric McBay

Those who bothered incessantly about security survived, but few of them had much beyond survival to their credit. To strike and then to survive was the real test.

—M. R. D. Foot, historian, on World War II resistance movements.

We live in an age of escalating political persecution, and we shouldn't expect that to go away. The more effective and serious a resistance movement becomes, the harsher the persecution of its members and their allies will be. Things will get worse before they get better, the Green Scare being a key example. Even people participating in outwardly innocuous actions are vulnerable to malicious persecution, as long as that action is effective or perceived as a threat by those in power. Those working aboveground have more to be concerned about than those working underground, because the people working aboveground are more accessible to those in power.

Fortunately, activists working against persecution by government and police have come up with ways to combat this problem through the use of a collective security culture. According to the must-read booklet *Security Culture: A Handbook for Activists*, security culture is "a culture where people know their rights and, more importantly, assert them. Those who belong to a security culture also know what behavior compromises security and they are quick to educate those people who, out of ignorance, forgetfulness, or personal weakness, partake in insecure behavior. This security consciousness becomes a culture when the group as a whole makes security violations socially unacceptable in the group."[1]

The handbook identifies six main topics that are inappropriate to discuss.[2] These are:

- Your involvement or someone else's involvement with an underground group.
- Someone else's desire to get involved with such a group.

- Asking others if they are a member of an underground group.
- Your or someone else's participation in an illegal action.
- Someone else's advocacy for such actions.
- Your or someone else's plans for a future action.

The key issue here comes from talking about specifics. Talking about particular people, groups, places, times, targets, events, and other specifics is a bad idea, even if it is a joke, gossip, or speculation. This is different from speaking about resistance or illegal activities in abstract or general terms. As the handbook states, "It is perfectly legal, secure, and desirable that people speak out in support of monkeywrenching and all forms of resistance."[3]

The authors write that there are three and only three specific exceptions to these general rules. The first is if you are planning an action with trusted members of your affinity group in a secure fashion. Even within the affinity group, critical discussion and information should be restricted to those actually participating in an action. The phrase to take away here is *need to know*. In good security culture, only people who need to know critical information have access to it.

The second exception is after a member of the resistance has been arrested, tried, and convicted. In this case, a person may speak about an action for which they've been convicted if the person chooses to. However, the individual must be careful to avoid giving away information that would implicate other people or cause a hazard to people still working underground. Since the *Security Culture* handbook was written, Rod Coronado was arrested and imprisoned for publically discussing details of a previous action he'd already served time for.[4] Take note: if those in power want to persecute you, they may still attack you for discussing previous actions. For this reason, the second exception is not universal.

The third exception noted in *Security Culture* is for anonymous letters and communiqués to the media. However, as the authors remark, this has to be done in a very careful way. The transmission of the communiqué itself must be secure and anonymous. Also, the communiqué should be carefully stripped of identifying information, dialect, or other clues.

There may be a fourth exception not discussed in *Security Culture*. Many of the most effective resistance movements and insurgencies, as discussed in the strategy chapter, work on an "open source" model in which effective attacks and tactics are quickly copied by many groups. In order to avoid reinventing the wheel, instructions and information on the specifics of these tactics must be disseminated to various cells. This may involve sharing information about the specifics of targets and other information that normally shouldn't be discussed. When this sharing is done, care should be taken to conceal identities and other operational details that are secret or not directly related to the topic, and the instructions should be circulated through secure means. The US military has an extensive collection of field manuals for training purposes; an analogous collection of underground resistance field manuals would be invaluable.

Security breaches, when they happen, occur for different reasons. Sometimes people gossip or speculate about who performed certain actions, or ask inappropriately. Sometimes people will lie and claim to have performed actions they have not in order to gain credibility. Others will brag, or hint heavily, about their involvement in underground or illegal activities. All of these behaviors are foolish if not downright stupid and dangerous. Some people in the Green Scare were arrested and put in jail because they or their comrades made security violations like these. Sometimes people do these things because they are being impulsive. Sometimes they do these things because they are using intoxicants. In any case, at the very least these security violations create rumors that can be passed on to listening informers, perhaps via gossip. People who do this act, in effect, as unwitting informers.

If you encounter these behaviors, the first response can be to educate. People aren't born knowing about security culture, and they simply may not have encountered good information or training. Make it clear, in private and tactfully if possible, but firmly, that their actions are violating good security culture. Explain what they did and why security culture is important, and point them toward further resources on the subject. Don't let violations pass or become habit.

Some people, unfortunately, are unable or unwilling to maintain good security culture, and may become chronic violators. They may not be doing it on purpose; you may like them, and they may be your friends. But they may also be acting as informers, either wittingly or unwittingly. The only effective way to deal with repeat violators is to cut them off from sources of information. This generally means asking them to leave your group, not to attend meetings or organizing spaces. To allow them to remain, as harsh as it may seem, is to invite security breaches. It would be far harsher, in the end, to allow potential informers to stay and put activists at risk of prosecution.

This can be very emotionally difficult, but it is necessary. It's well known that counterintelligence agents in government and corporations have surveilled, infiltrated, and sabotaged even mainstream antiwar and environmental groups like Greenpeace. Aboveground groups generally do not and should not have critical information that could end up putting people in jail if it got into the wrong hands. But it's not always clear-cut. Infiltrators who train in "safe" aboveground groups can go on to do more destructive work, including acting as agent provocateurs in your own community. Those infiltrators can also gather information about who sympathizes with militant or radical causes and learn about social networks and relationships. They can decide which revealing offhand comments and suspicious activists should potentially be investigated. Anyone who likes to ask inappropriate questions or gossip about illegal activities will eventually spill information to those in power, either directly or by discussing it electronically where it can be easily surveilled. Conversely, people who brag or lie about illegal or underground activities, or try to plan them with others in public, can draw unnecessary and unwanted attention to any resistance group.

People who cannot follow the simple and basic rules of security culture are either deliberate informers or fundamentally unsuitable for serious resistance. And even though it may be painful or unpleasant, such people need to be separated from groups and places where serious resistance is taking place.

◙ ◙ ◙

People involved in resistance must know their basic legal rights. There are many free pamphlets suited to many different countries. The booklet "If an Agent Knocks: Federal Investigators and Your Rights" from the Center for Constitutional Rights is a good start for the US. Another recent booklet is the National Lawyers Guild's "Operation Backfire: A Survival Guide for Environmental and Animal Rights Activists," available at www.nlg.org.

If you believe you are being followed or watched, or if you are contacted by the police, report this to others in your activist community. After you are contacted, write down the names of the agents who spoke to you, what they said, as many questions as you can remember, and anything else that seems important. This can be passed on to others in your community. In part this kind of transparency helps to maintain trust in a community (and to avoid rumors that someone saw so-and-so talking to the cops). It also helps warn others in the community that they should keep their guard up, refresh people on their rights, and perhaps initiate counterintelligence work.

Resisters who are involved in underground or illegal activities usually want to warn their comrades if they know or hear that police are poking around. The initial warning is sometimes a prearranged but outwardly innocuous code word or phrase which can be quickly relayed over phone, email, or other medium. (Of course, the police may also be tapping telephone or email, and waiting to see who a suspected resister contacts immediately after a probing visit.)

It's worth studying the investigative and interrogation techniques used by police. These techniques mostly aren't secrets, but can be found in books and other resources.

Firewalls

Good security isn't just about individual behavior. Good security includes everything we've talked about: decision making, recruitment, and overall structure.

As we discussed, it is crucial that a firewall exist between those carrying out underground activities and those doing aboveground work. Internal firewalls should also be in place between compartmentalized portions of an underground organization.

Information should only cross these firewalls under very narrow and circumscribed conditions. Groups need clearly stated internal policies about when and what information can cross, when contact can be made, how that information may be communicated, and so on. Generally speaking, it should be tightly controlled and very intentional; the vast majority of the time an underground group should maintain "radio silence" as far as discussion of activities outside the group are concerned.

There are three main reasons that information might pass through a firewall. The first is to gather information and reports from auxiliaries outside the immediate organization. The second is to send information such as proclamations or communiqués to the media or press office. The third is for internal communication within an underground resistance network. In all cases, identifying information should be stripped away from the communication. The time, place, and nature of the communication should be done according to a group's internal security policy. The people who bridge the gap between the aboveground and the underground group are also taking a definite risk, and should be aware of that.

The firewall also applies to other types of nonpolitical crime. Underground activists should avoid breaking other laws if only for reasons of self-preservation. This includes traffic laws. Breaking laws means risking the attention of police, adding an unnecessary risk for those working underground. Tre Arrow knows this, since while a fugitive he was arrested and imprisoned after being caught shoplifting bolt cutters. People who are underground must keep a low profile and at least *look* like regular, law-abiding citizens. Further, people who want to commit crimes for the sake of committing crimes are often not a good match for the underground. They may want to commit actions just for the rush, rather than for strategic or political reasons. And they often lack solidarity for others involved in such actions. (During the Green Scare many people were imprisoned because of informer Jacob Ferguson, a long-time petty criminal.)

In researching this book, we've encountered many examples of the inappropriate application of security culture; that is, some groups are applying draconian security culture measures where they are not appropriate or failing to apply security culture rules when they are appropriate. This confusion needs clearing up, because at best it can cause people to be hampered by unnecessary limitations, and, at worst, cause people to needlessly face charges or prison time.

It is important to understand what constitutes "security" and what does not. While I was writing this section, people contacted me with the complaint that some people in their aboveground groups wanted to institute excessive "security culture" measures. For example, a person in one group wanted to stop sending out minutes of the meetings to people who didn't attend meetings anymore. But this sort of thing doesn't increase security. They were still sending minutes out by email to those who attended meetings, and nothing sent by email can be considered secure. The effect was to unnecessarily exclude people, in basic contradiction of the aboveground need to maximize inclusiveness, outreach, and communication. And besides, an aboveground group should not be carrying out illegal activities that carry any serious risk of reprisal in the first place.

In contrast, some groups simply don't follow security culture measures when they should. One activist friend told me of a time in her youth when fellow militants built a barricade on a major street at 8:00 am one Monday as a gesture of solidarity for labor strikes going on elsewhere in the state. Then they set it on fire. This was not guerilla-style "hit and run"; groups of sign-waving protesters were marching across the street. When the police arrived, a number of people were arrested and faced serious charges.

Now, as much as we all may love flaming barricades, the subsequent legal troubles for arrestees far outweighed the benefits of the action. In some ways this is an issue of tactics and strategy, which we'll return to in the final part of this book. But it's also an organizational issue. Militant groups often carry out attacks that cause minor damage but don't shake the system itself. This happens often when a group is made up entirely of combatants, with the mentality common to combatants. They want to fight. They want to confront the cops, they want to confront those in power. They want to cause damage, to agitate, to shake things up. Not only are they unafraid of conflict, they seek it out. They're often young, and long-term strategy isn't on their minds.

These characteristics can be wonderful and admirable, and resistance movements cannot succeed without combatants who take to the front lines. But resistance movements also can't succeed without cadres. Cadres, as the backbone of an organization, are tasked with strategic and training concerns. They want to maximize a group's capacity and long-term success. It's their job to think strategically, to think long-term, to ensure combatants don't do reckless things that harm the organization.

There's no firm dividing line between combatants and cadres, and people aren't born into either role; they can move or change over time. But groups without a good proportion in each role are going to falter. You can't have an army that consists only of officers, and you can't have an army that consists only of foot soldiers.

New groups or those with high turnover often lack cadres. You can't read a how-to book and become a cadre (although cadres do study resistance intensively). Cadres need years or decades of experience, a solid political and organizational grounding, and mentorship or other training from existing cadres. If your group lacks cadres, you should either train them yourself, or borrow or recruit them from other groups.

◫ ◫ ◫

The way to avoid paranoia and an improperly applied security culture is to understand that different protective measures and security precautions are appropriate for different activities. If I were a knight on a medieval battlefield, I might wear a suit of armor; that would be an appropriate protective measure for that activity. On the other hand, if I were going to the swimming pool, I would wear a bathing suit; that would be appropriate protection there. Obviously it would be foolish to wear only swimming trunks to a castle siege. But it would be similarly foolish to wear a suit of armor to the swimming pool. It might make you *feel* more protected, but it would make swimming very difficult, if not fatal.

Underground groups protect themselves by keeping their location, identities, and activities secret. But aboveground groups lack clandestine mobility. If persecuted by the police or others, they use strength in numbers and a network of supporters to defend themselves. They use their communication and social networks to mobilize. An aboveground group that imposes needlessly restrictive "security" limitations isn't genuinely increasing its security at all. On the contrary, it's decreasing its security by alienating members and allies and by cutting itself off from its network of supporters.

The onus for keeping a low profile is on individual people and their chosen and trusted groups. Some people may join regular aboveground groups and push excessive security measures not suited for their group's activities. (This is a bit like signing up for the synchronized swimming team and then showing up for practice and complaining that people don't wear full-body chain mail.) If you need to keep a low profile, if you personally need a higher level of security restriction, then you shouldn't be part of such a group in the first place.

When people do try to use higher security measures with aboveground groups, they often do it inconsistently. A good friend of mine, Brent, used to work in an aboveground conservation group affiliated with a larger liberal foundation. Shortly after joining the group, he was added to their email listserv. At first Brent was very confused by the discussion, because he didn't recognize any of the people who were posting. Another person explained that the members were using false "code names" to protect themselves. When Brent inadvertently used someone's real name on the listserv, there was a flare-up. Brent then explained why he thought the code names were silly and alienating, and four people sent out angry emails attacking him—two sent from their work email addresses, and one using a school address, thus making their real names obvious. Can I convert my biodiesel van to run on irony instead?

If you want to use a higher level of security in an aboveground group that may be fine, but you can't just use some restrictions and ignore others. Security is only as strong as the weakest link. Trying to mix and match security measures is like wearing the top half of a suit of armor with a speedo; it's needlessly cumbersome, it offers no real protection, and it looks damned silly.

This is not to say that aboveground groups never do illegal actions (though organizing higher-risk illegal actions over an email listserv would be stupid no matter the group). Nor does it mean that aboveground groups have to let just anyone join up and start planning. But it is important to recognize that the purpose of security culture isn't to make people "safe" (since working against those in power never is), it's to make people more effective. People can't be very effective if they're in jail or caught up in the courts. But they also can't be effective aboveground if they shackle themselves with pointless "security" measures.

As M. R. D. Foot noted about resistance in occupied Europe: "in an excellent phrase of one of [British intelligence's] SOE's men in Stamboul, 'Caution axiomatic, but over-caution results in nothing done.' Those who bothered incessantly about security survived, but few of them had much beyond survival to their credit. To strike and then to survive was the real test."[5]

◙ ◙ ◙

There's no question that the firewall is crucial. Many successful resistance movements, including the French Resistance, used it. Other resistance movements did not, much to their detriment. But there are fuzzy zones. It's great to *say* that groups should avoid excessive security. But some aboveground groups do exist in an awkward middle ground; maybe they want to push the legal limit with their tactics. This makes basic security culture all the more important. Information about illegal actions needs to be limited to those directly involved. And those in power will often prosecute trivial crimes in order to attack people they couldn't get otherwise. (How many mob bosses were convicted for tax evasion or mail fraud?)

In other cases, people may be working aboveground but hope to go underground. Those people should consider keeping a low profile, especially as far as electronic means are concerned. Data mining and profiling can allow those in power to identify people who have radical sympathies or interests. Keeping a low profile means not leaving a "paper trail" (or, in the case of online records, a digital trail) which would make someone seem suspicious or of interest.

LEVELS OF SECURITY

Security is often organized at four different levels: individual, relational, operational, and organizational.

At the *individual* level, members of a resistance group are responsible for following good security practices. This includes the general precepts of security culture discussed above. Individual security is a group's first line of defense. Underground resisters generally stay inconspicuous. All resisters need training for specific dangerous situations, like arrest.

Relational security refers to the way in which members of a group relate to each other and people outside the group. Relational security measures maximize the benefits of collaboration and minimize the risk (especially by limiting the effects of someone else's individual security breach, which for some will mean hiding if a close ally is arrested). Good security culture, again, is important here. So are maintaining firewalls, either within or around an organization, and using secure communication.

Operational security measures are used during specific operations, missions, or tasks. These might include good reconnaissance and rehearsals before an operation, lookouts and tight communications during an operation, training in escape and evasion, and preplanned escape routes or safehouses for after an operation.

Organizational security measures are the highest level of security. A group's organizational security measures are reflected in its general organization structure, record keeping, recruitment practices, and so on. At an organizational level, a group (or its leadership) is responsible for determining the standards and group norms for security practices at all levels. Organizational security includes enforcement, perhaps carried out by security cadres within the organization. Such units are also responsible for identifying and suppressing collaborators or informers in the organization, and for counterintelligence measures like disinformation.

These measures apply not just to members of the group, but also to ex-members. When people leave a group, they must know what is expected of them. The essential rules of security culture apply even after a person has left a group. Members of resistance movements do not discuss illegal or underground activities except with the people in their group. If they leave the group, then they simply may not discuss group activities with *anyone*. Members of a disbanded underground group must refrain from chatting about past actions. People are either in an underground group or they are not. There is no middle ground. Discussing the "good old days" with former comrades has sent many people to prison during the Green Scare.

◫ ◫ ◫

Resistance movements organize themselves into groups based on their political means, and they organize those groups into networks that make them as effective as possible. Those groups make decisions well, they know how to recruit new members, and they can maintain their own security. That's no small task, but the next is bigger. The point of organizing is to fight, and fight to win. And to do that, they need real strategy.

◫ ◫ ◫

Q: What might distinguish an anticivilization resistance from other popular movements that those in power have successfully overpowered COINTELPRO-style? Do people have new strategies and tactics that can stand up to these new systems and technologies?

Derrick Jensen: Frankly, no. People now have a tremendous disadvantage over people in the past in that people now live inside a panopticon. The ability to surveil and to kill at a distance has greatly increased over what it was in times past. Contrast the powers of the state at present with those, say, in Nazi Germany. For the Nazis, fingerprint technology was still very new. They had nothing like the capacity to surveil that modern states have. They had only rudimentary computers. They didn't have voice recognition software. They didn't have *any* software. So those in power have a tremendous advantage over historical popular movements.

Indigenous and traditional resistance movements had villages where they could be safe. They had wild places where they could be safe. They had their own territory. People now don't have that. They do, however, have a significant advantage over the indigenous resistance movements of the last five hundred years in that they mix in. Tecumseh could not have walked into Philadelphia and not been recognized as different. People today have that advantage.

But the biggest advantage that people today have over people in times previous is that the age of exuberance is over. The age of cheap oil is over. The empires of today are on their way to collapse. It used to seem that as civilization dissolved, anyone who even remotely opposed it would be put up against a wall. But now it looks as though as civilization falls apart, its emperors may not even be able to deliver the mail, much less maintain the level of oppression that they have historically perpetrated on those who oppose empire. Think of the collapse of the Soviet Union; it just sort of fell apart instead of instigating purges or gulags. The Soviet Union didn't have the resources.

Even the United States is falling apart. The US government can't maintain the water systems in this country and it can't maintain the roads. State and federal governments can't pay for colleges anymore. Those in power don't have the money, and they don't have the resources, and those resources will never come back.

If someone would have taken out some important piece of infrastructure in years past, those in power would have been able to replace it. But now the governments of the world don't have the money. The more they spend on rebuilding, the less primary damage they can do.

PART III: STRATEGY AND TACTICS

Chapter 12

Introduction to Strategy

by Aric McBay

> I do not wish to kill nor to be killed, but I can foresee circumstances in which both these things would be by me unavoidable. We preserve the so-called peace of our community by deeds of petty violence every day. Look at the policeman's billy and handcuffs! Look at the jail! Look at the gallows! Look at the chaplain of the regiment! We are hoping only to live safely on the outskirts of this provisional army. So we defend ourselves and our hen-roosts, and maintain slavery.
>
> —Henry David Thoreau, "A Plea for Captain John Brown"

Anarchist Michael Albert, in his memoir *Remembering Tomorrow: From SDS to Life after Capitalism,* writes, "In seeking social change, one of the biggest problems I have encountered is that activists have been insufficiently strategic." While it's true, he notes, that various progressive movements "did just sometimes enact bad strategy," in his experience they "often had no strategy at all."[1]

It would be an understatement to say that this inheritance is a huge problem for resistance groups. There are plenty of possible ways to explain it. Because we sometimes don't articulate a clear strategy because we're outnumbered and overrun with crises or immediate emergencies, so that we can never focus on long-term planning. Or because our groups are fractured, and devising a strategy requires a level of practical agreement that we can't muster. Or it can be because we're not fighting to win. Or because many of us don't understand the difference between a *strategy* and a *goal* or a *wish*. Or because we don't teach ourselves and others to think in strategic terms. Or because people are acting like dissidents instead of resisters. Or because our so-called strategy often boils down to asking someone else to do something for us. Or because we're just not trying hard enough.

One major reason that resistance strategy is underdeveloped is because thinkers and planners who *do* articulate strategies are often attacked for doing so. People can always find something to disagree with. That's especially true when any one strategy is expected to solve all problems or address all causes claimed by progressives. If a movement depends more on ideological purity than it does on accomplishments, it's easy for internal sectarian arguments to take priority over getting things done. It's easier to attack resistance strategists in a burst of horizontal hostility than it is to get things together and attack those in power.

The good news is that we can learn from a few resistance groups with successful and well-articulated strategies. The study of strategy *itself* has been extensive for centuries. The fundamentals of strategy are foundational for military officers, as they must be for resistance cadres and leaders.

PRINCIPLES OF WAR AND STRATEGY

The US Army's field manual entitled *Operations* introduces nine "Principles of War." The authors emphasize that these are "not a checklist" and do not apply the same way in every situation. Instead, they are characteristic of successful operations and, when used in the study of historical conflicts, are "powerful tools for analysis." The nine "core concepts" are:

Objective. "Direct every military operation toward a clearly defined, decisive, and attainable objective." A clear goal is a prerequisite to selecting a strategy. It is also something that many resistance groups lack. The second and third requirements—that the objective be both decisive and attainable—are worth underlining. A decisive objective is one that will have a clear impact on the larger strategy and struggle. There is no point in going after one of questionable or little value. And, obviously, the objective itself must be attainable, because otherwise efforts toward that operation objective are a waste of time, energy, and risk.

Offensive. "Seize, retain, and exploit the initiative." To seize the initiative is to determine the course of battle, the place, and the nature of conflict. To give up or lose the initiative is to allow the enemy to determine those things. Too often resistance groups, especially those based on lobbying or demands, give up the initiative to those in power. Seizing the initiative positions the fight on our terms, forcing them to react to us. Operations that seize the initiative are typically offensive in nature.

Mass. "Concentrate the effects of combat power at the decisive place and time." Where the field manual says "combat power," we can say "force" more generally. When Confederate General Nathan Bedford Forrest summed up his military theory as "get there first with the most," this is what he was talking about. We must engage those in power where we are strong and they are weak. We must strike when we have overwhelming force, and maneuver instead of engaging when we are outmatched. We have limited numbers and limited force, so we have to use that when and where it will be most effective.

Economy of Force. "Allocate minimum essential combat power to secondary efforts." In order to achieve superiority of force in decisive operations, it's usually necessary to divert people and resources from less urgent or decisive operations. Economy of force requires that all personnel are performing important tasks, regardless of whether they are engaged in decisive operations or not.

Maneuver. "Place the enemy in a disadvantageous position through the flexible application of combat power." This hinges on mobility and flexibility, which are essential for asymmetric conflict. The fewer a group's numbers, the more mobile and agile it must be. This may mean concentrating forces, it may mean dispersing them, it may mean moving them, or it may mean hiding them. This is necessary to keep the enemy off balance and make that group's actions unpredictable.

Unity of Command. "For every objective, ensure unity of effort under one responsible commander." This is where some streams of anarchist culture come up against millennia of strategic advice. We've already discussed this under decision making and elsewhere, but it's worth repeating. No strategy can be implemented by consensus under dangerous or emergency circumstances. Participatory decision making is not compatible with high-risk or urgent operations. That's why the anarchist columns in the Spanish Civil War had officers even though they despised rulers. A group may arrive at a strategy by any decision-making method it desires, but when it comes to implementation, a hierarchy is required to undertake more serious action.

Security. "Never permit the enemy to acquire an unexpected advantage." When fighting in a panopticon, this principle becomes even more important. Security is a cornerstone of strategy as well as of organization.

Surprise. "Strike the enemy at a time or place or in a manner for which they are unprepared." This is key to asymmetric conflict—and again, not especially compatible with an open or participatory decision-making structures. Resistance movements are almost always outnumbered, which means they have to use surprise and swiftness to strike and accomplish their objective before those in power can marshal an overpowering response.

Simplicity. "Prepare clear, uncomplicated plans and clear, concise orders to ensure thorough understanding." The plan must be clear and direct so that everyone understands it. The simpler a plan is, the more reliably it can be implemented by multiple cooperating groups.

Many of these basic principles fall into conflict with the favored actions of dissidents. Protest marches, petitions, letter writing, and so on often lack a decisive or attainable objective, give the initiative to those in power, fail to concentrate force at a decisive juncture, put excessive resources into secondary efforts, limit maneuvering ability, lack unified command for the objective (such as there is), have mixed implementation of security, and typically offer no surprise. They are, however, simple plans, if that's any consolation.

In fact, these strategic principles might as well come from a different dimension as far as most (liberal) protest actions are concerned. That's because the military strategist has the same broad objective as the radical strategist: to use the decisive application of force to accomplish a task. Neither strategist is under the illusion that the opponent is going to correct a "mistake" if this enemy gets enough information or that success can occur by simple persuasion without the backing of political force. Furthermore, both are able to clearly identify their enemy. If you identify with those in power, you'll never be able to fight back. An oppositional culture has an identity that is distinct from that of those in power; this is a defining element of cultures of resistance. Without a clear knowledge of who your adversary is, you either end up fighting everyone (in classic horizontal hostility) or no one, and, in either case, your struggle cannot succeed.

☒ ☒ ☒

In the US Army's field manual on guerrilla warfare, entitled *Special Forces Operations*, the authors go further than the general principles of war to kindly describe the specific properties of successful asymmetric conflict. "Combat operations of guerilla forces"—and, I would add, resistance and asymmetric forces in general—"take on certain characteristics that must be understood."[2] Six key characteristics must be in place for resistance operations:

Planning. "Careful and detailed. . . . [p]lans provide for the attack of selected targets and subsequent operations designed to exploit the advantage gained. . . . Additionally, alternate targets are designated to allow subordinate units a degree of flexibility in taking advantage of sudden changes in the tactical situation." In other words, it is important to employ maneuvering and flexible application of combat power. It's important to emphasize that planning is *not* about coming up with a concrete or complex scheme. The point is to plan well enough that they have the flexibility to improvise. It might sound counterintuitive, but the goal is to create an adaptable plan that offers many possibilities for effective action that can be applied on the fly.

Intelligence. "The basis of planning is accurate and up-to-date intelligence. Prior to initiating combat operations, a detailed intelligence collection effort is made in the projected objective area. This effort supplements the regular flow of intelligence." That's strategic and operational intelligence. On a tactical level, "provisions are made for keeping the target or objective area under surveillance up to the time of attack."

Decentralized Execution. "Guerrilla combat operations feature centralized planning and decentralized execution." It is necessary to have a coherent plan, and in order for that plan to be a surprise, the details often have to be kept secret. A centralized plan allows separate cells to carry out their work independently but still accomplish something through coordination and building toward long-term objectives. Decentralized execution is needed to reach multiple targets for a group that lacks a command and control hierarchy.

Surprise. "Attacks are executed at unexpected times and places. Set patterns of action are avoided. Maximum advantage is gained by attacking enemy weaknesses." When planning a militant action, resisters don't announce when or where. The point is not to make a statement, but to make a decisive material impact on systems of power. This can again be enhanced by coordination between multiple cells. "Surprise may also be enhanced by the conduct of concurrent diversionary activities."

Short Duration Action. "Usually, combat operations of guerrilla forces are marked by action of short duration against the target followed by a rapid withdrawal of the attacking force. Prolonged combat action from fixed positions is avoided." Resistance groups don't have the numbers or logistics for sustained or pitched battles. If they try to draw out an engagement in one place, those in power can mobilize overwhelming force against them. So underground resistance groups appear, accomplish their objectives swiftly, and then disappear again.

Multiple Attacks. "Another characteristic of guerrilla combat operations is the employment of multiple attacks over a wide area by small units tailored to the individual missions." Again, coordination is required. "Such action tends to deceive the enemy as to the actual location of guerrilla bases, causes him to over-estimate guerrilla strength and forces him to disperse his rear area security and counter guerrilla efforts." That is, when those in power don't know where an attack will come, they must spend effort to defend every single potential target—whether that means guarding them, increasing insurance costs, or closing down vulnerable installations. And as forces become more dispersed in order to guard sprawling and vulnerable infrastructure, they become less concentrated and correspondingly make easier targets.

Other writers on resistance struggles have shared these understandings. Che Guevara outlined similar strategy and tactics in his book *Guerilla Warfare* (1961), which itself followed from Mao Tse-Tung's 1937 book on the subject. Colin Gubbins, former head of the British Special Operations Executive, wrote two pamphlets on the subject for use in Occupied Europe (written not long after Mao's book). These pamphlets—*The Partisan Leader's Handbook* and *The Art of Guerilla Warfare*—were based in part on what the British learned from T. E. Lawrence, but also from their attempts to quash resistance warfare in Ireland, Palestine, and elsewhere. In *The Partisan Leader's Handbook*,

Gubbins touched on the elements of surprise ("the most important thing in everything you undertake"), mobility, secrecy, and careful planning. "The whole object of this type of warfare is to strike the enemy, and disappear completely leaving no trace; and then to strike somewhere else and vanish again. By these means the enemy will never know where the next blow is coming," he wrote.

Gubbins also urged resistors to "never engage in any operation unless you think success is certain." Small resistance units don't have the numbers or morale to absorb unnecessary losses. Resistance groups should only engage the enemy at points and times where they can overwhelm. The first step to take before any action is to plan a safe line of retreat, and "break off the action as soon as it becomes too risky to continue." A newly founded resistance group often lacks the experience and training to accurately gauge how risky a situation is, which is why Gubbins recommends erring on the side of caution. It is better to learn iteratively and build up from a number of small successes than to get caught attempting operations that are too large and apt to end in failure. The takeaway message: successful resistance movements choose their battles carefully.

Just as asymmetric strategies require specific characteristics for success, they also have definite limitations.[3] Resistance forces typically have "limited capabilities for static defensive or holding operations." They often *want* to hold territory, to stand and fight. But when they try, it usually gets them killed, unless they've spent years developing extensive social and military groundwork and have a large force and popular support. Another limitation is that, especially in the beginning, resistance forces lack "formal training, equipment, weapons, and supplies" that would allow them to undertake large-scale operations. This can be gradually remedied through ongoing recruitment and training, good logistics, and the security and caution required to limit losses through attrition; however, resistance forces are often dependent on local supporters and auxiliaries—and perhaps an outside sponsoring power—for their supplies and equipment. If they can't find those supporters, they will probably lose.

Communications offer another set of limitations. Communications in underground groups are often difficult, limited, and slow. This also applies to organizational command; the more decentralized an organization is, the longer it takes to propagate decisions, orders, and other information. And because resistance groups have small numbers and finite resources, "the entire project is dependent upon precise, timely, and accurate intelligence."

DEVISING STRATEGY

Despite the limitations created by their smaller numbers, resistance movements do have real strategic choices, from the loftiest overarching strategy to the most detailed tactical level. Let's explore beyond the default palette of actions. Resisters can and must do far better than the strategy of the status quo.

There is a finite number of possible actions, and a finite amount of time, and resisters have finite resources. There are no perfect actions. Prevailing dogma puts the onus on dissenters to be "creative" enough to find a "win-win" solution that pleases those in power *and* those who disagree, that stops the destruction of the planet but permits the continuation of business as usual and lifestyles of conspicuous consumption. If resisters fall prey to this belief, if they accept its absurd and contradictory premises, they are engineering their own defeat before the fact. If resisters believe this, they are accepting all blame for the actions of those in power, accepting that the problems they face are *their* fault for not being "innovative" enough, rather than the fault of those in power for deliberately destroying the world to enrich themselves.

At the highest strategic level, any resistance movement has several general templates from which to choose. It may choose a war of containment, in which it attempts to slow or stop the spread of the opponent. It may choose a war of disruption, in which it targets *systems* to undermine their power. It may choose a war of public opinion, by which to win the populace over to their side. But the main strategy of the left, and of associated movements, has been a kind of war of attrition, a war in which the strategists hope to win by slowly eroding away the personnel and supplies of the other side, thus wearing down the omnicidal power structures and public opposition to change more quickly than those forces can destroy our communities, more quickly than they can gobble up biodiversity, more quickly than they can burn the remaining fossil fuels. Of course, this strategy has been an abysmal failure.

A strategy of attrition only works when there is an indefinite amount of time to maneuver, to prolong or delay conflict. Obviously that's not the case in the current situation, which is urgent and worsening. Furthermore, to achieve success in a war of attrition, the resistance must be able to wear down the enemy more quickly than it gets worn down; again, in the present case, those in power are not being worn down at all (except in the degree to which they are so rapidly consuming the commodities required for their own reign to continue).

Furthermore, a resistance movement fighting a war of attrition must reasonably expect that it will be in a better strategic position in the future than it is at the current time. But who genuinely believes that we—however you would define "we"—are moving toward a better strategic position? And in order to get ahead in a war of attrition, resisters would have to have more disposable resources than their opponent.

Another crucial element in a war of attrition is reliable recruitment and growth. It doesn't matter how many enemy bridges a group takes out if the adversary can build them faster than they can be destroyed. And on every level, civilization is recruiting and growing faster than resistance forces. To keep pace, resistance fighters would have to destroy dams more quickly than they are built, get people to hate capitalism faster than children are inculcated to love it, and so on. So far, at least, that's not happening.

Of course, we are not in a two-sided war of attrition. Those in power aren't holding back, but have been actively attacking. And those in the resistance haven't even been fighting a comprehensive war of attrition; it's more like a *moral* war of attrition. Rather than trying to erode the material basis of power, we've been hoping that eventually they'll run out of bad things to do, and perhaps then they'll come around to our way of thinking.

A movement that wanted to win would be smarter and more strategic than that. It would abandon the strategy of moral attrition. It would identify the most vulnerable targets those in power possess. It would strike directly and decisively at their infrastructure—physical, economic, political—and do it while there is still a planet left.

◻ ◻ ◻

Strategy and tactics form a continuum; there's no clear dividing line between them. So the tactics available, which will be discussed in the next chapter, guide strategy, and vice versa. But strategy forms the base. If resistance action is a tree, the tactics are spreading branches and leaves, finely divided and numerous, while the strategy is the trunk, providing stability, cohesion, and rootedness. If resisters ignore the necessity and value of strategy, as many would-be resistance groups do—they are all tactics, no strategy—then they don't have a tree, they have loose branches, tumbleweeds blowing this way and that with changing winds.

Conceptually, strategy is simple. First understand the context: where are we, what are our problems? Then, develop the goal(s): where do we want to be? Identify the priorities. Now figure out what actions are needed to get from point A to point B. Finally, identify the resources, people, and specific operations needed to carry out those activities.

Here's an example. Let's say you love salmon. Here's the context: salmon have been all but wiped out in North America, because of dams, industrial logging, industrial fishing, industrial agriculture, the murder of the oceans, and global warming. The goal is for the salmon population not only to stop declining, but to increase. The difference between a world in which salmon are being wiped out, and one in which they are thriving, comes down to those six obstacles. Over-coming them would be the priority in any successful strategy to save the salmon.

What actions must be taken to honor this priority? Remove the dams. Stop industrial forms of logging, fishing, and agriculture. Stop the massive production and dumping of plastics. Stop global warming, which means stop the burning of fossil fuels. In all these cases, existing structures and practices have to be demolished for salmon to survive, for the goal to be accomplished.[4]

Now it's time to proceed to the operational and tactical side of this strategy. According to the US Army field manual, all operations fit into one of three "all encompassing" categories: decisive, sustaining, or shaping.

Decisive operations "are those that directly accomplish the task" or objective at hand. In our salmon example, a decisive operation might be taking out a dam or preventing a clear-cut above a salmon spawning stream. Decisive operations are the centerpiece of strategy.

Sustaining operations "are operations at any echelon that enable shaping and decisive operations" by offering direct support to those other operations. These supporting operations might include funding or logistical support, communications, security, or other aid and services. In the salmon example, this might mean providing transportation to people taking out a dam, bringing food to tree-sitters, or helping to research timber sale appeals. It might mean running an escape line or safehouse, or providing prisoner support.

Shaping operations "create and preserve conditions for the success of the decisive operation." They alter the circumstances of the conflict and help bring about the conditions required for victory. Shaping operations could include carrying out a campaign on the importance of removing dams, undermining a particular logging company, or helping to develop a culture of resistance that values effective action and refuses to collaborate. However, shaping operations are not necessarily broad-based or indirect. If an allied underground cell were to attack a nearby pipeline as a distraction, allowing the main group to take out a dam, that diversionary measure would be considered a shaping operation. The lobby effort that created the Clean Water Act could even be considered a shaping operation, because it helps to preserve the conditions necessary for victory.

If you look at the taxonomy of action chart on page 243, you'll see that the actions on the left consist mostly of shaping operations, the actions along the center-right consist mostly of sustaining operations, and the right-most actions are generally decisive.

These categories are used for a reason. Every effective operation— and hence every effective tactic—must fall into one or more of these categories. It must do one of those things. If it doesn't—if that operation's or tactic's contribution to the end goal is undefined or inexpressible—then successful resisters don't waste time on that tactic.

◙ ◙ ◙

LEARNING FROM NONVIOLENT STRATEGY

It's also worth looking at the principles that guide strategic nonviolence. Effective nonviolent organizing is not a pacifist attempt to convince the state of the error of its ways, but a vigorous, aggressive application of force that uses a subset of tactics different from those of military engagements.

Gene Sharpe recognized this, and Peter Ackerman and Christopher Kruegler followed Sharpe's strategic tradition in their book *Strategic Nonviolent Conflict: The Dynamics of People Power in the Twentieth Century*. They understand that there is no dividing line between "violent" and "nonviolent" tactics, but rather a continuum of action. Furthermore, they also understand the need for tactical flexibility; sticking to only one tactic, such as mass demonstrations, gives those in power a chance to anticipate and neutralize the resistance strategy. In terms of strategy, they argue "that most mass nonviolent conflicts to date have been largely improvised" and could greatly benefit from greater preparation and planning.[5] I would argue that the same applies to any resistance movement, regardless of the particular tactics it employs.

Having assessed the history of nonviolent resistance strategy in the twentieth century, Ackerman and Kruegler offer twelve strategic principles "designed to address the major factors that contribute to success or failure" in nonviolent resistance movements. They class these as principles of development, principles of engagement, and principles of conception.

Their principles of development are as follows:

Formulate functional objectives. The first principle is clearly important in any resistance movement using any tactics. "All competent strategy derives from objectives that are well chosen, defined, and understood.

Yet it is surprising how many groups in conflict fail to articulate their objectives in anything but the most abstract terms."[6]

Ackerman and Kruegler also observe that "[m]ost people will struggle and sacrifice only for goals that are concrete enough to be reasonably attainable." As such, if the ultimate strategic goal is something that would require a prolonged and ongoing effort, the strategy should be subdivided into multiple intermediate goals. These goals help the resistance movement to evaluate its own success, grow support and improve morale, and keep the movement on course in terms of its overall strategy. This is especially important when the dominant power structure has been in control for a long time (as opposed to a recent occupier). "The tendency to view the dominant power as omnipotent can best be undermined by a steady stream of modest, concrete achievements."[7] This is especially relevant to groups that have very large, ambitious goals like abolishing capitalism, ending racism, or bringing down civilization.

Develop organizational strength. Ackerman and Kruegler write that "to create new groups or turn preexisting groups and institutions into efficient fighting organizations" is a key task for strategists.[8] They also note that the "operational corps"—who we've been calling cadres—have to organize themselves effectively to deal with threats to organizational strength, specifically "opportunists, free-riders, collaborators, misguided enthusiasts who break ranks with the dominant strategy, and would-be peacemakers who may press for premature accommodation."[9] These threats damage morale and undermine the effectiveness of the strategy.

Secure access to critical material resources. They identify two main reasons for setting up effective logistical systems: for physical survival and operations of the resisters, and to enable the resistance movement to disentangle itself from the dominant culture so that various noncooperation activities can be undertaken. "Thought should be given, at an early stage, to controlling sufficient reserves of essential materials to see the struggle through to a successful conclusion. While basic goods and services are used primarily for defensive purposes, such other assets as communications infrastructure and transportation equipment form the underpinnings of offensive operations."[10] In particular, they suggest stockpiling communications equipment.

Cultivate external assistance. The benefits of cultivating external assistance and allies should be clear. Combating an enemy with global power requires as many allies and as much solidarity as resisters can rally.

Expand the repertoire of sanctions. The fifth principle is key because it is highly transferable. By "expand the repertoire of sanctions," they simply mean to expand the diversity of tactics the movement is capable of carrying out effectively. They also encourage strategists to evaluate the risk versus return of various tactics. "Some sanctions can be very inexpensive to wield or can operate at very low risk. Unfortunately, such sanctions may also have a correspondingly low impact. A minute of silence at work to display resolve is a case in point. Other sanctions are grand in design, costly, and replete with risk. They also may have the greatest impact."[11]

Their second group of principles consists of principles of engagement:

Attack the opponents' strategy for consolidating control. This is specifically intended for mass movements, but essentially the authors mean to undermine the control structure of those in power, to generally subvert them, and to ensure that any repression or coercion those in power attempt to carry out is made difficult and expensive by the resistance.

Mute the impact of the opponents' violent weapons. "The corps [or cadres] cannot prevent the adversaries' deployment and use of violent methods, but it can implement a number of initiatives for muting their impact. We can see several ways of doing this: get out of harm's way, take the sting out of the agents of violence, disable the weapons, prepare people for the worst effects of violence, and reduce the strategic importance of what may be lost to violence."[12] These options—mobility, the use of intelligence for maneuver, and so on—are basic resistance approaches to any attack by those in power, and not limited to nonviolent activists.

Alienate opponents from expected bases of support. Ackerman and Kruelger suggest using "political jiujitsu" so that the violent actions of those in power are used to undermine their support. Of course, we could extend this to generally undermining all kinds of support structures that those in power rely on—social, political, infrastructural, and so on.

Maintain nonviolent discipline. Interestingly, the key word in their discussion seems to be not "nonviolence," but "discipline." "Keeping non-violent discipline is neither an arbitrary nor primarily a moralistic choice. It advances the conduct of strategy."[13] They compare this to soldiers in an army firing only when ordered to. Regardless of what tactics are used, it's clear that they should be used only when appropriate in the larger strategy."

Their third and final group is the principles of conception:

Assess events and options in light of levels of strategic decision making. Planning should be done on the basis of context and the big picture to identify the strategy and tactics used. Often, as we have discussed, this is simply not done. The failure to have a long-term operational plan with clear steps makes it impossible to measure success. "Lack of persistence, a major cause of failure in nonviolent conflict, is often the product of a short-term perspective."[14]

Adjust offensive and defensive operations according to the relative vulnerabilities of the protagonists. Strategists need to analyze and fluidly react to the changing tactical and strategic situation in order to shift to offensive or defensive postures as appropriate.

Sustain continuity between sanctions, mechanisms, and objectives. There must be a sensible continuum from the goals, to the strategy, to the tactics used.

There are clearly elements of this that are less appropriate for taking down civilization. For reasons we've already discussed—lack of numbers chief among them—a strategy of strict nonviolence isn't going to succeed in stopping this culture from killing the planet. And there are many things about which I would disagree with Ackerman and Kruegler. But they aren't dogmatic in their approach; they view the use of nonviolence (which for them includes sabotage) as a tactical and strategic measure rather than a purely moral or spiritual one. What I take away from their principles—and what I hope you'll take away, too—is that effective strategy is guided by the same general principles regardless of the particular tactics it employs. Both require the aggressive use of a well-planned offensive. Strategy inevitably changes depending on the subset of tactics that are relevant and available, and a strategy that does not employ violent tactics is simply one example of that. The main strategic difference between resistance forces and military forces in history is not that military forces use violence and resistance forces don't, but that military officers are trained to develop an effective strategy, while resistance forces too often simply stumble along toward a poorly defined objective.

GRAND STRATEGY

There's one nagging thought that always returns to me when I'm studying WWII resistance strategy: resisters in Occupied Europe were brave, even heroic, but their actions alone did not bring down the Nazis. Resisters weakened the Nazis, hampered their actions, disrupted their logistics, and destroyed materiel. But they lacked the resources and organization to decisively engage and defeat Hitler's forces. It took a conventional military assault by the Allies to finish the job. And the overwhelming majority of this was done by the Russians, with their large army relying heavily on infantry tactics. We can speculate about whether guerrilla uprisings in occupied countries would have eventually developed and ended Nazi rule, but that's not what happened during the actual years of occupation.

For those of us who want to stop this culture from killing the planet, there are no capital "A" Allies with vast resources and armies. That's the nature of our predicament. We may be able to ally ourselves with powers of lesser evil, the way that Spanish Anarchists allied themselves with Spanish Republicans and Soviets in Spain, or the way antebellum abolitionists allied themselves with Union Republicans against the Confederate South. But that will only get us so far, and joining the lesser evil can be dangerous.

How, then, would a successful resistance movement expand its actions beyond resistance that merely *hampers* to that which decisively dismantles civilization's centralized systems of power, those that are allowing it to steal from the poor and destroy the planet? We'll return to this in the Core Strategy chapter, but there are three main answers in terms of any theoretical deep green resistance movement's "allies." One is that the depletion of finite resources, along with the dead-ending of that pyramid scheme called industrial capitalism, will provoke a cascading industrial and economic collapse. Indeed, just during the time we've been writing this book, we've seen a banking crisis turn into a major credit crisis, which has cascaded into a recession and simmering global economic crisis. That disruption will undermine the ability of those in power to exercise their influence and concentrate wealth, and generally throw industrial civilization into a state of disarray.

A second answer is ecological and climate collapse. Cheap oil has so far insulated urban industrial people from most effects of increasing and catastrophic damage to the biosphere. But industrial collapse will mean the end of that insulation, and will mean that thousands of years of civilization's "ecological debt" will come due. Furthermore, the earth is not just a passive battlefield—it's alive, and it's fighting on the side of the living.

A third, more tentative answer is that as all of this transpires, less overtly militant aboveground forces may fight against those in power out of self-interest. Once those in power no longer have the "energy slaves" offered by cheap oil and industry, they will (once again) increasingly try to extract that labor from human beings, from literal slaves. Hopefully people in the minority world, where the rich and powerful minority live, will have the good sense to see that and fight back against this enslavement, as so many people in the majority world, where the impoverished minority live, have already been doing for so long. But this is a more tenuous proposition. Popular resentment may be quick to build against a particular head of state or particular political party. Developing a mass culture of resistance against an entire economic or political system, however, can take decades. People who are privileged and entitled take a long time to change, if they change at all. More likely they will side with someone who makes big but ultimately empty promises.

☒ ☒ ☒

Good strategy is part planning and part opportunity, and success depends on the effective use of both. In his book *Guerrilla Strategies: An Historical Anthology*, Gérard Chaliand suggests that the lessons of revolutionary warfare in the mid-twentieth century boil down to two key points. First, he writes, "The conditions for the insurrection must be as ripe as possible, the most favorable situation being one in which foreign domination or aggression makes it possible to mobilize broad support for a goal that is both social and national. Failing this, the ruling stratum should be in the middle of an acute political crisis and popular discontent should be both intense and wide ranging." Second, he suggests, "The most important element in a guerrilla campaign is the *underground political infrastructure*, rooted in the population itself and coordinated by middle-ranking cadres. Such a structure is a prerequisite for growth and will provide the necessary recruits, information, and local logistics."[15]

We're clearly heading into a period of prolonged emergency, although the crisis will vary between chronic and acute over time. That increases the prospects for revolutionary—or rather, devolutionary— struggle, especially if radical organizations are able to anticipate and effectively seize opportunities offered by particular crises. It's unlikely that mass support will be rallied for anticivilizational causes in the foreseeable future, because most people are happy to get the material benefits of this culture and ignore the consequences. However, an increase in political discontent can be beneficial even if it doesn't create a majority.

Chaliand's second conclusion is key, and even I find it a bit surprising that he would rank underground development so highly. But it makes sense; aboveground organizational infrastructure, though it may be hard work, is comparatively easy to expand. Underground infrastructure seems troublesome or irrelevant in times where resistance movements are too marginal or inactive to pose a threat. But as soon as they become successful enough to provoke significant repression, the underground becomes indispensible, and creating it at that point is extremely difficult.

The use of a crisis as an opportunity isn't a new idea, but it has played a key role in strategic theory. Napoleon Bonaparte said that "the whole art of war consists of a well-reasoned and extremely circumspect defensive followed by rapid and audacious attack." A similar opinion was shared by British strategist Basil Liddell Hart. As a foot soldier in World War I, Liddell Hart was injured in a gas attack and became horrified by the needless bloodshed. After the war he tried to develop strategy that would avoid the kind of carnage he'd been part of. In his book *Strategy: The Indirect Approach* (first published in 1941), he argued for a military strategy that has a lot in common with asymmetric strategy. Rather than attempting to carry out a direct assault on enemy military forces, he recommended making an indirect and unexpected attack on the adversary's support systems, to decisively end the war and avoid prolonged and bloody battles.

Resisters can learn from this kind of approach. Often, because of the disorganized nature of many resistance movements, initial offensive actions are tentative and poorly coordinated. Sometimes these are celebrated because, well, at least they're something. But they are rarely effective in and of themselves, and they may tip the hand of the resistance and allow those in power to seize the initiative.

When I'm looking for an analogy for civilization, I often think of the Borg from *Star Trek*. Relentlessly expansionist and essentially colonial, they insist that every indigenous culture they encounter "adapt to service" them—that every individual either assimilate to their basic imperative or die. Like any coercive hegemony, they insist that resistance is futile. They're fundamentally industrial. They have overwhelming military force, and they're very good at adapting to resistance. The good guys only get a few shots with their phasers before the Borg adapt, making the weapons virtually useless. Then the good guys have to rejig their tactics or run away until they have a better chance.

That's basically what happens when a resistance group makes a token attack at the wrong time. If, instead of being "rapid and audacious," an operation is slow and timid, the effect may be to point out the enemy's weakness and allow them to shore it up. It removes the element of surprise. And that applies whether the resistance movement is using armed tactics, sabotage, or nonviolence.

SUCCESS AND FAILURE

The key problem with identifying successful strategies is that the context of historical resistance is different from the present. Their goals were often different as well. There's a difference between destroying or expelling a foreign power, and forcing a power to negotiate or offer concessions, and dismantling a domestic system of power or economics. Such differences are the reason we've used relatively few anti-colonial movements as case studies; their context and strategy are too different.

Resistance groups often fall prey to several major strategic failures. We'll discuss five big ones here:

- A failure to adhere to the principles of asymmetric struggle.
- A failure to devise a consistent strategy and goal.
- An inappropriate excess of hope; ignoring the scope of the problem.
- A failure to adequately negotiate the relationship between aboveground and underground operations.
- An unwillingness or inability to use the required tactics.

The first of these is a failure to adhere to the principles of asymmetric struggle. Yes, most resisters want to fight the good fight, and an out-and-out fight can be tempting. But that can only happen where resisters have superior forces on their side, which is almost never. The original IRA engaged in and lost pitched battles on more than one occasion.

In occupied Europe, writes M. R. D. Foot, "whenever there was a prospect that a large partisan force could be set up, people started asking for heavy weapons" instead of the submachine guns they were usually delivered. But artillery was always short on the front lines of conventional conflict, its presence drastically cut the mobility of a resistance group, and ammunition was hard to come by. "Bodies of resisters who clamoured for artillery were victims of the fallacy of the national redoubt . . . and of the old-fashioned idea that a soldier should stand and fight. The irregular soldier is usually much more use to his cause if he runs away, and fights in some other time and place of his own choosing."[16]

Former Black Panthers have identified a similar problem with BPP strategy, specifically with their habit of equipping offices and houses to use as pseudofortresses. Explains Curtis Austin, "Using offices inside the ghetto as bases of operations was also a mistake. As a paramilitary organization, it should not have made defending clearly vulnerable offices a matter of policy. Sundiata Acoli echoed these sentiments when he noted this policy 'sucked the BPP into taking the unwinnable position of making stationary defenses of BPP offices. . . . small military forces should never adopt as a general action the position of making stationary defences of offices, homes, buildings, etc.' The frequency and quickness with which they were surrounded and attacked should have led them to develop a policy that would have allowed them to move from one headquarters to another with speed and stealth. Instead, the fledgling group constantly found itself defending sandbagged and otherwise well-fortified offices until their limited supplies of ammunition expired."[17]

Early Weather Underground and SDS strategy similarly ignored the importance of surprise in planning actions by advertising and promoting open conflicts with the state and police in advance. This was criticized by other groups at the time. Writes Ron Jacobs, "From the Yippies' vantage point, the idea of setting a date for a battle with the state was ridiculous: it provided the police with a greater capacity to counter-attack, and it also took away the element of surprise, the activists' only advantage. . . . Pointing out the differences between the planned, offensive violence of Weatherman and Yippie's spontaneous, defensive version, Abbie Hoffman termed Weatherman's confrontations 'Gandhian violence for the element of purging guilt through moral witness.'"[18] (This analysis is interesting, if perhaps surprising and a little ironic, given the Yippies' propensity for symbolic and theatrical actions.)

A most notable example of this problem was the "Days of Rage" gathering in Chicago in 1969. According to Weatherman John Jacobs, the intent of the Days of Rage was to confront the forces of the state and "shove the war down their dumb, fascist throats and show them, while we were at it, how much better we were than them, both tactically and strategically, as a people."[19] Jacobs told the Black Panthers that 25,000 protesters would be present.[20] However, only about 200 showed up, met by more than a 1,000 trained and well-equipped police. In a speech the day of the event, Jacobs changed tack and argued for the importance of fighting for righteous and moral (rather than tactical or strategic) reasons: "We'll probably lose people today . . . We don't really have to win here . . . just the fact that we are willing to fight the police is a political victory."[21] The protesters then started something of a riot, smashing some police cars and luxury businesses, but also miscellaneous cars, a barbershop, and the windows of lower- and middle-class homes[22]—not a great argument for superior strategy and tactics. The police quickly dispatched the protesters with tear gas, batons, and bullets. In the following days, almost 300 people were arrested, including most of the Weather Underground and SDS leadership. The Black Panthers—who were not afraid of political violence or of fighting the police—denounced the action as foolish and counterproductive. The Weather Underground, at least, did seem to learn from this when they went underground and used tactics better suited to an asymmetric conflict. (How effective their tactics were while underground is another question.)

All of this brings us to the second common strategic problem of resistance groups. Although their drive and values may be laudable— and although their revolutionary commitment is not in question—many resistance groups have simply failed to devise a consistent strategy and goal. In order for a strategy to be verifiably feasible, it has to have an endpoint that can be described as well as a clear and reasonable path or steps that connect the implementation of the strategy to the endpoint.

Some people call this the "A to B" factor. Does a proposed strategy actually lay out a reasonable path between point A and point B? If you can't explain how the strategy might work or how you can implement it, you certainly can't evaluate the strategy effectively.

It seems dead obvious when put in these terms, but a real A to B strategy is often missing in resistance groups. The problems may seem so insurmountable, the risk of group schisms so concerning, that many movements just stagger along, driven by a deep desire for justice and in some cases a need to fight back. But this leads to short-term, small-scale thinking, and soon the resisters can't see the strategic forest for the tactical trees.

This problem is not a new one. M. R. D. Foot describes it in his writings about resistance against the Nazis in Occupied Europe. "Less well-trained clandestines were more liable to lose sight of their goal in the turmoil of subversive work, and to pursue whatever was most easy to do, and obviously exasperating to the enemy, without making sure where that most easy course would lead them."[23]

It's good and courageous to want to fight injustice, but resisters who *only* fight back on a piecemeal basis without a long-term strategy will lose. Often the question of real strategy doesn't even enter into discussion. Jeremy Varon wrote in his book on the Weather Underground and the German Red Army Faction that "1960s radicals were driven by an apocalyptic impulse resting on a chain of assumptions: that the existing order was thoroughly corrupt and had to be destroyed; that its destruction would give birth to something radically new and better; and that the transcendent nature of this leap rendered the future a largely blank or unrepresentable utopia."[24] Certainly they were correct that the existing order was (and still is) thoroughly corrupt and deeply destructive. The idea that destroying it would inevitably lead to something better by conventional human standards is more slippery. But the main problem is the profound gap in terms of their strategy and objective. They had virtually no plan beyond their choice of tactics which, in the case of the Weather Underground, became largely symbolic in nature despite their use of explosives. Their uncritical "apocalyptic" beliefs about the nature of revolution—something shared by many other militant groups—almost guaranteed that they would fail to develop an effective long-term strategy, a problem to which we'll return later on.

It's very interesting—and hopefully illuminating—that a group like the Weather Underground did so many things right but completely fell down strategically. We keep coming back to them and criticizing them not because their actions were necessarily wrong, but because they were on the right track in so many ways. The internal organization of the Weather Underground as a clandestine group was highly developed and effective, for example. And their desire to bring the war home, their commitment to action, far surpassed that of most leftists agitating against the Vietnam War.

But as Varon observed, "The optimism of American and West German radicals about revolution was based in part on their reading of events, which seemed to portend dramatic change. They debated revolutionary strategy, and their activism in a general way suggested the nature of the liberated society to come. But they never specified how turmoil would lead to radical change, how they would actually seize power, or how they would reorganize politics, culture, and the economy after a revolution. Instead, they mostly rode a strong sense of outrage and an unelaborated faith that chaos bred crisis, and that from crisis a new society would emerge. In this way, they translated their belief that revolution was politically and morally necessary into the mistaken sense that revolution was therefore likely or even inevitable."[25]

All of this brings us to a third common flaw in resistance strategy—an excess of hope. Obviously, we now know that a 1960s American revolution was far from inevitable. So why did the Weather Underground and others believe that it was? To some degree, this sort of anchorless optimism is a coping mechanism. Resistance groups are up against powerful foes, and believing that your desired victory is somehow inevitable can help morale. It can also be wrong. We should remember former prisoner of war James Stockdale's "very important lesson": "You must never confuse faith that you will prevail in the end—which you can never afford to lose—with the discipline to confront the most brutal facts of your current reality, whatever they might be."[26]

Another factor is what you might call the bubble or silo effect. People tend to self-sort into groups of people they have something in common with. This can lead to activists being surrounded by people with similar beliefs, and even becoming socially isolated from those who don't share their ideas. Eventually, groupthink occurs, and people start to believe that far more people share their perspective than actually do. It's only a short step to feeling that vast change is imminent. This is especially true if the goal is nebulous and difficult to evaluate.

The false belief that "the revolution is nigh" is hardly limited to '60s or leftist groups, of course. Even World War II German dissidents like Carl Friedrich Goerdeler, a conservative but anti-Nazi politician, fell prey to the same misapprehension. Writes Allen Dulles: "Despite Goerdeler's realization of the Nazi peril, he greatly overestimated the strength of the relatively feeble forces in Germany which were opposing it. Optimistic by temperament, he was often led to believe that plans were realities, that good intentions were hard facts. As a revolutionary he was possibly naïve in putting too much confidence in the ability of others to act."[27]

Significantly, but perhaps not surprisingly, his naïveté extended not just to potential resisters but even to Hitler. Prior to the July 20 plot, he firmly believed that if only he could sit down and meet with Hitler, he could rationally convince him to admit the error of his ways and to resign. His friends were barely able to stop him from trying on more than one occasion, which would have obviously been foolish and dangerous to the resistance because of their planned assassination.[28] Of course, Nazi Germany was not just a big misunderstanding, and after the failed *putsch*, Goerdeler was arrested, tortured for months by the Gestapo, and then executed.

The fourth common strategic flaw is a failure to adequately negotiate the relationship between aboveground and underground operations. We touched on this on a number of occasions in the organization section. Many groups—notably the Black Panthers—failed to implement an adequate firewall between the aboveground and underground. But we aren't just talking about organizational partitions and separation; the history of resistance has showed again and again the larger strategic challenge of coordinating cooperative aboveground and underground action.

This has a lot to do with building mutual support and solidarity. The Weather Undeground in its early years was notably abysmal at this. Their attitude and rhetoric was aggressively militant. The organization, in the words of its own members (written after the fact), had a "tendency to consider only bombings or picking up the gun as revolutionary, with the glorification of the heavier the better," an attitude which even alienated other armed revolutionary organizations like the BPP.[29] Indeed, the Weather Underground would deliberately seek confrontation for the sake of confrontation even with people with whom it professed alignment. For example, in one action during the Vietnam War, Weather Underground members went to a working-class beach in Boston and erected a Vietcong flag, knowing that many on the beach had family in the US armed forces. When encircled, instead of discussing the war, they aggressively ratcheted up the tension, idealistically believing that after a brawl both sides could head over to the bar for a serious chat. Instead, the Weather Underground got their asses kicked.[30]

Now, there's something to be said for pushing the limits of "legitimate" resistance. There's something to be said for giving hesitant resisters a kick in the pants—or at least a good example—when they should be doing better. But that's not what the Weather Underground did. In part the problem was their lack of a clear and articulable strategy. In his memoir, anarchist Michael Albert relates a story about being asked to attend an early Weather Underground action so that he could see what they do. "About ten of us, or thereabouts, piled into a subway car heading for the stop nearest a large dorm at Boston University. While in the subway, trundling along underground, one of the Weathermen, according to prearranged agreement, stood up on his seat to give a speech to his captive audience of other subway riders. He nervously yelled out 'Country Sucks, Kick Ass,' and promptly sat down. That was their entire case. It was their whole damn enchilada."[31] What are people supposed to get from that? By contrast, no one reading the Black Panther Party's Ten Point Plan would be confused about their strategy and goals.

But the Weather Underground's most ineffective actions in the aboveground vs. underground department were those that actually harmed aboveground organizations. Their actions in Students for a Democratic Society (SDS) are a prime example. SDS was a broad-based organization with wide support, which focused on participatory democracy, direct action, and nonviolent civil disobedience for civil rights and against the war. Before the formation of the Weather Underground, a group called the Revolutionary Youth Movement (RYM), led by Bernardine Dohrn, later a leader of the Weather Underground, essentially hijacked SDS. They gained power at a 1969 national SDS convention and expelled members of a rival faction (the Progressive Labor Party and Worker Student Alliance). They hoped to push the entire organization into more militant action, but their coup caused a split in the organization, which rapidly disintegrated in the following years. In the decades since, no leftist student organization has managed to even approach the scale of SDS.

The bottom line is that RYM took a highly functional aboveground group and destroyed it. The Weather Underground's exaltation of militancy got in the way of radical change and caused a permanent setback in popular leftist organizing. What the Weather Underground members failed to realize is that not everyone is going to participate in underground or armed resistance, and that everyone does not *need* to participate in those things. The civil rights and antiwar movements were appropriate places for actionists to try to build nonviolent mass movements, where very important work was being done, and SDS was a crucial group doing that work. Aboveground and underground groups need each other, and they must work in tandem, both organizationally and strategically. It's a major strategic error for any faction—aboveground or underground—to dismiss the other half of their movement. To arrogantly destroy a functioning organization is even worse.

▨ ▨ ▨

There is a fifth common strategic failure, which in some ways is the most important of them all: the unwillingness or inability to apply appropriate tactics to carry out the strategy. Is your resistance movement using its entire tool chest? A resistance movement that is fighting to win considers every operation and every tactic it can possibly employ. That doesn't mean that it actually uses every tool or tactic. But nothing is simply dismissed without consideration.

The Weather Underground, to return again to their example, was a group which began with an earnest desire to fight back, to "bring the war home," and express genuine solidarity with the people of Vietnam and other countries under American attack by taking up arms. Initially, this was meant to include attacks on human beings in key positions in the military-industrial complex. Indeed, before they went underground, as we've already discussed, the Weather Underground was eager to attack even low-level representatives of the state hierarchy, specifically police. Shortly after going underground, they changed their strategy.

The turning point in the Weather Underground's strategy of violence versus nonviolence was the Greenwich Village townhouse explosion. In the spring of 1970, an underground cell there was building bombs in preparation for a planned attack on a social event for noncommissioned officers at a nearby army base. However, a bomb detonated prematurely in the basement, killing three people, injuring two others (who fled), and destroying the house. After the explosion, the Weather Underground took what you could call a nonviolent approach to bombings—they attacked symbols of power like the Pentagon and the

Capitol building, but went out of their way to case the scenes before detonation to ensure that there were no human casualties.

Rather ironically, their post–Greenwich Village tactical approach again became largely symbolic and nonviolent, much like the aboveground groups they criticized. Lacking connections to other movements and organizations, and lacking a clear strategic goal, the Weather Underground's efforts were doomed to be ineffective.

ABOLITIONIST STRATEGY

The Weather Underground was far from the only group that had difficulty implementing necessary tactics. The story of abolitionists prior to the Civil War gives us one of the best examples of this, in part because of the length and breadth of their struggle. Starting from a marginalized position in society, the struggle over slavery eventually inflamed an entire culture and provoked the bloodiest war in American history.

We'll begin the story in the 1830s when several different currents of antislavery activism were growing rapidly. One of these currents was the Underground Railroad, run by both black and white people. Another current consisted of what you might call liberal abolitionists, predominantly white with a few black participants as well.

The general story of the Underground Railroad has become well-known, but there are many common misconceptions. Black slave escapes date back to the 1500s (when escapes south to Spanish Florida were rather more common), although some aspects of the nineteenth century Railroad were more systematically organized. One common but incorrect belief about the Underground Railroad is that it was run by magnanimous whites in order to aid black people otherwise unable to help themselves. In fact, this revisionist mythology is quite far from the truth.[32] Until the 1840s, it was primarily run by and for black people who distrusted the involvement of whites. Escaped blacks were always in much greater danger than whites, and had to possess a great deal of skill, knowledge, and bravery in order to escape. The great majority of escapes were orchestrated by the slaves themselves, who spent months or years planning and reconnoitering escape routes and hiding places.

Indeed, some historians have calculated that by the 1850s about 95 percent of escaping slaves were alone or with one or two companions.[33]

Furthermore, although the Underground Railroad is now recognized as a heroic and important part of the history of slave resistance, not all abolitionists of the time participated. In fact, some actually opposed the Underground Railroad. According to one history, "Abolitionists were divided over strategy and tactics, but they were very active and very visible. Many of them were part of the organized Underground Railroad that flourished between 1830 and 1861. Not all abolitionists favored aiding fugitive slaves, and some believed that money and energy should go to political action."[34]

There's no question that those who participated in the Underground Railroad were very brave, regardless of the color of their skin, and the importance of the Railroad to escaped slaves and their families cannot be overstated. The problem was that the Railroad just wasn't enough to pose a threat to the institution of slavery itself. In 1830, there were around two million slaves in the United States. But at its peak, the Underground Railroad freed fewer than 2,000 slaves each year, less than one in one thousand. This escape rate was much lower than the rate of increase of the enslaved population through birth. Of course, many fugitive slaves worked to save money and buy their families out of slavery, which meant that the Railroad freed more people than just those who physically travelled it.

Tactical Development: From Moral Suasion to Political Confrontation[35]

While the Underground Railroad was growing in the 1830s, another antislavery current was growing as well. This one consisted mostly of white abolitionists, driven by Christian principles and a desire to convince slave owners to stop sinning and release their slaves. These early white Christian abolitionists recognized the horrors of slavery, but adopted an approach of pacifist moral exhortation. Historian James Brewer Stewart discusses their approach: "Calling this strategy 'moral suasion,' these neophyte abolitionists believed that theirs was a message of healing and reconciliation best delivered by Christian peacemakers, not by divisive insurgents. . . . They appealed directly to the (presumably) guilty and therefore receptive consciences of slaveholders with cries for immediate emancipation." They believed, as liberals usually do, that the oppressive horrors perpetrated by those in power were mostly a misunderstanding (rather than an interlocking system of power that rewarded the oppressors for evil). So, of course, they believed that they could correct the mistake by politely arguing their case.

Stewart continues: "This would inspire masters to release their slaves voluntarily and thereby lead the nation into a redemptive new era of Christian reconciliation and moral harmony. . . . immediate abolitionists saw themselves as harmonizers, not insurgents, because the vast majority of them forswore violent resistance. . . . 'Immediatists,' in short, saw themselves not as resisting slavery by responding to it reactively, but instead as uprooting it by spiritually revolutionizing the corrupted values of its practitioners and supporters." In other words, they fell prey to four of the strategic failings we've discussed so far. They didn't use asymmetric strategic principles, largely because they weren't using a resistance strategy at all. They were essentially lobbying, and their "morally superior" approach meant that, as a minority faction, they had no political force to bring to bear on those whom they lobbied. Furthermore, they were hopelessly naïve (or to state the problem more precisely, they were hope*fully* naïve) about the nature of power and the slave economy. As a result, they were unable to concoct a reasonable A to B strategy. Their so-called strategy, though well-meaning and moral, was more akin to a collective fantasy that overlooked the nature and extent of violence that slave culture would bring to bear on its adversaries.

Stewart recognizes this problem as well. "By adopting Christian pacifism and regarding themselves as revolutionary peacemakers, these earliest white immediatists woefully underestimated the power of the forces opposing them. Well before they launched their crusade, slavery had secured formidable dominance in the nation's economy and political culture. To challenge so deeply entrenched and powerful an institution meant adopting postures of intransigence for which these abolitionists were, initially, wholly unprepared."[36]

Need I spell out the parallels to our current situation? Pick any liberal or mainstream environmental or social justice movement. Mainstream environmentalism has been particularly naïve in this regard, largely ignoring the deeply entrenched nature of ecocidal activities in the capitalist economy, in industry, in daily life, and in the psychology of the civ-ilized. Furthermore, mainstream environmentalists—who often do not come out of a long tradition of resistance—utterly ignore the force that those in power will bring to bear on any threat to that power. By assuming that society will adopt a sustainable way of life if only individual people can be persuaded, mainstream environmentalists ignore the rewards offered for unsustainability, and too often ignore those who pay the costs for such rewards.

Of course, mainstream environmentalism is hardly unique in this. Indeed, this basic trajectory is so common that it is nearly archetypal. Again and again, whenever privileged people have tried to ally them-selves with oppressed people, we have seen this phenomenon at work. Seemingly ignorant of the daily violence perpetrated by the dominant culture, many people of privilege have wandered off into a strategic and tactical Neverland, which is based on their own personal wishes about how resistance ought to be, rather than a hard strategy that is designed to be effective and that draws on the experience of oppressed peoples and their long history of resistance. Sometimes the people of privilege listen and learn, and sometimes they don't.

Of course, these early white abolitionists were on the right side, and, of course, their response to slavery was, morally speaking, far above that of the majority of white people's. But, writes Stewart: "With the nation's most powerful institutions so tightly aligned in support of slavery and white supremacy, it is clear that young white abolitionists were pro-foundly self-deceived when they characterized their work as 'the destruction of error by the potency of truth—the overthrow of prejudice by the power of love—the abolition of slavery by the spirit of repentance.' When so contending, they were deeply sincere and grievously wrong. To crusade for slavery's rapid obliteration was, in truth, to stimulate not 'the power of love' and 'repentance,' but instead to promote the opposition of not only an overwhelming number of powerful enemies—the entire political system—but also the nation's most potent economic interests—society's most influential elites—and a popular political culture in the North that was more deeply suffused with racial bigotry than at previous times in the nation's history." This is a lesson we must remember.

They were highly optimistic about their chances. After increasing racial tensions and a series of violent uprisings in the early 1830s, one immediatist predicted that "the whole system of slavery will fall to pieces with a rapidity that will astonish."[37] This attitude is again reminiscent of the excess of hope we discussed earlier.

We should note that it was not just white abolitionists who were opposed to serious resistance at this stage, but some people of color as well. Historian Lois E. Horton writes that one black editor of a newspaper "penned an article addressed 'To the Thoughtless part of our Colored Citizens,' in which he admonished readers to act with more dignity and self-restraint when fugitive slaves were captured. [The editor] urged African Americans to leave the defense of fugitives to the lawyers . . . Public protest, even public assembly, [he] warned, would risk the loss of support from respectable allies. He was especially shocked by the involvement of Black women in this protest, singling them out for 'everlasting shame' and charging that they 'degraded' themselves by their participation."[38]

But more militant abolitionists continued to gain prominence. Former fugitive slave Henry Highland Garnet rejected the pacifism of both white and black abolitionists, saying "There is not much hope of Redemption without the shedding of blood."

Many white abolitionists retained their pacifist beliefs and practices, but as the abolition movement grew, it was increasingly perceived as a threat by slaveholders and those in power. An escalating wave of violent repression occurred, in which abolitionists and their allies were attacked, and their mailings and offices were burned. Many white abolitionists abandoned pacifism after white newspaper editor and abolitionist Elijah Lovejoy was gunned down in his office by proslavery thugs. William Lloyd Garrison, publisher of the foundational abolitionist paper the *Liberator*, wrote: "When we first unfurled the banner of the *Liberator* . . . we did not anticipate that, in order to protect southern slavery, the free states would voluntarily trample under foot all law and order, and government, or brand the advocates of universal liberty as incendiaries and outlaws. . . . It did not occur to us that almost every religious sect, and every political party would side with the oppressor."[39] Of course, they did not consider and dismiss the idea—it simply didn't occur to them. This repression did, however, induce increasing numbers of Northerners to join with the abolitionists out of concern for the violations of law by the government and antiabolitionists.

The good news was that by the 1850s, more and more abolitionists were defying fugitive slave laws and even taking up arms to aid escaped slaves inside and outside of the Underground Railroad. Violent confrontations began to occur in a scattershot fashion or, to be more precise, *defensive* violence carried out *by abolitionists* became more common, since slavery had been based on violent confrontations since the beginning, and none of that was new to black people. It was soon not unheard of in the North for slaveholders or slave catchers to be shot—on one occasion in Boston in 1854, a crowd even stormed a courthouse where a fugitive slave was being held and overpowered the guards. Writes Stewart, "And even when physical violence did not result . . . oratorical militants increasingly urged their audiences to resort to physical destruction if more peaceable methods failed to stop federal slave catchers. On several occasions well-organized groups of abolitionists overwhelmed the marshals and spirited fugitives to safety. At other times they stored weapons, planned harassing manoeuvres, and massed as intimidating mobs."[40] Though only a decade earlier they were taking oaths never to use force, white abolitionists came to agree that use of lethal force against slave catchers, in self-defense, was morally justified. Armed defiance of slave catchers was a long tradition for black activists at that time, but a considerable change for white abolitionists. Many Christian abolitionists changed their tactics, arguing that not only was pacifism *not* required by God, but that it was a Christian's *duty* and the "Law of God" to shoot a slave catcher.

JOHN BROWN AND THE HARPER'S FERRY RAID

This shift toward more militant defiance of slavery was all wonderful, of course. And it certainly increased the success of the Underground Railroad if a slave catcher knew that a trip into strongly abolitionist areas might end with a bullet in his chest. But, again, there was a problem. Even this rapidly growing and increasingly defiant abolitionist movement had not been able to successfully challenge the institution of slavery itself. The situation continued to get worse. Writes Stewart,

"More than two decades of peacefully preaching against the sin of slavery had yielded not emancipation but several new slave states and an increase of over half a million held in bondage, trends that seemingly secured a death grip by the 'slave power' on American life."[41] Cotton agriculture in the South often destroyed the landbase, and that, combined with a growing slave population, meant that it was profitable—according to some historians, even imperative—that slaveholders expand westward in order to maintain the slave economy. Each new slave state shifted the balance of political power in the Union even more toward slavery.

Enter John Brown, an ardent abolitionist and deeply moral man who had clashed with proslavery militants on several occasions before. Brown, a wool grower by trade, had fought in the struggle to make the new state of Kansas an antislavery state. He was apparently not much interested in making speeches, and thought little of rhetoric alone given the seriousness of the situation. Brown was frustrated with mainstream abolitionists, reportedly exclaiming, "These men are all talk. What we need is action—action!"

And action was exactly what he had in mind. In 1858, Brown ran a series of small raids from Kansas into Missouri, liberating slaves and stealing horses and wagons. He helped bring the liberated slaves to Canada, but his main plan was much more daring. Secretly raising funds from wealthy abolitionist donors, buying arms, and training a small group of paramilitary recruits, Brown planned a raid on the armory at Harpers Ferry, West Virginia. The plan was simple. Brown and his troops would raid the armory, which contained tens of thousands of small arms. They would steal as many arms as they could, then liberate and arm the slaves in the area. They would head south, operating as guerrillas, liberating and arming slaves and fighting only in self-defense. Brown hoped for a movement that would grow exponentially as they moved into the Deep South, a cascade of action that would unravel and destroy the institution of slavery itself.

Although some historians—especially those impugning Brown—have considered him an insurrectionist, that's not an accurate reflection of his intended strategy. Brown's biographer, Louis A. DeCaro, has discussed this very fact: "Brown nowhere planned insurrection, which is essentially an armed uprising with the intention of eliminating slave masters. Brown planned an armed defensive campaign. His intention was to lead enslaved people away from slavery, arm them to fight defensively while they liberated still more people, fighting in small groups in the mountains, until the economy of slavery collapsed. Brown did not believe in killing unless it was absolutely necessary."[42]

Tragically, things were not to go as planned. Part of the problem was numbers. While a draft plan for the Harpers Ferry raid called for thousands of men, on the day of the raid Brown had only twenty-one, both white and black. In an unusual situation for resistance fighters, Brown had far more guns than men. From Northern abolition societies, Brown had received about ten carbines (short rifles) for each fighter available. Nonetheless, Brown, deeply driven, decided to proceed.

At first the raid went smoothly. They easily entered the town of Harpers Ferry, cut the telegraph wires, and captured the armory. But Brown made a tactical error—the worst tactical error a guerrilla can make—by failing to seize the arms and move on as soon as possible. As a result, local militia were soon firing on the armory from the town while the militants remained inside. After continuing exchanges of fire and several deaths, US Marines under the command of Robert E. Lee arrived, surrounding and then storming the armory. Five of Brown's fighters escaped, ten were killed, and the rest captured. Those captured were imprisoned and stood trial. John Brown and five others were subsequently hanged.

It's extremely important to understand why the raid failed. The problem was tactical, rather than strategic in nature. Although he was unsuccessful, even his enemies at the time said that "it was among the best planned and executed conspiracies that ever failed."[43] In fact, even on the tactical level Brown's *planning* was excellent. But instead of employing the hit-and-run tactics asymmetric forces depend on, Brown got bogged down in the armory. According to biographer Louis A. DeCaro, "The reason the raid did not succeed was because he paid too much concern to his hostages, including some whining slave masters, and undermined himself in trying to negotiate with them." Furthermore, rather astonishingly, DeCaro notes that Brown even allowed "his prisoners to go home and see their families under guard and send out for their breakfast."[44] Indeed, Brown was in Harpers Ferry for almost two days before the marines arrived. According to DeCaro: "Had he kept to his own plan and schedule, he and his fugitive allies would have almost walked away from Harper's Ferry without facing any significant opposition, and could have easily retreated to the mountains as planned. Contrary to the notion that he was a crazy man and a killer, it seems that John Brown was actually too tender-hearted and still hoped to resolve some of the issue by negotiation. This was his greatest error."[45]

News of the raid spread swiftly. The knee-jerk response among many abolitionists and their sympathizers was one of contempt for Brown's actions. Even Lincoln (perhaps afraid of offending the South) called him a "misguided fanatic." Henry David Thoreau, notably, was one of the few who immediately sprang to Brown's defense. He begged his fellow citizens to listen: "I hear many condemn these men because they were so few. When were the good and the brave ever in a majority?"[46] (Now is a good time to ask that question of ourselves and our allies, especially if we are waiting for someone else to act.)

DeCaro notes that Brown's reputation in history has been consistently attacked and "the 'facts' of his case have been mediated from slave masters, pro-slavery people, and pacifists."[47] (Those in the latter category will hopefully find it relevant, if embarrassing, that they are lumped in with such dreadful company.) But not everyone has been so easily convinced that Brown was wrongheaded. Malcolm X, not surprisingly, had great respect for John Brown and little patience for white liberals who criticized his methods. "John Brown . . . was a white man who went to war against white people to help free slaves. And any white man who is ready and willing to shed blood for your freedom—in the sight of other whites, he's nuts." In other words, those who hate Brown do so in large part because he was a "race traitor."

The raid on Harpers Ferry increased tensions between the North and South. Some historians rank it among the proximal causes of the Civil War. This is ironic, as Brown despised unnecessary bloodshed, and, like many at the time, was aware that a war between North and South was very likely looming. It was his hope that his strategy of guerrilla warfare would end the slave economy while *averting* a civil war, which could be even bloodier. It's possible that, had he been more ruthless, he might have succeeded. His hesitation to be ruthless, then, may have resulted in a much greater number of deaths. Brown's problem, as with many of those who fight injustice, was that he was simply too nice, even when dealing with vicious oppressors. Brown himself realized this too late. On the day he was hanged he wrote the following: "I, John Brown, am now quite certain that the crimes of this guilty land will never be purged away but with blood. I had, as I now think, vainly flattered myself that without very much bloodshed it might be done."[48]

OUTRIGHT CIVIL WAR

Brown's failed attack was a flashpoint for the rising strain between North and South, and outright Civil War shortly followed. This is not the place to discuss the full history of the Civil War or all its causes, but there are a few points that are relevant to understanding how outright Civil War impacted resistance. Many people have been taught that the Civil War was "fought to end slavery," but this is not true. Social justice was not a main driving force behind the Civil War, and prior to the outbreak of hostilities, Abraham Lincoln insisted that slavery was a choice for each state to make. It might be more accurate to say that the Civil War was precipitated by the growth of "slave power" (that is, the power of slaving-holding states) and by the tensions between conflicting economic and political institutions. The immediate cause of the Civil War was the secession of slave-holding states into the Confederacy, which Abraham Lincoln would not allow.

The outbreak of Civil War (and especially the invasion of the Confederacy by Union forces) resulted in two distinct changes for abolitionists. First, slave resistance in the South was vastly increased, and second, many Northerners who were not abolitionists were forced to come face to face with slavery.

The impact of the Civil War on slave resistance was extensive even where armed conflict was not yet occurring. Many slaves attempted escape to get across Union lines where they would be ostensibly free, and many of those escapees joined the Union army to fight for the end of the Confederacy and the end of slavery. But even those slaves who did not run were roused to active resistance—or at least withdrawal of their labor. As in France in 1943, more and more slaves began to resist when it became clear that the slave owners might *lose*.

Historian Bruce Levine notes that:

The wartime breakdown of slavery became apparent beyond those Southern districts actually penetrated by Union troops. In still-unoccupied parts of the Confederacy, masters, army officers, and government officials clashed repeatedly over which of them had the greater need for and claim to the labor of remaining slaves. This process eroded the real power of Rebel masters—and emboldened those still under their formal control. A South Carolina overseer bemoaned the "goodeal of obstanetry" he faced among "Some of the Peopl" working on his plantation, "mostly amongst the Woman a goodeal of Quarling and disputing and telling lies." James Alcorn, a Mississippi planter, found that Union raids in his area had "thoroughly demoralized" his slaves. (This phrase was common planter parlance for saying that power over a slave—and a slave's fear of a master—had faded.) That change, moaned Alcorn, had rendered his human property "no longer of any practical value." Even among those field laborers who had not fled, a Louisiana overseer reported to his employer, "but very few are faithful—Some of those who remain are worse than those who have gone." In one district after another, bondspeople began to call for improvements in their conditions as well as implicit but no less momentous alterations in their status—and they withheld their labor until such demands were met. . . . "Their condition is one of perfect anarchy and rebellion," Georgia plantation mistress Mary Jones confided in her journal. "They have placed themselves in perfect antagonism to their owners and to all government and control. We dare not predict the end of all this."[49]

The nature of slave resistance changed as well, with organizers shifting from the survival-orientated operations of the Underground

Railroad to decisive military operations. Many former slaves worked with the Union forces, including Harriet Tubman, who worked as a scout and led raids and mass liberations of slaves.

The war also forced nonabolitionist northerners to confront the nature of slavery head-on. Writes Levine, "The wartime crisis of slavery left a deep imprint not only on southern whites but also on Union troops. As Lincoln and others had feared, and as the 1862 elections made clear, the decision to add the destruction of slavery to the North's war aims at first provoked fierce opposition in parts of the Union. Few Union soldiers had gone to war committed to abolition . . . the Union soldier's firsthand exposure to the real nature of slavery did much, however, to change minds and soften hearts."[50]

When a destructive system is deeply entrenched, and when average people are isolated from the costs of that system, real change doesn't come just from speeches. Real change happens—and only can happen—when that system is broken down by force. Then the oppressed gain the breathing room needed to fight back, and the apathetic can get their first look at that system's real face.

EVALUATING STRATEGY

Resistance is not one-sided. For any strategy resisters can come up with, those in power will do whatever they can to disrupt and undermine it. Any strategic process—for either side—will change the context of the strategy. A strategic objective is a moving target, and there is an intrinsic delay in implementing any strategy. The way to hit a moving target is by "leading" it—by looking slightly ahead of the target. Don't aim for where the target *is;* aim for where it's going to be.

Too often we as activists of whatever stripe don't do this. We often follow the target, and end up missing badly. This is especially clear when dealing with issues of global ecology, which often involve tremendous lag time. We're worried about the global warming that's happening now, but to avert current climate change, we should have acted thirty years ago. Mainstream environmentalism in particular is decades behind the target, and the movement's priorities show it. The most serious mainstream environmental efforts are for tiny changes that don't reflect the seriousness of our *current* situation, let alone the situation thirty years from now. They've got us worried about hybrid cars and changing lightbulbs, when we should be trying to head off runaway global warming, cascading ecological collapses, the creation of hundreds of millions of ecological refugees or billions of human casualties, and the social justice disasters that accompany such phenomena. If we can't avert global ecological collapse, then centuries of social justice gains will go down the toilet.

It's worth spelling this out. There have been substantial improvements in humans rights in recent decades, along with major social justice concessions in many parts of the world. Much of this progress can be rightly attributed to the tireless work of social justice advocates and extensive organized resistance. But look at, for example, the worsening ratio between the income of the average employee and the average CEO. The economy has become less equitable, even though the middle rungs of income now have a higher "standard of living." And all of this is based on a system that systematically destroys natural biomes and rapidly draws down finite resources. It's not that everyone is getting an equal slice of the pie, or even that the pie is bigger now. If we're getting more pie, it's largely because we're eating tomorrow's pie today. And next week's pie, and next month's pie.

For example, the only reason large-scale agriculture even functions is because of cheap oil; without that, large-scale agriculture goes back to depending on slavery and serfdom, as in most of the history of civilization. In the year 1800, at the dawn of the industrial revolution, close to 80 percent of the human population of this planet was in some form of serfdom or slavery.[51] And that was with a fraction of the current human population of seven billion. That was with oceans still relatively full of fish, global forests still relatively intact, with prairie and agricultural lands in far better condition than they are now, with water tables practically brimming by modern standards. What do you think is going to happen to social justice concessions when cheap oil—and hence, almost everything else—runs out? Without a broad-based and militant resistance movement that can focus on these urgent threats, the year 1800 is going to look downright cheerful.

If we want to be effective strategists, we must be capable of planning for the long term. We must anticipate changes and trends that affect our struggle. We must plan and prepare for the changing nature of our fight six months down the road, two years down the road, ten years down the road, and beyond.

We need to look ahead of the target, but we also need to plan for setbacks and disruptions. That's one of the reasons that the strategy of protracted popular warfare was so effective for revolutionaries in China and Vietnam. That strategy consisted of three stages: the first was based on survival and the expansion of revolutionary networks; the second was guerrilla warfare; and the third was a transition to conventional engagements to decisively destroy enemy forces. The intrinsic flexibility of this strategy meant that revolutionaries could seamlessly move along that continuum as necessary to deal with a changing balance of power. It was almost impossible to derail the strategy, since even if the revolutionaries faced massive setbacks, they could simply return to a strategy of survival.

◙ ◙ ◙

How does anyone evaluate a particular strategy? There are several key characteristics to check, based on everything we've covered in this chapter.

Objective. Does the strategy have a well-defined and attainable objective? If there is no clear objective there *is* no strategy. The objective doesn't have to be a static end point—it can be a progressive change or a process. However, it should not be a "blank or unrepresentable utopia."

Feasibility. Can the organization get from A to B? Does the strategy have a clear path from the current context to the desired objective? Does the plan include contingencies to deal with setbacks or upsets? Does the strategy make use of appropriate strategic precepts like the nine principles of war? Is the strategy consonant with the nature of asymmetric conflict?

Resource Limitations. Does the movement or organization have the number of people with adequate skills and competencies required to carry out the strategy? Does it have the organizational capacity? Does the movement or organization have the number of people with adequate skills and competencies required to carry out the strategy? Does it have the organizational capacity? If not, can it scale up in a reasonable time?

Tactics. Are the required tactics available? Are the tactics and operations called for by the plan adequate to the scale, scope, and seriousness of the objective? If the required tactics are not available or not being implemented currently, why not? Is the obstacle organizational or ideological in nature? What would need to happen to make the required tactics available, and how feasible are those requirements?

Risk. Is the level of risk required to carry out the plan acceptable given the importance of the objective? Remember, this goes both ways. It is important to ask, *what is the risk of acting?* as well as *what is the risk of not acting?* A strategy that overreaches based on available resources and tactics might be risky. And, although it may seem counterintuitive at first, a strategy that is too hesitant or conservative may be even more risky, because it may be unable to achieve the objective. If the objective of the strategy is to prevent catastrophic global warming, taking serious action may indeed seem risky—but the consequences of insufficient action are far more severe.

Timeliness. Can the plan accomplish its objective within a suitable time frame? Are events to happen in a reasonable sequence? A strategy that takes too long may be completely useless. Indeed, it may be worse than useless, and become actively harmful by drawing people or resources from more effective and timely strategic alternatives.

Simplicity and Consistency. Is the plan simple and consistent? The plan should not depend on a large number of prerequisites or complex chains of events. Only simple plans work in emergencies. The plan itself must be explained in a straightforward manner without the use of weasel words or vague or mystical concepts. The plan must also be internally consistent—it must make sense and be free of serious internal contradictions.

Consequences. What are the other consequences or effects of this strategy beyond the immediate objective and operations? Might there be unintended consequences, reprisals, or effects on bystanders? Can such undesirable effects be limited by adjusting the strategy? Does the value of the objective outweigh the cost of those consequences?

◙ ◙ ◙

A solid grand strategy is essential, but it's not enough. Any strategy is made out of smaller tactical building blocks. In the next chapter, "Tactics and Targets," I outline the tactics that an effective resistance movement to stop this culture from killing the planet might use, and discuss how such a movement might select targets and plan effective actions.

◙ ◙ ◙

Q: How can I accept the risks of being caught when that could mean never being able to see or help my family/lover/children in these difficult times?

Derrick Jensen: Nothing in this book is meant to exhort people to do things they don't want to do. In fact, nothing in this book is meant to exhort people to do anything illegal (recognizing that innocence of actual criminal activity is no guarantee that one will not be punished by those in power). We've said numerous times that there are plenty of ways that a culture of resistance can manifest, any number of activities that you can participate in that are not as immediately risky as belowground actions. If your primary concern is the risk of being caught, there are plenty of other things you can do.

But remember that when state repression gets really bad, being aboveground does not mean that the state won't come for you. It's often the public intellectuals, the organizers, and the writers who are thrown in jail. The people underground, without a public profile, are sometimes safer.

Perhaps, though, we should turn the question around. "Are you willing to risk not having fish in the oceans?" If things continue the way they are, by 2050 there will be no fish in the oceans. Amphibians are already dying. Migratory songbirds are already dying. The planet is dying. Are you willing to risk that?

None of this is theoretical. When the industrial system starts to collapse, I will be dead. I am reliant upon high-tech medicine for my life. But there is something larger and more important than my life.

Q: Civilization is the only thing keeping violent criminals from raping/killing people like in those horrible places far away. Who will protect my family if we dismantle civilization?

Derrick Jensen: A couple of years ago, I got an email from a policeman in Chicago. He was reading *Endgame* and liking it except that he thought I came down too hard on cops. He said, "Our job is to protect people from sociopaths and that's what I do every day. I protect people from sociopaths." I wrote back, "I think that's really great that you protect us from sociopaths. When my mom's house got burgled, the first thing we did was call the cops. When my house got burgled, I turned it over to the cops. It's great that you protect us from sociopaths. My problem is that you really only protect us from poor sociopaths, not rich sociopaths."

After Bhopal, Warren Anderson was tried and found guilty in absentia for the atrocities of running Union Carbide. He was sentenced to hang. And the United States refuses to extradite him. If it were up to me, all the people associated with the Gulf oil spill, which is murdering the Gulf, would be executed. That would be part of the function of a state. Instead, one of the primary functions of government is to protect the rich sociopaths from the outrage of the rest of us. Who is protecting the farmers in India from Monsanto? Who is protecting the farmers in the United States from Cargill and ADM?

I did a benefit for a group of Mexican Americans who were attempting to stop yet another toxic waste dump from being placed in their neighborhood. The toxic waste was, of course, from somewhere far away. The conversation turned to what it would be like if police and prosecutors were not enforcing the dictates of distant corporations instead of the wishes of the local communities. What if they were enforcing cancer-free zones? Or clear-cut-free zones? Or rape-free zones, for that matter? And then everyone laughed, because everyone knows it's not going to happen. But what if we in our communities started to form community defense groups and said, "This is going to be a cancer-free zone. This will be a clear-cut-free zone. This will be a rape-free zone. This will be a dam-free zone." What would happen if we did that?

That's exactly what we're talking about in this book. We want our communities to be cancer-free. We want them to be clear-cut-free. We want them to be rape-free. We want them to be dam-free. And we need to stop the sociopaths who are hurting us.

As civic society collapses in a patriarchy, things can become much worse. Look at the Democratic Republic of Congo, where there are organized mass rapes. What do we do about that? One of the things we need to do is to prepare now. That's why we've emphasized in this book so often that the revolutionaries need to be of good character. A friend of mine says that he does the environmental work he does because as things become increasingly chaotic, he wants to make sure that some doors remain open. If the grizzly bears are gone in twenty years, they'll be gone forever. But if they are there in twenty years, they may be able to be there forever. It's the same for the bull trout, the same with the redwoods—if you cut down this forest, it's gone, but if it's standing, who knows what will happen in the future? And it's the same for people's social attitudes; as things become increasingly chaotic, events become increasingly uncontrollable. We must make sure that certain ideas are in place before that happens. That's why we have emphasized zero tolerance for horizontal hostility, zero tolerance for violence against women, zero tolerance for racism. Because as civic society collapses—no matter the cause of this collapse—men will rape more, and the time to defend against that is not then, but now.

There are two approaches to the problem of men assaulting women. One of them is in a line by Andrea Dworkin, "My prayer for women of the twenty-first century: harden your hearts and learn to kill." Women need to learn self-defense, and they need to form self-defense organizations, and they need to be feminists. And men must make their allegiance to women absolute. They must have a zero tolerance policy for the abuse of women.

The same is true for race-based hate crimes. As the economic system collapses, those whose entitlement has put them at the top of the heap are going to start blaming everyone else (witness the Tea Party, for example). As Nietzsche wrote, "One does not hate what one can despise." And so long as your entitlement is in place and so long as your entitlement isn't threatened, you can despise those whom you're exploiting. But as soon as that entitlement is threatened, that contempt turns over into outright hatred and violence. As civilization collapses, we will see an increase in male-pattern violence. We will see an increase in violence against those who resist. We will see an increase in violence against people of color. We are already seeing this.

My answer for people of color is, learn to defend yourself and form self-defense organizations. And the job of white allies is to make our allegiance to the victims of white oppression absolute.

There have been many resistance movements who have formed self-defense organizations and their own police forces. The IRA acted as neighborhood police, the Spanish Anarchists organized their own police force in some of the bigger cities, and we will talk about the Gulabi Gang in Chapter 13. We need something similar. We need to form self-defense organizations to defend those humans and nonhumans who are assaulted and violated. Those assaults will continue to happen until we stop them.

To be clear, civilization is not the same as society. Civilization is a specific, hierarchical organization based on "power over." Dismantling civilization, taking down that power structure, does not mean the end of all social order. It should ultimately mean more justice, more local control, more democracy, and more human rights, not less.

Chapter 13

Tactics and Targets

by Aric McBay

For me, nonviolence was not a moral principle but a strategy; there is no moral goodness in using an ineffective weapon.

—Nelson Mandela

Deeds, not words!

—Slogan of the Women's Social and Political Union

Recall that all operations—and hence all tactics—can be divided into three categories:

- Decisive operations, which directly accomplish the objective.
- Sustaining operations, which directly assist and support those carrying out decisive operations.
- Shaping operations, which help to create the conditions necessary for success.

Where tactics fall depends on the strategic goal. If the strategic goal is to be self-sufficient, then planting a garden may very well be a decisive operation, because it directly accomplishes the objective, or part of it. But if the strategic goal is bigger—say, stopping the destruction of the planet—then planting a garden cannot be considered a decisive operation, because it's not the absence of gardens that is destroying the planet. It's the presence of an omnicidal capitalist industrial system.

If one's strategic goal is to dismantle that system, then one's tactical categories would reflect that. The only decisive actions are those that directly accomplish that goal. Planting a garden—as wonderful and important as that may be—is not a decisive operation. It may be a shaping or sustaining operation under the right circumstances, but nothing about gardening will directly stop this culture from killing the planet, nor dismantle the hierarchical and exploitative systems that are causing this ecocide. Remember, the world used to be filled with indigenous societies which were sustainable and enduring. Their sustainability did not prevent civilization from decimating them again and again.

In this chapter we'll break down aboveground and underground tactics into the three operational categories. For each class of operations, we'll further break tactics down by scale for individuals, affinity groups, and larger organizations. This is summarized in the illustration on the following pages (Figures 13-1 and 13-2). As a rule, any tactic an individual can carry out can also be accomplished by a larger organization. So the tactics for each scale can nest into the next, like Russian matryoshka dolls.

Every resistance movement has certain basic activities it must carry out: things like supporting combatants, recruitment, and public education. These activities may be decisive, sustaining, or shaping, as shown in the illustration. And they may be carried out at different scales. Operations like education, awareness raising, and propaganda (shown under aboveground shaping) may occur across the range from the individual to large organizations. The scope of education may change as larger and larger groups take it on, but the basic activities are the same.

Other operations change as they are undertaken by larger groups and networks. Look in the underground tactics under sustaining. Individuals may use escape and evasion themselves, to start with. Once a cell is formed, they can actually run their own safehouse. And once cells form into networks, they can combine their safehouses to form escape lines or an entire Underground Railroad. The basic operation of escape and evasion evolves into a qualitatively different activity when taken on by larger networks. A similar dynamic is at work in recruitment; individuals are limited to mutual recruitment, but established groups can carry out organizational recruitment and training.

And, of course, some resistance units are too small to take on certain tasks, as we shall discuss. Individuals have few options for decisive action aboveground. Underground, they are limited in their sustaining operations, because secrecy demands that they limit contact with other actionists whom they could support. But once organizations become large enough, they can embrace new operations that would otherwise

Figure 13-1

Aboveground Operations

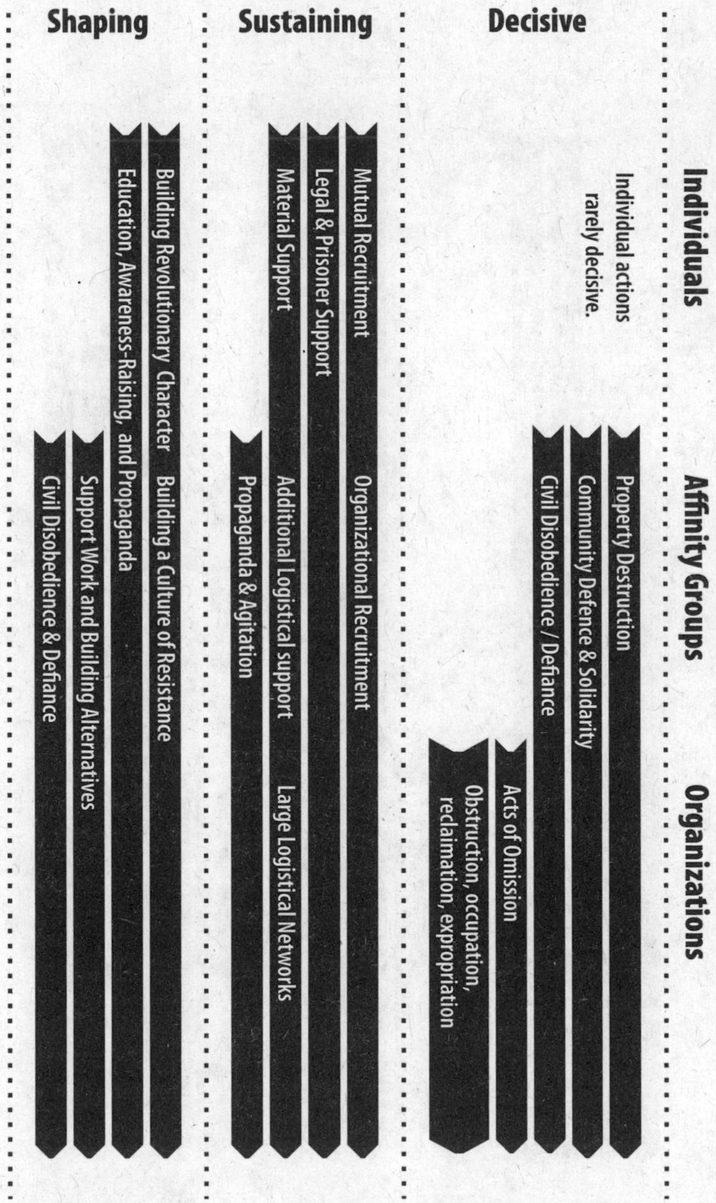

	Shaping	Sustaining	Decisive	
Individuals			Individual actions rarely decisive.	
Affinity Groups	Building Revolutionary Character Education, Awareness-Raising, and Propaganda	Mutual Recruitment Legal & Prisoner Support Material Support	Property Destruction Community Defence & Solidarity Civil Disobedience / Defiance	
Organizations	Building a Culture of Resistance Support Work and Building Alternatives Civil Disobedience & Defiance	Organizational Recruitment Additional Logistical support Propaganda & Agitation Large Logistical Networks	Acts of Omission Obstruction, occupation, reclaimation, expropriation	

Figure 13-2

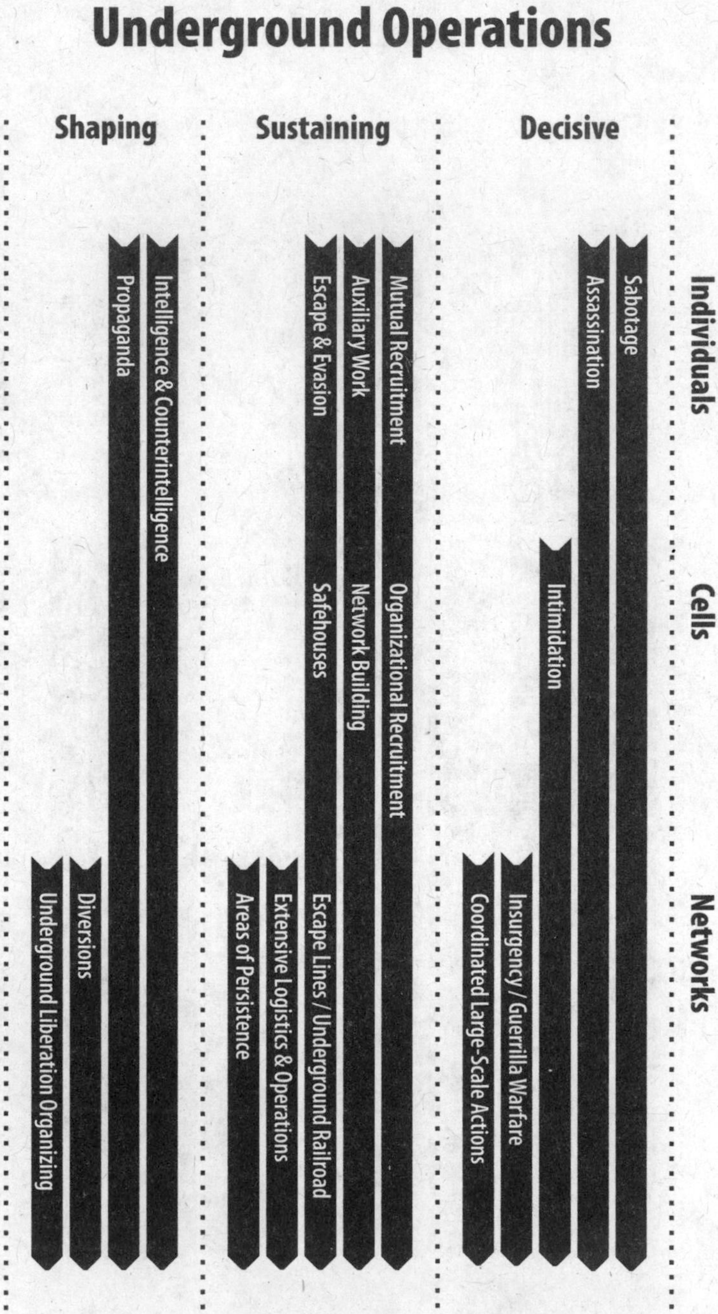

Underground Operations

	Shaping	Sustaining	Decisive
Individuals	Propaganda Intelligence & Counterintelligence	Mutual Recruitment Auxiliary Work Escape & Evasion	Sabotage Assassination
Cells		Organizational Recruitment Network Building Safehouses	Intimidation
Networks	Underground Liberation Organizing Diversions	Areas of Persistence Extensive Logistics & Operations Escape Lines / Underground Railroad	Coordinated Large-Scale Actions Insurgency / Guerrilla Warfare

be out of their reach. Aboveground, large movements can use acts of omission like boycotts or they can occupy and reclaim land. And underground networks can use their spread for coordinated large-scale actions or even guerrilla warfare.

ABOVEGROUND TACTICS

Broadly speaking, aboveground tactics are those that can be carried out openly—in other words, where the gain in publicity or networking outweighs the risk of reprisals. Underground tactics, in contrast, are those where secrecy is needed to carry out the actions to avoid repression or simply to do the actions. The dividing line between underground and aboveground can move. Its position depends on two things: the social and political context, and the audacity of the resisters.

There have been times when sabotage and property destruction have been carried out openly. Conversely, there have been times when even basic education and organizing had to happen underground to avoid repression or reprisals. This means, explicitly, that when we use the term *underground* we do not necessarily mean acts of sabotage or violence: smuggling Jews out of Nazi Germany was an underground activity, and the Underground Railroad was by definition, er, underground. One of the most important jobs of radicals is to push actions across the line from underground to aboveground. That way, more people and larger organizations are able to use what was once a fringe tactic.[1]

Provoking open defiance of the laws or rules in question also impairs the ability of elites to exercise their power. The South African government, for example, was terrified that people of color in South Africa would simply stop obeying the law of the apartheid government. In even the most openly fascist state, the police force is still a minority of the population. If enough people disobey as part of their daily activities, then the country becomes ungovernable; there aren't enough police to force everyone to perform their jobs at gunpoint.

When enough serious people have gathered to push a tactic back into the aboveground arena, those in power have few choices. If they continue to insist that the law be obeyed, resistance sympathizers may increasingly disregard *any* laws as dissidents begin to view the government as generally illegitimate—often a government's worst nightmare. Or the government may offer concessions or change the law. Any of the above could be considered a victory. Usually governments strive to retain the image of control through selective concessions or legislation because the other road ends with civil unrest, revolution, or anarchy.

The cases of Dr. Martin Luther King Jr. and Malcolm X exemplifies how a strong militant faction can enhance the effectiveness of less militant tactics. In his book *Pure Fire: Self-Defense as Activism in the Civil Rights Era*, Christopher B. Strain explains that Martin Luther King Jr. pushed his agenda by using Malcolm X "to illustrate the alternative to legislative reform: chaos. . . . King would usually present the matter in terms of a choice: 'We can deal with [the problem of second-class citizenship] now, or we can drive a seething humanity to a desperation it tried, asked, and hoped to avoid.' . . . [He] suggested if white leaders failed to heed him 'millions of Negroes, out of frustration and despair' will 'seek solace' in Malcolm X, a development that 'will lead inevitably to a frightening racial nightmare.'"[2] But Strain emphasizes that King and Malcolm X were by no means enemies. "Despite their differing opinions, both men recognized that their brands of activism were complementary, serving to shore up the other's weaknesses."[3]

Some presume that Malcolm X's "anger" was ineffective compared to King's more "reasonable" and conciliatory position. That couldn't be further from the truth. It was Malcolm X who *made* King's demands seem eminently reasonable, by pushing the boundaries of what the status quo would consider extreme.

Pushing boundaries doesn't have to involve underground property destruction or violence. Breaking antisegregation laws through lunch counter sit-ins, for example, pushed the limits of acceptability during the civil rights struggle. The second generation of suffragists, too, got tired of simply asking for what they wanted and started breaking the law. In both cases, the old guard activists were leery at first.

To be perfectly explicit: it isn't just militants who can push the boundaries; even nonviolent groups can and should be pushing the envelope for militancy—vocally and through their actions—wherever and whenever possible. It's hard to overstate the importance of this for any grand strategy of resistance. In this way, and many others, aboveground and underground activists are mutually supportive and work in tandem.

DECISIVE OPERATIONS ABOVEGROUND

Open property destruction is not always decisive. Take the Plowshares Movement activists, who break into military installations and use hammers and other tools to attack everything from soldiers' personal firearms to live nuclear weapons, after which they wait and accept personal legal responsibility for their actions. There's no doubt that this involves bravery—obviously it requires a lot of guts to take a sledgehammer to a hydrogen bomb—but these acts are not intended to be decisive. They are chiefly symbolic actions; neither the intent nor the effect of the action is to cause a measurable decrease in the military arsenal. (Presumably they could accomplish this if they really wanted to; anyone with the wherewithal to bypass military security and get within arm's reach of a live nuclear warhead could probably do it more than once.)

In fact, open property destruction as a decisive aboveground tactic is historically rare. Remember, those in power view their property as being more important than the lives of those below them on civilization's hierarchy. If large amounts of their property are being destroyed openly, they have few qualms about using violent retaliation. Because of this, situations where property can be destroyed openly tend to be very unstable. If those in power retaliate, the resistance movement either falters, shifts underground, or escalates. The Boston Tea Party is an excellent example. After the dumping of tea in December 1773, a boycott was imposed on British tea imports. In October 1774, the ship *Peggy Stewart* was caught attempting to breach the boycott while landing in Annapolis, Maryland. Protesters burned the ship to the waterline, a considerable escalation from the earlier dumping of tea. Within a year, mere property destruction segued into armed conflict and the Revolutionary War broke out.

Aboveground acts of omission are the more common tactical choice.

An individual's reduced consumption is not decisive, for reasons already discussed; in a society running short of finite resources like petroleum, well-meaning personal conservation may simply make supplies more available to those who would put them to the worst use, like militaries and corporate industry. But large-scale conservation could reduce the rate of damage slightly, and buy us more time to enact decisive operations, or, at least, when civilization does come down, leave us with slightly more of the world intact.

The expropriation or reclamation of land and materiel can be very effective decisive action when the numbers, strategy, and political situation are right. The Landless Workers Movement in Latin America has been highly successful at reclaiming "underutilized" land. Their large numbers (around two million people), proven strategy of reclaiming land, and political and legal framework in Brazil enable their strategy.

Many indigenous communities around the world engage in direct reoccupation and reclamation of land, especially after prolonged legal land claims, with mixed success. There are enough examples of success to suggest that direct reclamation can be successful, especially with wider support from both indigenous and settler communities. The specifics of conflicts like those at Kanehsatake and Oka, Caledonia, Gustafsen Lake, Ipperwash, and Wounded Knee (1973), are too varied to get into here. But it's clear that indigenous land reclamations attack the root of the legitimacy—even the existence—of colonial states, which is why those in power respond so viciously to them, and why those struggles are so critical and pivotal for broader resistance in general.

SUSTAINING OPERATIONS ABOVEGROUND

Sustaining operations directly support resistance. For individuals aboveground, that means finding comrades through mutual recruitment or offering material or moral support to other groups. But individual mutual recruitment can be difficult (although this is easier if the recruiter in question is strongly driven, charismatic, well organized, persuasive, and so on). Affinity groups, with more people available to prospect, screen, and train new members, are able to recruit and enculturate very effectively. Individual recruiters have personality, but a group, even a small one, has a culture—hopefully a healthy culture of resistance.

Aboveground sustaining operations mostly revolve around solidarity, both moral and material. Legal and prisoner support are important ways of supporting direct action. So are other kinds of material support, fund raising, and logistical aid. The hard part is often building a relationship between supporters and combatants. There can be social and cultural barriers between supporters (say, settler solidarity activists) and those on the front lines (say, indigenous resisters). Indigenous activists may be tired of white people telling them how to defend themselves or perhaps simply wary of people whom they don't know whether they should trust.

Propaganda and agitation supporting a particular campaign or struggle are other important sustaining actions. Liberation struggles like those in South Africa and Palestine have been defended internationally by vocal activists and organizers over decades. This propaganda has increased support for those struggles (both moral and material) and made it more difficult for those in power to repress resisters.

Larger organizations can undertake sustaining operations like fund raising and recruitment on a larger scale. They may also do a better job of training or enculturation. A single affinity group has many benefits, but can also be a bubble, a cultural fishbowl of people who come together *because* they believe the same thing. Being part of a larger network can mean that a new member gets a more well-rounded experience. Of course, the opposite can happen—dysfunctional large groups can quash ideological diversity. Often in "legitimate" groups that means quashing more radical, militant, or challenging beliefs in favor of an inoffensive liberal approach.

The converse problem is factionalism. There's a difference between allowing internal dialogue and dissent, on one hand, and having acrimonious internal conflicts (like in the Black Panthers or the Students for a Democratic Society), on the other. The larger an organization is the harder it is to walk the line between unity and splintering (especially when the COINTELPRO types are trying hard to destroy any effective operation).

Larger organizations have a better capacity for sustaining operations (and decisive operations, for that matter) than individuals and small groups, but they rarely apply it effectively. Internal conflicts limit operations to the lowest common denominator: the lowest risk, the lowest level of internal controversy, and the lowest level of effectiveness. The big green and big leftist organizations will only go as far as holding press conferences and waving signs. Meanwhile, indigenous people who are struggling (often at gunpoint) to defend and reclaim their lands are ignored if they act outside the government land claims process. Tree sitters, even those who are avowedly nonviolent, get ignored by the big green organizations when police and loggers come in to attack them. The big organizations almost always fail to deploy their resources for sustaining operations when and where they are needed most. On a moral level, that's deeply deplorable. On a strategic level, it's unspeakably stupid. On a species and planetary level, it's simply suicidal.

Of course, it doesn't have to be that way. Effective resistance movements in history are usually composed of a cross section of many different organizations on many different scales, performing the different tasks best suited to them, and larger organizations are an important part of that. History has shown that it's possible for large organizations to operate in solidarity and with foresight. Even if they don't actually carry out decisive operations themselves, large aboveground organizations can offer incredibly important support.

SHAPING OPERATIONS ABOVEGROUND

Most day-to-day aboveground resistance actions are shaping operations of one kind or another. But many actions could be sustaining *or* shaping operations, depending on the context. Building a big straw-bale house out in the country would be considered a shaping operation if the house were built simply for the purpose of building a straw-bale house. But if that building were used as a retreat center for resistance training, it might then become part of sustaining operations. Consider the Black Panthers. A free breakfast program for children that was devoid of political content would have been a charity or perhaps mutual aid. A breakfast program integrated within a larger political strategy of education, agitation, and recruitment became a sustaining operation (as well as a threat to the state).

One of the most important shaping operations is building a culture of resistance. On an individual level, this might mean cultivating the revolutionary character—learning from resisters of the past, and turning their lessons into habits to gain the psychological and analytical tools needed for effective action. Building a culture of resistance goes hand in hand with education, awareness raising, and propaganda. It also ties into support work and building alternatives, especially concrete political and social alternatives to the status quo. As always, every action must be tied into the larger resistance strategy.

Most large organizations focus on shaping operations without making sure they are tied to a larger strategy. They try to raise awareness in the hopes that it will lead indirectly to change. This can be a fine choice if made deliberately and intelligently. But I think that most progressive organizations eschew decisive or sustaining operations because they simply don't consider themselves to be resistance organizations; they identify strongly with those in power and with the culture that is destroying the planet. They keep trying to convince those in power to please change, and it doesn't work, and they fail to adjust their tactics accordingly. The planet keeps dying, and people drop out of doing progressive work by the thousands, because it so often *doesn't* work. We simply don't have time for that anymore. We need a livable planet, and at this point a livable planet requires a resistance movement.

UNDERGROUND TACTICS

Some tactics *can* be carried out underground—like general liberation organizing and propaganda—but are more effective aboveground. Where open speech is dangerous, these types of tactics may move underground to adapt to circumstances. The African National Congress, in its struggle for basic human rights, should have been allowed to work aboveground, but that simply wasn't possible in repressive apartheid South Africa.

And then there are tactics that are *only* appropriate for the underground, obligate underground operations that depend on secrecy and security. Escape lines and safehouses for persecuted persons and resistance fugitives are example of those operations. There's a reason it's called the "underground" railroad—it's not transferable to the aboveground, because the entire operation is completely dependent on secrecy. Clandestine intelligence gathering is another case; the French Resistance didn't gather enemy secrets by walking up to the nearest SS office and asking for a list of their troop deployments.

Some tactics are almost always limited to the underground:

- Clandestine intelligence
- Escape
- Sabotage and attacks on materiel
- Attacks on troops
- Intimidation
- Assassination

As operational categories, intelligence and escape are pretty clear, and few people looking at historical struggles will deny the importance of gathering information or aiding people to escape persecution. Of course, some abolitionists in the antebellum US didn't support the Underground Railroad. And many Jewish authorities tried to make German Jews cooperate with registration and population control measures. In hindsight, it's clear to us that these were huge strategic and moral mistakes, but at the time it may only have been clear to the particularly perceptive and farsighted.

Sabotage and attacks on materiel are overlapping tactics. Oftentimes, sabotage is more subtle; for example, machinery may be disabled without being recognized as sabotage. Attacks on materiel are often more overt efforts to destroy and disable the adversary's equipment and supplies. In any case, they form an inclusive continuum, with sabotage on the more clandestine end of the scale.

It's true that harm can be caused through sabotage, and that sabotage can be a form of violence. But allowing a machine to operate can also be more violent than sabotaging it. Think of a drift net. How many living creatures does a drift net kill as it passes through the ocean, regardless of whether it's being used for fishing or not? Destroying a drift net—or sabotaging a boat so that a drift net cannot be deployed—would save countless lives. Sabotaging a drift net is clearly a nonviolent act. However, you could argue conversely that not sabotaging a drift net (provided you had the means and opportunity) is a profoundly violent act—indeed, violent not just for individual creatures, but violent on a massive, ecological scale. The drift net is an obvious example, but we could make a similar (if longer and more roundabout) argument for most any industrial machinery.

You're opposed to violence? So where's your monkey wrench?

Sabotage is not categorically violent, but the next few underground categories may involve violence on the part of resisters. Attacks on troops, intimidation, assassination, and the like have been used to great effect by a great many resistance movements in history. From the assassination of SS officers by escaping concentration camp inmates to the killing of slave owners by revolting slaves to the assassination of British torturers by Michael Collins's Twelve Apostles, the selective use of violence has been essential for victory in a great many resistance and liberation struggles.

Attacks on troops are common where a politically conscious population lives under overt military occupation. In these situations, there is often little distinction between uniformed militaries, police, and government paramilitaries (like the Black and Tans or the miliciens). The violence may be secondary. Sometimes the resistance members are trying to capture equipment, documents, or intelligence; how many guerrillas have gotten started by killing occupying soldiers to get guns? Sometimes the attack is intended to force the enemy to increase its defensive garrisons or pull back to more defended positions and abandon remote or outlying areas. Sometimes the point is to demonstrate the strength or capabilities of the resistance to the population and the occupier. Sometimes the point is actually to kill enemy soldiers and deplete the occupying force. Sometimes the troops are just sentries or guards, and the primary target is an enemy building or facility.

Of course, for these attacks to happen successfully, they must follow the basic rules of asymmetric conflict and general good strategy. When raiding police stations for guns, the IRA chose remote, poorly guarded sites. Guerrillas like to go after locations with only one or two sentries, and any attack on those small sites forces the occupier to make tough choices: abandon an outpost because it can't be adequately defended or increase security by doubling the number of guards. Either benefits the resistance and saps the resources of the occupier.

And although in industrial conflicts it's often true that destroying materiel and disrupting logistics can be very effective, that's sometimes not enough. Take American involvement in the Vietnam War. The American cost in terms of materiel was enormous—in modern dollars, the war cost close to $600 billion. But it wasn't the cost of replacing helicopters or fuelling convoys that turned US sentiment against the war. It was the growing stream of American bodies being flown home in coffins.

There's a world of difference—socially, organizationally, psychologically—between fighting the occupation of a foreign government and the occupation of a domestic one. There's something about the psychology of resistance that makes it easier for people to unite against a foreign enemy. Most people make no distinction between the people living in their country and the government of that country, which is why the news will say "America pulls out of climate talks" when they are talking about the US *government*. This psychology is why millions of Vietnamese people took up arms against the American invasion, but only a handful of Americans took up arms against that invasion (some of them being soldiers who fragged their officers, and some of them being groups like the Weather Underground who went out of their way *not* to injure the people who were burning Vietnamese peasants alive by the tens of thousands). This psychology explains why some of the patriots who fought in the French Resistance went on to torture people to repress the Algerian Resistance. And it explains why most Germans didn't even support *theoretical* resistance against Hitler a decade after the war.

This doesn't bode well for resistance in the minority world, where the rich and powerful minority live. People in poorer countries may be able to rally against foreign corporations and colonial dictatorships, but those in the center of empire contend with power structures that most people consider natural, familiar, even friendly. But these domestic institutions of power—be they corporate or governmental—are just as foreign, and just as destructive, as an invading army. They may be based in the same geographic region as we are, but they are just as alien as if they were run by robots or little green men.

◻ ◻ ◻

Intimidation is another tactic related to violence that is usually conducted underground. This tactic is used by the "Gulabi Gang" (also called the Pink Sari Gang) of Uttar Pradesh, a state in India.[4] Leader Sampat Pal Devi calls it "a gang for justice." The Gulabi Gang formed as a response to deeply entrenched and violent patriarchy (especially domestic and sexual violence) and caste-based discrimination. The members use a variety of tactics to fight for women's rights, but their "vigilante violence" has gained global attention. With over 500 members, they can exert considerable force. They've stopped child marriages. They've beaten up men who perpetrate domestic violence. The gang forced the police to register crimes against Untouchables by slapping police officers until they complied. They've hijacked trucks full of food that were going to be sold for a profit by corrupt officials. Their hundreds of members practice self-defense with the *lathi* (a traditional Indian stick or staff weapon). It's no surprise their ranks are growing.

Many of these examples tread the boundary of our aboveground-underground distinction. When struggling against systems of patriarchy that have closely allied themselves with governments and police (which is to say, virtually all systems of patriarchy), women's groups that have been forced to use violence or the threat of violence may have to operate in a clandestine fashion at least some of the time. At the same time, the effects of their self-defense must be prominent and publicized. Killing a rapist or abuser has the obvious benefit of stopping any future abuses by that individual. But the larger beneficial effect is to intimidate other would-be abusers—to turn the tables and prevent other incidents of rape or abuse by making the consequences for perpetrators known. The Gulabi Gang is so popular and effective in part because they openly defy abuses of male power, so the effect on both men and women is very large. Their aboveground defiance rallies more support than they could by causing abusive men to die in a series of mysterious accidents. The Black Panthers were similarly popular because they publically defied the violent oppression meted out by police on a daily basis. And by openly bearing arms, they were able to intimidate the police (and other people, like drug dealers) into reducing their abuses.

There are limits to the use of intimidation on those in power. The most powerful people are the most physically isolated—they might have bodyguards or live in gated houses. They have far more coercive force at their fingertips than any resistance movement. For that reason, resistance groups have historically used intimidation primarily on low-level functionaries and collaborators who give information to those in power when asked or who cooperate with them in a more limited way.

It's important to acknowledge the distinction between intimidation and terrorism. Terrorism consists of violent attacks on civilians. Resistance intimidation directly targets those responsible for oppressive and exploitative acts and power structures, and lets those people know that there are consequences for their actions. The reason it gets people so riled up is because it involves violence (or the threat of violence) going up the hierarchy. But resistance intimidation is ultimately, of course, an attempt to reduce violence. Groups like the Gulabi Gang beat abusive men instead of just killing them. There's a reasonable escalation that gives men a chance to stop their wrongdoing and also makes the consequences for further wrongdoing clear. Rape and domestic abuse are terrorism; they're senseless and unprovoked acts of violence against unarmed civilians, designed to threaten and terrorize women (and men) into compliance. The intimidation of rapists or domestic abusers is one tactic that can be used to stop their violence while employing the minimum amount of violence possible.

N N N

No resistance movement wants to engage in needless cycles of violence and retribution with those in power. But a refusal to employ violent tactics when they *are* appropriate will very likely lead to more violence. Many abolitionists did not support John Brown because they considered his plan for a defensive liberation struggle to be too violent—but Brown's failure led inevitably to a lengthy and gruesome Civil War (as well as continued years of bloody slavery), a consequence that was orders of magnitude more violent than Brown's intended plan.

◙ ◙ ◙

This leads us to the last major underground tactic: assassination.

In talking about assassination (or any attack on humans) in the context of resistance, two key questions must be asked. First, is the act strategically beneficial, that is, would assassination further the strategy of the group? Second, is the act morally just, given the person in question? (The issue of justice is necessarily particular to the target; it's assumed that the broader strategy incorporates aims to increase justice.)

As is shown on my two-by-two grid of all combinations (see Figure 13-3), an assassination may be strategic and just, it may be strategic and unjust, it may be unstrategic but just, or it may be both unstrategic and

Figure 13-3

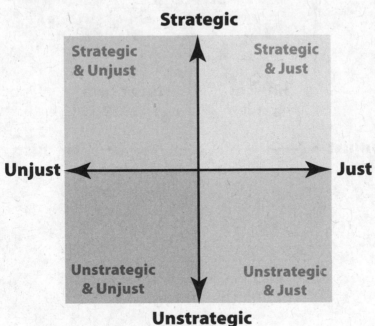

unjust. Obviously, any action in the last category would be out of the question. Any action in the strategic and just category could be a good bet for an armed resistance movement. The other two categories are where things get complex.

Hitler exemplified a number of different strategy vs. justice combinations at different points in time. It's a common moral quandary to ask whether it would be a good idea to go back in time and kill Hitler as a child, provided time travel were possible. There's a good bet that this would have averted World War II and the Holocaust, which would have been a good thing, so put a check mark in the "strategic" column. The problem is that most people would consider it unjust to murder an innocent child who had yet to commit any crimes, so it would be difficult to call that action just in the immediate sense.

Once Hitler had risen to power in the late 1930s, though, his aim

Figure 13-4

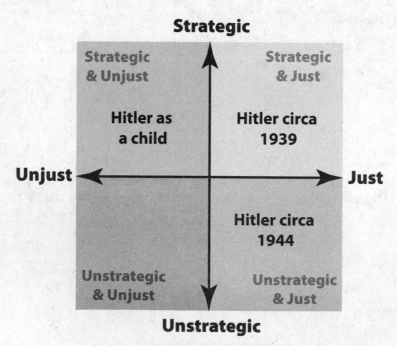

was clear, as he had already been whipping up hate and expanding his control of Nazi Germany. At that point, it would have been both strategic and just to assassinate him. Indeed, elements in the Wehrmacht (army) and the Abwehr (intelligence) considered it, because they knew what Hitler was planning to do. Unfortunately, they were indecisive, and did not commit to the plan. Hitler soon began invading Germany's neighbors, and as his popularity soared, the assassination plan was shelved. It was years before inside elements would actually stage an assassination attempt.

That famous attempt took place—and failed—on July 20, 1944.[5] What's interesting is that the Allies were also considering an attempt on Hitler's life, which they called Operation Foxley. They knew that Hitler routinely went on walks alone in a remote area, and devised a plan to parachute in two operatives dressed as German officers, one of them a sniper, who would lay in wait and assassinate Hitler when he walked by. The plan was never enacted because of internal controversy. Many in the SOE and British government believed that Hitler was a poor strategist, a maniac whose overreach would be his downfall. If he were assassinated, they believed, his replacement (likely Himmler) would be a more competent leader, and this would draw out the war and increase Allied losses. In the opinion of the Allies it was unquestionably *just* to kill Hitler, but no longer strategically beneficial (Figure 13-4).

There is no shortage of situations where assassination would have been just, but of questionable strategic value. Resistance groups pondering assassination have many questions to ask themselves in deciding whether they are being strategic or not. What is the value of this potential target to the enemy? Is this an exceptional person or does his or her influence come from his or her role in the organization? Who would replace this person, and would that person be better or worse for the struggle? Will it make any difference on an organizational scale or is the potential target simply an interchangeable cog? Uniquely valuable individuals make uniquely valuable targets for assassination by resistance groups.

Of course, in a military context (and this overlaps with attacks on troops), snipers routinely target officers over enlisted soldiers. In theory, officers or enlisted soldiers are standardized and replaceable, but, in practice, officers constitute more valuable targets. There's a difference between theoretical and practical equivalence; there might be other officers to replace an assassinated one, but the replacement might not arrive in a timely manner nor would he have the experience of his predecessor (experience being a key reason that Michael Collins assassinated intelligence officers). That said, snipers don't *just* target officers. Snipers target any enemy soldiers available, because war is essentially about destroying the other side's ability to wage war.

The benefits must also outweigh costs or side effects. Resistance members may be captured or killed in the attempt. Assassination also provokes a major response—and major reprisals—because it is a direct attack on those in power. When SS boss Reinhard Heydrich ("the butcher of Prague") was assassinated in 1942, the Nazis massacred more than 1,000 Czech people in response. In Canada, martial law (via the War Measures Act) has only ever been declared three times—during WWI and WWII, and again after the assassination of the Quebec Vice Premier of Quebec by the Front de Libération du Québec. Remember, aboveground allies may bear the brunt of reprisals for assassinations, and those reprisals can range from martial law and police crackdowns to mass arrests or even executions.

There's an important distinction to be made between assassination as an ideological tactic versus as a military tactic. As a military tactic, employed by countless snipers in the history of war, assassination decisively weakens the adversary by killing people with important experience or talents, weakening the entire organization. Assassination as an ideological tactic—attacking or killing prominent figures because of ideological disagreements—almost always goes sour, and quickly. There are few more effective ways to create martyrs and trigger cycles of violence without actually accomplishing anything decisive. The assassination of Michael Collins, for example, by his former allies led only to bloody civil war.

DECISIVE OPERATIONS UNDERGROUND

Individuals working underground focus mostly on small-scale acts of sabotage and subversion that make the most of their skill and opportunity. Because they lack escape networks, and because they must be opportunistic, it's ideal for their actions to be what French resisters called *insaisissable*–untraceable or appearing like an accident—unless the nature of the action requires otherwise.

Individual saboteurs are more effective with some informal coordination—if, for example, a general day of action has been called. It also helps if the individuals seize an opportunity by springing into action when those in power are already off balance or under attack, like the two teenaged French girls who sabotaged trains carrying German tanks after D-Day, thus hampering the German ability to respond to the Allied landing.

One individual resister who attempted truly decisive action was Georg Elser, a German-born carpenter who opposed Hitler from the beginning. When Hitler started the World War II in 1939, Elser resolved to assassinate Hitler. He spent hours every night secretly hollowing out a hidden cavity in the beer hall where Hitler spoke each year on the anniversary of his failed coup. Elser used knowledge he learned from working at a watch factory to build a timer, and planted a bomb in the hidden cavity. The bomb went off on time, but by chance Hilter left early and survived. When Elser was captured, the Gestapo tortured him for information, refusing to believe that a single tradesperson with a grade-school education could come so close to killing Hitler without help. But Elser, indeed, worked entirely alone.

Underground networks can accomplish decisive operations that require greater coordination, numbers, and geographic scope. This is crucial. Large-scale coordination can turn even minor tactics—like simple sabotage—into dramatically decisive events. Underground saboteurs from the French Resistance to the ANC relied on simple techniques, homemade tools, and "appropriate technology." With synchronization between even a handful of groups, these underground networks can make an entire economy grind to a halt.

The change is more than quantitative, it's qualitative. A massively coordinated set of actions is fundamentally different from an uncoordinated set of the same actions. Complex systems respond in a nonlinear fashion. They can adapt and maintain equilibrium in the face of small insults, minor disruptions. But beyond a certain point, increasing attacks undermine the entire system, causing widespread failure or collapse.

Because of this, coordination is perhaps the most compelling argument for underground networks over mere isolated cells. I'll discuss coordinated actions in more detail in the next chapter.

SUSTAINING OPERATIONS UNDERGROUND

Since individuals working underground are pretty much alone, they have very few options for sustaining operations. They may potentially recruit or train others to form an underground cell. Or they may try to make contact with other people or groups (either underground or aboveground) to work as an auxiliary of some kind, such as an intelligence source, especially if they are able to pass on information from inside a government or corporate bureaucracy. But making this connection is often very challenging.

Individual escape and evasion may also be a decisive or sustaining action, at least on a small scale. Antebellum American slavery offers some examples. In a discussion of slave revolts, historian Deborah Gray White explains, "[I]ndividual resistance did not overthrow slavery, but it might have encouraged masters to make perpetual servitude more tolerable and lasting. Still, for many African Americans, individual rebellions against the authority of slaveholders fulfilled much the same function as did the slave family, Christianity, and folk religion: it created the psychic space that enabled Black people to survive."[6]

Historian John Michael Vlach observes: "Southern plantations actually served as the training grounds for those most inclined to seek their freedom." Slaves would often escape for short periods of time as a temporary respite from compelled labor before returning to plantations, a practice often tolerated by owners. These escapes provided opportunities to build a camp or even steal and stock up on provisions for another escape. Sometimes slaves would use temporary escapes as attempts to compel better behavior from plantation owners.[7] In any case, these escapes and minor thefts helped to build a culture of resistance by challenging the omnipotence of slave owners and reclaiming some small measure of autonomy and freedom.

Individuals have some ability to assert power, but recruitment is key in underground sustaining operations. A single cell can gather or steal equipment and supplies for itself, but it can't participate in wider sustaining operations unless it forms a network by recruiting organizationally, training new members and auxiliaries, and extending into new cells. One underground cell is all you need to create an entire network. Creating the first cell—finding those first few trusted comrades, developing communications and signals—is the hardest part, because other cells can be founded on the same template, and the members of the existing cell can be used to recruit, screen, and train new members.

Even though it's inherently difficult for an underground group to coordinate with other distinct underground groups, it is possible for an underground cell to offer supporting operations to aboveground campaigns. It was an underground group—the Citizen's Commission to Investigate the FBI—that exposed COINTELPRO, and allowed many aboveground groups to understand and counteract the FBI's covert attacks on them. And the judicious use of sabotage could buy valuable time for aboveground groups to mobilize in a given campaign.

There are clearly campaigns in which aboveground groups have no desire for help from the underground, in which case it's best for the underground to focus on other projects. But the two can work together on the same strategy without direct coordination. If a popular aboveground campaign against a big-box store or unwanted new industrial site fails because of corrupt politicians, an underground group can always pick up the slack and damage or destroy the facility under construction. Sometimes people argue that there's no point in sabotaging anything, because those in power will just build it again. But there may come a day when those in power start to say "there's no point in building it—they'll just burn it down again."

Underground cells may also run a safehouse or safehouses for themselves and allies. Single cells can't run true underground railroads, but even single safehouses are valuable in dealing with repression or persecution. A key challenge in underground railroads and escape lines is that the escapees have to make contact with underground helpers without exposing themselves to those in power. Larger, more "formalized" underground networks have specialized methods and personnel for this, but a single cell running a safehouse may not. If an underground cell is conscientious, its members will be the only ones aware that the safehouse exists at all, which puts the burden on them to contact someone who requires refuge.

Mass persecution and repression has happened enough times in history to provide a wealth of examples where this would be appropriate. The internment of Japanese Canadians during World War II is quite well-known. Less well-known is the internment of hundreds of leftist radicals and labor activists starting in 1940. Leading activists associated with certain other ethnic organizations (especially Ukrainian), the labor movement, and the Communist party were arrested and sent to isolated work camps in various locations around Canada. A few managed to go into hiding, at least temporarily, but the vast majority were captured and sent to the camps, where a number of them died.[8] In a situation like that, an underground cell could offer shelter to a persecuted aboveground activist or activists on an invitational basis without having to expose themselves openly.

Many of these operations work in tandem. Resistance networks from the SOE to the ANC have used their escape lines and underground railroads to sneak recruits to training sites in friendly areas and then infiltrated those people back into occupied territory to take up the fight.

Underground networks may be large enough to create "areas of persistence" where they exert a sizeable influence and have developed an underground infrastructure rooted in a culture of resistance. If an underground network reaches a critical mass in a certain area, it may be able to significantly disrupt the command and control systems of those in power, allowing resisters both aboveground and underground a greater amount of latitude in their work.

There are a number of examples of resistance movements successfully creating areas of persistence. The Zapatistas in Mexico exert considerable influence in Chiapas, so much so that they can post signs to that effect. "You are in Zapatista rebel territory" proclaims one typical sign (translated from Spanish). "Here the people give the orders and the government obeys." The posting also warns against drug and alcohol trafficking or use and against the illegal sale of wood. "No to the destruction of nature."⁹ Other Latin American resistance movements, such as the FMLN in El Salvador and the Sandinistas in Nicaragua, created areas of persistence in Latin America in the late twentieth century. Hamas in Palestine and Hezbollah in Lebanon have similarly established large areas of persistence in the Middle East.

SHAPING OPERATIONS UNDERGROUND

Because working underground is dangerous and difficult, effective resisters mostly focus on decisive and sustaining operations that will be worth their while. That said, there are still some shaping operations for the underground.

This includes general counterintelligence and security work. Ferreting out and removing informers and infiltrators is a key step in allowing resistance organizations of every type to grow and resistance strategies to succeed. Neither the ANC nor the IRA were able to win until they could deal effectively with such people.

Underground cells can also carry out some specialized propaganda operations. For reasons already discussed, propaganda in general is best carried out by aboveground groups, but there are exceptions. In particularly repressive regimes, basic propaganda and education projects must move underground to continue to function and protect identities. Underground newspapers and forms of pirate radio are two examples. Entire, vast underground networks have been built on this principle. In Soviet Russia, *samizdat* was the secret copying of and distribution of illegal or censored texts. A person who received a piece of illegal literature—say, Vaclav Havel's *Power of the Powerless*—was expected to make more copies and pass them on. In a pre–personal computer age, in a country where copy machines and printing presses were under state control, this often meant laboriously copying books by hand or typewriter.

Underground groups may also want to carry out certain high-profile or spectacular "demonstration" actions to demonstrate that underground resistance is possible and that it is happening, and to offer a model for a particular tactic or target to be emulated by others. Of course, demonstrative actions may be valuable, but they can also degrade into symbolism for the sake of symbolism. Plenty of under-ground groups, the Weather Underground included, hoped to use their actions to "ignite a revolution." But, in general—and especially when "the masses" can't be reasonably expected to join in the fight—under-ground groups must get their job done by being as decisive as possible.

TARGET SELECTION

A good tactic used on a poor target has little effect.

The *Field Manual on Guerrilla Warfare* identifies four "important fac-tors related to the target which influence its final selection."[10] These criteria are meant specifically for targets to be disrupted or destroyed, not necessarily when choosing potential targets for intelligence gath-ering or further investigation. The four criteria are as follows:

Criticality. How important is this target to the enemy and to enemy operations? "A target is critical when its destruction or damage will exercise a significant influence upon the enemy's ability to conduct or support operations. Such targets as bridges, tunnels, ravines, and mountain passes are critical to lines of communication; engines, ties, and POL [petroleum, oil, and lubricant] stores are critical to trans-portation. Each target is considered in relationship to other elements of the target system." Resistance movements (and the military) look for bottlenecks when selecting a target. And they make sure to think in big picture terms, rather than just in terms of a specific individual target. What target(s) can be disrupted or destroyed to cause maximum damage to the entire enemy system? Multiple concurrent surprise attacks are ideal for resistance movements, and can cause cascading failures.

Vulnerability. How tough is the target? "Vulnerability is a target's sus-ceptibility to attack by means available to [resistance] forces. Vulnerability is influenced by the nature of the target, i.e., type, size, disposition and composition." In military terminology, a "soft target" is one that is relatively vulnerable, while a "hard target" is well defended or fortified. A soft target could be a sensitive electrical component, a flammable storage shed, or a person. A hard target might be a roadway, a concrete bunker, or a military installation. Hard targets require more capacity or armament to disable. A battle tank might have lower vulnerability when face with a resister armed with a Molotov cocktail, but high vulnerability against someone armed with a rocket-propelled grenade.

Accessibility. How easy is it to get near the target? "Accessibility is measured by the ability of the attacker to infiltrate into the target area. In studying a target for accessibility, security controls around the target area, location of the target, and means of infiltration are considered." It's important to make a clear distinction between accessibility and vulnerability. For a resister in Occupied France, a well-guarded fuel depot might be explosively vulnerable, but not very accessible. For resisters in German-occupied Warsaw, the heavy wall surrounding the Warsaw Ghetto might be easily accessible, but not very vulnerable unless they carried powerful explosives. Good intelligence and reconnaissance are key to identifying and bypassing obstacles to access.

Recuperability. How much effort would it take to rebuild or replace the target? "Recuperability is the enemy's ability to restore a damaged facility to normal operating capacity. It is affected by the enemy capability to repair and replace damaged portions of the target." Specialized installations, hard-to-find parts, or people with special unique skills are difficult to replace. Targets with very common or mass-produced and stockpiled components would be poorer targets in terms of recuperability. Undermining enemy recuperability can be done with good planning and multiple attacks: SOE saboteurs were trained to target the same important parts on every machine. If they were to sabotage all of the locomotives in a stockyard, they would blow up the same part on each train, thus preventing the engineers from cannibalizing parts from other trains to make a working one.

From this perspective the ideal target would be highly critical (such that damage would cause cascading systems failures), highly vulnerable, very accessible, and difficult and time-consuming to repair or replace. The poorest target would be of low importance for enemy operations, hardened, inaccessible, and easily replaced. You'll note that there's no category for "symbolic value" to the enemy, because the writers of the manual weren't interested in symbolic targets. They consistently emphasize that successful operations will undermine the morale of the adversary, while increasing morale of the resisters and their supporters. The point is to carry out decisively effective action with the knowledge that such action will have emotional benefits for your side, not to carry out operations that seem emotionally appealing in the hopes that those choices will lead to effective action.

An additional criterion not discussed above would be *destructivity*. How damaging is the existence of the target to people and other living creatures? A natural gas–burning power plant might be more valuable based on the four first criteria, but a coal-fired power plant could be more destructive, making it a higher priority from a practical and symbolic perspective.

It's rare to find a perfect target. It's more likely that choosing among targets will require certain trade-offs. A remote enemy installation might be more vulnerable, but it could also be more difficult to access and possibly less important to the adversary. Larger, more critical installations are often better guarded and less vulnerable. Target decisions have to be made in the context of the larger strategy, taking into account tactics and organizational capability.

One of the reasons that the Earth Liberation Front (ELF) has had limited decisive success so far is that its targets have had low criticality and high recuperability. New suburban subdivisions are certainly crimes against ecology, but partially constructed homes are not very important to those in power, and they are relatively replaceable. The effect is primarily symbolic, and it's hard to find a case in which a construction project has actually been given up because of ELF activity—although many have certainly been made more expensive.

Most often, it seems that resistance targets in North America are chosen on the basis of vulnerability and accessibility, rather than on criticality. It's easy to walk up to a Walmart window and smash it in the middle of the night or to destroy a Foot Locker storefront during a protest march. Aggressive symbolic attacks do get attention, and if a person's main indicator of success is a furor on the 10:00 pm news, then igniting the local Burger King is likely to achieve that. But making a decisive impact on systems of power and their basis of support is more difficult to measure. If those in power are clever, they'll downplay the really damaging actions to make themselves seem invulnerable, but scream bloody murder over a smashed window in order to whip up public opinion. And isn't that what often happens on the news? If a biotech office is smashed and not a single person injured, the corporate journalists and pundits start pontificating about "violence" and "terrorism." But if a dozen US soldiers are blown up by insurgents in Iraq, the White House press secretary will calmly repeat over and over that "America" is winning and that these incidents are only minor setbacks.

The Black Liberation Army (BLA) is an example of a group that chose targets in alignment with its goals. The BLA formed as an offshoot (or, some would argue, as a parallel development) of the Black Panther Party. The BLA was not interested in symbolic targets, but in directly targeting those who oppressed people of color. Writes historian Dan Berger: "The BLA's Program included three components: retaliation against police violence in Black communities; elimination of drugs and drug dealers from Black communities; and helping captured BLA members escape from prison."[11] The BLA essentially believed that aboveground black organizing was doomed because of violent COINTELPRO-style tactics, and that the BPP had become a reformist organization. They argued that "the character of reformism is based on unprincipled class collaboration with our enemy."[12] In part because of their direct personal experience of violent repression at the hands of the state, they did not hesitate to kill white police officers in retaliation for attacks on the black community.

The IRA was also ruthless in their target selection, though they had limited choices in terms of attacking their occupiers. By the time WWII rolled around, resisters in Europe had a wide variety of potential and critical targets for sabotage, such as rail and telegraph lines, and further industrialization has only increased the number of critical mechanical targets, but a century ago, Ireland was hardly mechanized at all. That is why Michael Collins correctly identified British intelligence agents as the most critical and least recuperable targets available. Furthermore, his networks of spies and assassins made those agents—already soft targets—highly accessible. They were a perfect match for all four target selection criteria.

It's worth noting that these four criteria are not just applicable to targets that are going to be destroyed. The same criteria are used to select "pressure points" on which to exert political force for any strategy of resistance, even one that is explicitly nonviolent. Effective strikes or acts of civil disobedience can exert more political force by disrupting more critical and vulnerable targets—the more accessible, the better.

These criteria for target selection go both ways. Our own resistance movements are targets for those in power, and it's important to understand our organizations as potential targets. Leaders have often been attacked because they were crucial to the organization. Underground leaders are less accessible, but potentially more vulnerable if they can be isolated from their base of support. And aboveground groups often have better recuperability, because they have a larger pool to draw from and fewer training requirements; recall the waves after waves of civil rights activists willing to be arrested in Birmingham, Alabama.

❒ ❒ ❒

Anyone who casts their lot with a resistance movement must be prepared for reprisals. Those reprisals will come whether the actionists are aboveground or underground, choosing violence or nonviolence. Many activists, especially from privileged backgrounds, naïvely assume that fighting fair will somehow cause those in power to do the same. Nothing could be further from the truth. The moment that any power structure feels threatened, it will retaliate. It will torture Buddhists and nuns, turn fire hoses on school children, and kill innocent civilians. A brief perusal of Amnesty International's website will acquaint you with nonviolent protestors around the globe currently being detained and tortured or who have disappeared for simple actions like letter writing or peaceably demonstrating.

This is a reality that privileged people must come to terms with or else any movement risks a rupture when power comes down on actionists. Those retaliations are not anyone's fault; they are to be expected. Any serious resistance movement should be intellectually and emotionally prepared for the power structure's response. People are arrested, detained, and killed—often in large numbers—when power strikes back. Those who provide a challenge to power will be faced with consequences, some of them inhumanly cruel. The sooner everyone understands that, the better prepared we all will be to handle it.

◫ ◫ ◫

Now, having discussed what makes good strategy, how resistance groups organize effectively, and what sort of culture resistance groups need to support them, it is time to take a deep breath. A real deep breath.

This culture is killing the planet. It systematically dispossesses sustainable indigenous cultures. Runaway global warming (and other toxic effects of this culture) could easily lead to billions of human deaths, and indeed the murder of the oceans, and even more, the effective destruction of this planet's capacity to support life.

The question becomes: what is to be done?

◫ ◫ ◫

Q: What has happened to those who have tried to use violence? Fred Hampton, Laura Whitehorn, and Susan Rosenberg are just a few of the many who have tried to use force and have ended up dead, framed, or in jail. You say we all have a role; how do you feel about proposing that others do what you will not do?

Derrick Jensen: It's not a question of taking more or less risks by going aboveground or underground. As repression becomes more open, it is the people who are *aboveground* who are often first targeted by those in power. Erich Mühsam was aboveground. So was Ken Saro-Wiwa. Many writers have been. *That is our role.* Our role is to put big bull's-eye targets on our chests so that we can help to form a culture of resistance. Our role is to be public. And, of course, if you are public, you cannot also be underground; there must be an absolute firewall between aboveground and underground activities and organizations. This is basic security culture.

We are not asking anyone else to do things we aren't willing to do. In fact, we aren't asking anyone to do anything in specific. We all need to find our own roles, based on our personal assessment of what risks we can take and what our gifts are.

Those in power will come down on us if we resist. It doesn't matter if that resistance is violent or nonviolent. It's resistance that brings the risk and retaliation, and it's resistance that our planet needs.

Q: If we dismantle civilization, won't that kill millions of people in cities? What about them?

Derrick Jensen: No matter what you do, your hands will be blood red. If you participate in the global economy, your hands are blood red because the global economy is murdering humans and nonhumans the planet over. A half million children die every year as a direct result of so-called debt repayment from nonindustrialized nations to industrialized nations. Sixty thousand people die every day from pollution. And what about all the people who are being forced off their land? There are a lot of people dying already. Failing to act in the face of atrocity is no answer.

The grim reality is that both energy descent and biotic collapse will be more and more severe the more the dominant culture continues to destroy the basis for life on this planet. And yet some people will say that those who propose dismantling civilization are, in fact, suggesting genocide on a mass scale.

Polar bears and coho salmon would disagree. Traditional indigenous peoples would disagree. The humans who inherit what is left of this world when the dominant culture finally comes down would disagree.

I disagree.

My definition of dismantling civilization is depriving the rich of their ability to steal from the poor and depriving the powerful of their ability to destroy the planet. Nobody but a capitalist or a sociopath (insofar as there is a difference) could disagree with that.

Years ago I asked Anuradha Mittal, former director of Food First, "Would the people of India be better off if the global economy disappeared tomorrow?" And she said, "Of course." She said the poor the world over would be better off if the global economy collapsed. There are former granaries of India that now export dog food and tulips to Europe. The rural poor the world over are being exploited by this system. Would they be better off? What about the farmers in India who are being forced off their land so that Coca-Cola can have their water? What about those who are committing suicide because of Monsanto? A significant portion of people in the world do not have access to electricity. Would they be worse off after the grid crashes? No, they'd be better off immediately. What about the indigenous peoples of Peru who are fighting to stop oil exploration by Hunt Oil on their land, allowed because of United States–Peruvian trade agreements?

When someone says, "A lot of people are going to die," we've got to talk about which people. People all over the world are already enduring famines, but for the most part they are not dying of starvation; they're dying of colonialism, because their land and their economies have been stolen. We hear all the time that the world is running out of water. There is still as much water as there ever was, but 90 percent of the water used by humans is being used for agriculture and industry. People are dying of thirst because the water is being stolen.

When I asked a member of the Peruvian rebel group MRTA, the Tupacameristas, "What do you want for the people of Peru?" his response was, "What we want is to be able to grow and distribute our own food. We already know how to do that. We merely need to be allowed to do so." That's the entire struggle right there.

It is true that the urban poor would be worse off at first, because the dominant culture, like any good abusive system, has made its victims dependent upon it for their lives. That's what abusers do, whether they are domestic violence abusers, or whether they are larger-scale perpetrators. That's how slavers work: they make enslaved people dependent upon them for their lives. One of the brilliant things this culture has done has been to insert itself between us and our self-sufficiency, us and the source of all life. So we come to believe that the system provides our sustenance, not that the real world does. Yes, life would be much harder at first.

But in the long run, the urban poor would be better off. Most of the urban poor are people who live in third-world slums. That's more than a billion people, and, if trends continue, that will double in two decades. Many of these are people who have been forced off their traditional land. The poor will be able to take back this land if the governments of the world are no longer capable of propping up colonial arrangements of exploitation.

I have another answer, too. As this culture collapses, much of the misery will be caused by the wealthy attempting to maintain their lifestyles. As this culture continues to collapse, those who are doing the exploiting will continue to do the exploiting. Don't blame those who want to stop that exploitation. Instead, help to stop the exploitation that is killing people in the first place.

The authors of this book are not blithely asking who will die. In at least one of our cases, the answer is "I will." I have Crohn's disease, and I am reliant on high-tech medicines for my life. Without these medicines, I will die. But my individual life is not what matters. The survival of the planet is more important than the life of any single human being, including my own.

Since industrial civilization is systematically dismantling the ecological infrastructure of the planet, the sooner civilization comes down, the more life will remain afterwards to support both humans and non-humans. We can provide for the well-being of those humans who will be alive during and immediately after energy and ecological descent by preparing people for a localized future. We can rip up asphalt in vacant parking lots to convert them to neighborhood gardens, go teach people how to identify local edible plants so that people won't starve when they can no longer head off to the store for groceries. We can start setting up neighborhood councils to make decisions, settle conflicts, and provide mutual aid.

Chapter 14

Decisive Ecological Warfare

by Aric McBay

There's a time when the operation of the machine becomes so odious, makes you so sick at heart, that you can't take part, you can't even passively take part, and you've got to put your bodies upon the gears and upon the wheels, upon the levers, upon all the apparatus, and you've got to make it stop!

—Mario Savio, Berkeley Free Speech Movement

To gain what is worth having, it may be necessary to lose everything else.

—Bernadette Devlin, Irish activist and politician

BRINGING IT DOWN: COLLAPSE SCENARIOS

At this point in history, there are no good short-term outcomes for global human society. Some are better and some are worse, and in the long term some are very good, but in the short term we're in a bind. I'm not going to lie to you—the hour is too late for cheermongering. The only way to find the best outcome is to confront our dire situation head on, and not to be diverted by false hopes.

Human society—because of civilization, specifically—has painted itself into a corner. As a species we're dependent on the draw down of finite supplies of oil, soil, and water. Industrial agriculture (and annual grain agriculture before that) has put us into a vicious pattern of population growth and overshoot. We long ago exceeded carrying capacity, and the workings of civilization are destroying that carrying capacity by the second. This is largely the fault of those in power, the wealthiest, the states and corporations. But the consequences—and the responsibility for dealing with it—fall to the rest of us, including nonhumans.

Physically, it's not too late for a crash program to limit births to reduce the population, cut fossil fuel consumption to nil, replace agricultural monocrops with perennial polycultures, end overfishing, and

425

cease industrial encroachment on (or destruction of) remaining wild areas. There's no physical reason we couldn't start all of these things tomorrow, stop global warming in its tracks, reverse overshoot, reverse erosion, reverse aquifer drawdown, and bring back all the species and biomes currently on the brink. There's no physical reason we couldn't get together and act like adults and fix these problems, in the sense that it isn't against the laws of physics.

But socially and politically, we know this is a pipe dream. There are material systems of power that make this impossible as long as those systems are still intact. Those in power get too much money and privilege from destroying the planet. We aren't going to save the planet—or our own future as a species—without a fight.

What's realistic? What options are actually available to us, and what are the consequences? What follows are three broad and illustrative scenarios: one in which there is no substantive or decisive resistance, one in which there is limited resistance and a relatively prolonged collapse, and one in which all-out resistance leads to the immediate collapse of civilization and global industrial infrastructure.

NO RESISTANCE

If there is no substantive resistance, likely there will be a few more years of business as usual, though with increasing economic disruption and upset. According to the best available data, the impacts of peak oil start to hit somewhere between 2011 and 2015, resulting in a rapid decline in global energy availability.[1] It's possible that this may happen slightly later if all-out attempts are made to extract remaining fossil fuels, but that would only prolong the inevitable, worsen global warming, and make the eventual decline that much steeper and more severe. Once peak oil sets in, the increasing cost and decreasing supply of energy undermines manufacturing and transportation, especially on a global scale.

The energy slide will cause economic turmoil, and a self-perpetuating cycle of economic contraction will take place. Businesses will be unable to pay their workers, workers will be unable to buy things, and more companies will shrink or go out of business (and will be unable to pay their workers). Unable to pay their debts and mortgages, homeowners, companies, and even states will go bankrupt. (It's possible that this process has already begun.) International trade will nosedive because of a global depression and increasing transportation and manufacturing costs. Though it's likely that the price of oil will increase over time, there will be times when the contracting economy causes falling demand for oil, thus suppressing the price. The lower cost of oil may, ironically but beneficially, limit investment in new oil infrastructure.

At first the collapse will resemble a traditional recession or depression, with the poor being hit especially hard by the increasing costs of basic goods, particularly of electricity and heating in cold areas. After a few years, the financial limits will become physical ones; large-scale energy-intensive manufacturing will become not only uneconomical, but impossible.

A direct result of this will be the collapse of industrial agriculture. Dependent on vast amounts of energy for tractor fuel, synthesized pesticides and fertilizers, irrigation, greenhouse heating, packaging, and transportation, global industrial agriculture will run up against hard limits to production (driven at first by intense competition for energy from other sectors). This will be worsened by the depletion of groundwater and aquifers, a long history of soil erosion, and the early stages of climate change. At first this will cause a food and economic crisis mostly felt by the poor. Over time, the situation will worsen and industrial food production will fall below that required to sustain the population.

There will be three main responses to this global food shortage. In some areas people will return to growing their own food and build sustainable local food initiatives. This will be a positive sign, but public involvement will be belated and inadequate, as most people still won't have caught on to the permanency of collapse and won't want to have to grow their own food. It will also be made far more difficult by the massive urbanization that has occurred in the last century, by the destruction of the land, and by climate change. Furthermore, most subsistence cultures will have been destroyed or uprooted from their land—land inequalities will hamper people from growing their own food (just as they do now in the majority of the world). Without well-organized resisters, land reform will not happen, and displaced people will not be able to access land. As a result, widespread hunger and starvation (worsening to famine in bad agricultural years) will become endemic in many parts of the world. The lack of energy for industrial agriculture will cause a resurgence in the institutions of slavery and serfdom.

Slavery does not occur in a political vacuum. Threatened by economic and energy collapse, some governments will fall entirely, turning into failed states. With no one to stop them, warlords will set up shop in the rubble. Others, desperate to maintain power against emboldened secessionists and civil unrest, will turn to authoritarian forms of government. In a world of diminishing but critical resources, governments will get leaner and meaner. We will see a resurgence of authoritarianism in modern forms: technofascism and corporation feudalism. The rich will increasingly move to private and well-defended enclaves. Their country estates will not look apocalyptic—they will look like eco-Edens, with well-tended organic gardens, clean private lakes, and wildlife refuges. In some cases these enclaves will be tiny, and in others they could fill entire countries.

Meanwhile, the poor will see their own condition worsen. The millions of refugees created by economic and energy collapse will be on the move, but no one will want them. In some brittle areas the influx of refugees will overwhelm basic services and cause a local collapse, resulting in cascading waves of refugees radiating from collapse and disaster epicenters. In some areas refugees will be turned back by force of arms. In other areas, racism and discrimination will come to the fore as an excuse for authoritarians to put marginalized people and dissidents in "special settlements," leaving more resources for the privileged.[2] Desperate people will be the only candidates for the dangerous and dirty manual labor required to keep industrial manufacturing going once the energy supply dwindles. Hence, those in power will consider autonomous and self-sustaining communities a threat to their labor supply, and suppress or destroy them.

Despite all of this, technological "progress" will not yet stop. For a time it will continue in fits and starts, although humanity will be split into increasingly divergent groups. Those on the bottom will be unable to meet their basic subsistence needs, while those on the top will attempt to live lives of privilege as they had in the past, even seeing some technological advancements, many of which will be intended to cement the superiority of those in power in an increasingly crowded and hostile world.

Technofascists will develop and perfect social control technologies (already currently in their early stages): autonomous drones for surveillance and assassination; microwave crowd-control devices; MRI-assisted brain scans that will allow for infallible lie detection, even mind reading and torture. There will be no substantive organized resistance in this scenario, but in each year that passes the technofascists will make themselves more and more able to destroy resistance even in its smallest expression. As time slips by, the window of opportunity for resistance will swiftly close. Technofascists of the early to mid-twenty-first century will have technology for coercion and surveillance that will make the most practiced of the Stasi or the SS look like rank amateurs. Their ability to debase humanity will make their predecessors appear saintly by comparison.

. Not all governments will take this turn, of course. But the authoritarian governments—those that will continue ruthlessly exploiting people and resources regardless of the consequences—will have more sway and more muscle, and will take resources from their neighbors and failed states as they please. There will be no one to stop them. It won't matter if you are the most sustainable eco-village on the planet if you live next door to an eternally resource-hungry fascist state.

Meanwhile, with industrial powers increasingly desperate for energy, the tenuous remaining environmental and social regulations will be cast aside. The worst of the worst, practices like drilling offshore and in wildlife refuges, and mountaintop removal for coal will become commonplace. These will be merely the dregs of prehistoric energy reserves. The drilling will only prolong the endurance of industrial civilization for a matter of months or years, but ecological damage will be long-term or permanent (as is happening in the Arctic National Wildlife Refuge). Because in our scenario there is no substantive resistance, this will all proceed unobstructed.

Investment in renewable industrial energy will also take place, although it will be belated and hampered by economic challenges, government bankruptcies, and budget cuts.[3] Furthermore, long-distance power transmission lines will be insufficient and crumbling from age. Replacing and upgrading them will prove difficult and expensive. As a result, even once in place, electric renewables will only produce a tiny fraction of the energy produced by petroleum. That electric energy will not be suitable to run the vast majority of tractors, trucks, and other vehicles or similar infrastructure.

As a consequence, renewable energy will have only a minimal moderating affect on the energy cliff. In fact, the energy invested in the new infrastructure will take years to pay itself back with electricity generated. Massive infrastructure upgrades will actually *steepen* the energy cliff by decreasing the amount of energy available for daily activities. There will be a constant struggle to allocate limited supplies of energy under successive crises. There will be some rationing to prevent riots, but most energy (regardless of the source) will go to governments, the military, corporations, and the rich.

Energy constraints will make it impossible to even attempt any full-scale infrastructure overhauls like hydrogen economies (which wouldn't solve the problem anyway). Biofuels will take off in many areas, despite the fact that they mostly have a poor ratio of energy returned on energy invested (EROEI). The EROEI will be better in tropical countries, so remaining tropical forests will be massively logged to clear land for biofuel production. (Often, forests will be logged *en masse* simply to burn for fuel.) Heavy machinery will be too expensive for most plantations, so their labor will come from slavery and serfdom under authoritarian governments and corporate feudalism. (Slavery is currently used in Brazil to log forests and produce charcoal by hand for the steel industry, after all.)[4] The global effects of biofuel production will be increases in the cost of food, increases in water and irrigation drawdown for agriculture, and worsening soil erosion. Regardless, its production will amount to only a small fraction of the liquid hydrocarbons available at the peak of civilization.

All of this will have immediate ecological consequences. The oceans, wracked by increased fishing (to compensate for food shortages) and warming-induced acidity and coral die-offs, will be mostly dead. The expansion of biofuels will destroy many remaining wild areas and global biodiversity will plummet. Tropical forests like the Amazon produce the moist climate they require through their own vast transpiration, but expanded logging and agriculture will cut transpiration and tip the balance toward permanent drought. Even where the forest is not actually cut, the drying local climate will be enough to kill it. The Amazon will turn into a desert, and other tropical forests will follow suit.

Projections vary, but it's almost certain that if the majority of the remaining fossil fuels are extracted and burned, global warming would become self-perpetuating and catastrophic. However, the worst effects will not be felt until decades into the future, once most fossil fuels have already been exhausted. By then, there will be very little energy or industrial capacity left for humans to try to compensate for the effects of global warming.

Furthermore, as intense climate change takes over, ecological remediation through perennial polycultures and forest replanting will become impossible. The heat and drought will turn forests into net carbon emitters, as northern forests die from heat, pests, and disease, and then burn in continent-wide fires that will make early twenty-first century conflagrations look minor.[5] Even intact pastures won't survive the temperature extremes as carbon is literally baked out of remaining agricultural soils.

Resource wars between nuclear states will break out. War between the US and Russia is less likely than it was in the Cold War, but ascending superpowers like China will want their piece of the global resource pie. Nuclear powers such as India and Pakistan will be densely populated and ecologically precarious; climate change will dry up major rivers previously fed by melting glaciers, and hundreds of millions of people in South Asia will live bare meters above sea level. With few resources to equip and field a mechanized army or air force, nuclear strikes will seem an increasingly effective action for desperate states.

If resource wars escalate to nuclear wars, the effects will be severe, even in the case of a "minor" nuclear war between countries like India and Pakistan. Even if each country uses only fifty Hiroshima-sized bombs as air bursts above urban centers, a nuclear winter will result.[6] Although lethal levels of fallout last only a matter of weeks, the ecological effects will be far more severe. The five megatons of smoke produced will darken the sky around the world. Stratospheric heating will destroy most of what remains of the ozone layer.[7] In contrast to the overall warming trend, a "little ice age" will begin immediately and last for several years. During that period, temperatures in major agricultural regions will routinely drop below freezing in summer. Massive and immediate starvation will occur around the world.

That's in the case of a small war. The explosive power of one hundred Hiroshima-sized bombs accounts for only 0.03 percent of the global arsenal. If a larger number of more powerful bombs are used—or if cobalt bombs are used to produce long-term irradiation and wipe out surface life—the effects will be even worse.[8] There will be few human survivors. The nuclear winter effect will be temporary, but the bombing and subsequent fires will put large amounts of carbon into the atmosphere, kill plants, and impair photosynthesis. As a result, after the ash settles, global warming will be even more rapid and worse than before.

Nuclear war or not, the long-term prospects are dim. Global warming will continue to worsen long after fossil fuels are exhausted. For the planet, the time to ecological recovery is measured in tens of millions of years, if ever.[9] As James Lovelock has pointed out, a major warming event could push the planet into a different equilibrium, one much warmer than the current one.[10] It's possible that large plants and animals might only be able to survive near the poles.[11] It's also possible that the entire planet could become essentially uninhabitable to large plants and animals, with a climate more like Venus than Earth.

All that is required for this to occur is for current trends to continue without substantive and effective resistance. All that is required for evil to succeed is for good people to do nothing. But this future is not inevitable.

LIMITED RESISTANCE

What if some forms of limited resistance were undertaken? What if there was a serious aboveground resistance movement combined with a small group of underground networks working in tandem? (This still would not be a majority movement—this is extrapolation, not fantasy.) What if those movements combined their grand strategy? The above-grounders would work to build sustainable and just communities wherever they were, and would use both direct and indirect action to try to curb the worst excesses of those in power, to reduce the burning of fossil fuels, to struggle for social and ecological justice. Meanwhile, the undergrounders would engage in limited attacks on infrastructure (often in tandem with aboveground struggles), especially energy infrastructure, to try to reduce fossil fuel consumption and overall industrial activity. The overall thrust of this plan would be to use selective attacks to accelerate collapse in a deliberate way, like shoving a rickety building.

If this scenario occurred, the first years would play out similarly. It would take time to build up resistance and to ally existing resistance groups into a larger strategy. Furthermore, civilization at the peak of its power would be too strong to bring down with only partial resistance. The years around 2011 to 2015 would still see the impact of peak oil and the beginning of an economic tailspin, but in this case there would be surgical attacks on energy infrastructure that limited new fossil fuel extraction (with a focus on the nastier practices like mountain-top removal and tar sands). Some of these attacks would be conducted by existing resistance groups (like MEND) and some by newer groups, including groups in the minority world of the rich and powerful. The increasing shortage of oil would make pipeline and infrastructure attacks more popular with militant groups of all stripes. During this period, militant groups would organize, practice, and learn.

These attacks would not be symbolic attacks. They would be serious attacks designed to be effective but timed and targeted to minimize the amount of "collateral damage" on humans. They would mostly constitute forms of sabotage. They would be intended to cut fossil fuel consumption by some 30 percent within the first few years, and more after that. There would be similar attacks on energy infrastructure like power transmission lines. Because these attacks would cause a significant but incomplete reduction in the availability of energy in many places, a massive investment in local renewable energy (and other measures like passive solar heating or better insulation in some areas) would be provoked. This would set in motion a process of political and infrastructural decentralization. It would also result in political repression and real violence targeting those resisters.

Meanwhile, aboveground groups would be making the most of the economic turmoil. There would be a growth in class consciousness and organization. Labor and poverty activists would increasingly turn to community sufficiency. Local food and self-sufficiency activists would reach out to people who have been pushed out of capitalism. The unemployed and underemployed—rapidly growing in number—would start to organize a subsistence and trade economy outside of capitalism. Mutual aid and skill sharing would be promoted. In the previous scenario, the development of these skills was hampered in part by a lack of access to land. In this scenario, however, aboveground organizers would learn from groups like the Landless Workers Movement in Latin America. Mass organization and occupation of lands would force governments to cede unused land for "victory garden"–style allotments, massive community gardens, and cooperative subsistence farms.

The situation in many third world countries could actually improve because of the global economic collapse. Minority world countries would no longer enforce crushing debt repayment and structural adjustment programs, nor would CIA goons be able to prop up "friendly" dictatorships. The decline of export-based economies would have serious consequences, yes, but it would also allow land now used for cash crops to return to subsistence farms.

Industrial agriculture would falter and begin to collapse. Synthetic fertilizers would become increasingly expensive and would be carefully conserved where they are used, limiting nutrient runoff and allowing oceanic dead zones to recover. Hunger would be reduced by subsistence farming and by the shift of small farms toward more traditional work by hand and by draft horse, but food would be more valuable and in shorter supply.

Even a 50 percent cut in fossil fuel consumption wouldn't stave off widespread hunger and die-off. As we have discussed, the vast majority of all energy used goes to nonessentials. In the US, the agricultural sector accounts for less than 2 percent of all energy use, including both direct consumption (like tractor fuel and electricity for barns and pumps) and indirect consumption (like synthetic fertilizers and pesticides).[12] That's true even though industrial agriculture is incredibly inefficient and spends something like ten calories of fossil fuel energy for every food calorie produced. Residential energy consumption accounts for only 20 percent of US total usage, with industrial, commercial, and transportation consumption making up the majority of all consumption.[13] And most of that residential energy goes into household appliances like dryers, air conditioning, and water heating for inefficiently used water. The energy used for lighting and space heating could be itself drastically reduced through trivial measures like lowering thermostats and heating the spaces people actually live in. (Most don't bother to do these now, but in a collapse situation they will do that and more.)

Only a small fraction of fossil fuel energy actually goes into basic subsistence, and even that is used inefficiently. A 50 percent decline in fossil fuel energy could be readily adapted to form a subsistence perspective (if not financial one). Remember that in North America, 40 percent of all food is simply wasted. Of course, poverty and hunger have much more to do with power over people than with the kind of power measured in watts. Even now at the peak of energy consumption, a billion people go hungry. So if people are hungry or cold because of selective militant attacks on infrastructure, that will be a direct result of the actions of those in power, not of the resisters.

In fact, even if you want humans to be able to use factories to build windmills and use tractors to help grow food over the next fifty years, forcing an immediate cut in fossil fuel consumption should be at the top of your to-do list. Right now most of the energy is being wasted on plastic junk, too-big houses for rich people, bunker buster bombs, and predator drones. The only way to ensure there is some oil left for basic survival transitions in twenty years is to ensure that it isn't being squandered now. The US military is the single biggest oil user in the world. Do you want to have to tell kids twenty years from now that they don't have enough to eat because all the energy was spent on pointless neo-colonial wars?

Back to the scenario. In some areas, increasingly abandoned suburbs (unlivable without cheap gas) would be taken over, as empty houses would become farmhouses, community centers, and clinics, or would be simply dismantled and salvaged for material. Garages would be turned into barns—most people couldn't afford gasoline anyway—and goats would be grazed in parks. Many roads would be torn up and returned to pasture or forest. These reclaimed settlements would not be high-tech. The wealthy enclaves may have their solar panels and electric windmills, but most unemployed people wouldn't be able to afford such things. In some cases these communities would become relatively autonomous. Their social practices and equality would vary based on the presence of people willing to assert human rights and social justice. People would have to resist vigorously whenever racism and xenophobia are used as excuses for injustice and authoritarianism.

Attacks on energy infrastructure would become more common as oil supplies diminish. In some cases, these attacks would be politically motivated, and in others they would be intended to tap electricity or pipelines for poor people. These attacks would steepen the energy slide initially. This would have significant economic impacts, but it would also turn the tide on population growth. The world population would peak sooner, and peak population would be smaller (by perhaps a billion) than it was in the "no resistance" scenario. Because a sharp collapse would happen earlier than it otherwise would have, there would be more intact land in the world per person, and more people who still know how to do subsistence farming.

The presence of an organized militant resistance movement would provoke a reaction from those in power. Some of them would use resistance as an excuse to seize more power to institute martial law or overt fascism. Some of them would make use of the economic and social crises rippling across the globe. Others wouldn't need an excuse.

Authoritarians would seize power where they could, and try to in almost every country. However, they would be hampered by aboveground and underground resistance, and by decentralization and the emergence of autonomous communities. In some countries, mass mobilizations would stop potential dictators. In others, the upsurge in resistance would dissolve centralized state rule, resulting in the emergence of regional confederations in some places and in warlords in others. In unlucky countries, authoritarianism would take power. The good news is that people would have resistance infrastructure in place to fight and limit the spread of authoritarians, and authoritarians would have not developed as much technology of control as they did in the "no resistance" scenario.

There would still be refugees flooding out of many areas (including urban areas). The reduction in greenhouse gas emissions caused by attacks on industrial infrastructures would reduce or delay climate catastrophe. Networks of autonomous subsistence communities would be able to accept and integrate some of these people. In the same way that rooted plants can prevent a landslide on a steep slope, the cascades of refugees would be reduced in some areas by willing communities. In other areas, the numbers of refugees would be too much to cope with effectively.[14]

The development of biofuels (and the fate of tropical forests) is uncertain. Remaining centralized states—though they may be smaller and less powerful—would still want to squeeze out energy from wherever they could. Serious militant resistance—in many cases insurgency and guerilla warfare—would be required to stop industrialists from turning tropical forests into plantations or extracting coal at any cost. In this scenario, resistance would still be limited, and it is questionable whether that level of militancy would be effectively mustered.

This means that the long-term impacts of the greenhouse effect would be uncertain. Fossil fuel burning would have to be kept to an absolute minimum to avoid a runaway greenhouse effect. That could prove very difficult.

But if a runaway greenhouse effect could be avoided, many areas could be able to recover rapidly. A return to perennial polycultures, implemented by autonomous communities, could help reverse the greenhouse effect. The oceans would look better quickly, aided by a reduction in industrial fishing and the end of the synthetic fertilizer runoff that creates so many dead zones now.

The likelihood of nuclear war would be much lower than in the "no resistance" scenario. Refugee cascades in South Asia would be diminished. Overall resource consumption would be lower, so resource wars would be less likely to occur. And militaristic states would be weaker and fewer in number. Nuclear war wouldn't be impossible, but if it did happen, it could be less severe.

There are many ways in which this scenario is appealing. But it has problems as well, both in implementation and in plausibility. One problem is with the integration of aboveground and underground action. Most aboveground environmental organizations are currently opposed to any kind of militancy. This could hamper the possibility of strategic cooperation between underground militants and aboveground groups that could mobilize greater numbers. (It would also doom our aboveground groups to failure as their record so far demonstrates.)

It's also questionable whether the cut in fossil fuel consumption described here would be sufficient to avoid runaway global warming. If runaway global warming does take place, all of the beneficial work of the abovegrounders would be wiped out. The converse problem is that a steeper decline in fossil fuel consumption would very possibly result in significant human casualties and deprivation. It's also possible that the mobilization of large numbers of people to subsistence farming in a short time is unrealistic. By the time most people are willing to take that step, it could be too late.

So while in some ways this scenario represents an ideal compromise—a win-win situation for humans and the planet—it could just as easily be a lose-lose situation without serious and timely action. That brings us to our last scenario, one of all-out resistance and attacks on infrastructure intended to guarantee the survival of a livable planet.

ALL-OUT ATTACKS ON INFRASTRUCTURE

In this final scenario, militant resistance would have one primary goal: to reduce fossil fuel consumption (and hence, all ecological damage) as immediately and rapidly as possible. A 90 percent reduction would be the ballpark target. For militants in this scenario, impacts on civilized humans would be secondary.

Here's their rationale in a nutshell: Humans aren't going to do anything in time to prevent the planet from being destroyed wholesale. Poor people are too preoccupied by primary emergencies, rich people benefit from the status quo, and the middle class (rich people by global standards) are too obsessed with their own entitlement and the technological spectacle to do anything. The risk of runaway global warming is immediate. A drop in the human population is inevitable, and fewer people will die if collapse happens sooner.

Think of it like this. We know we are in overshoot as a species. That means that a significant portion of the people now alive may have to die before we are back under carrying capacity. And that disparity is growing by the day. Every day carrying capacity is driven down by hundreds of thousands of humans, and every day the human population increases by more than 200,000.[15] The people added to the overshoot each day are needless, pointless deaths. *Delaying* collapse, they argue, is itself a form of mass murder.

Furthermore, they would argue, humans are only one species of millions. To kill millions of species for the benefit of one is insane, just as killing millions of people for the benefit of one person would be insane. And since unimpeded ecological collapse would kill off humans anyway, those species will ultimately have died for nothing, and the planet will take millions of years to recover. Therefore, those of us who care about the future of the planet have to dismantle the industrial energy infrastructure as rapidly as possible. We'll all have to deal with the social consequences as best we can. Besides, rapid collapse is ultimately good for humans—even if there is a partial die-off—because at least some people survive. And remember, the people who need the system to come down the most are the rural poor in the majority of the world: the faster the actionists can bring down industrial civilization, the better the prospects for those people and their landbases. Regardless, without immediate action, everyone dies.

In this scenario, well-organized underground militants would make coordinated attacks on energy infrastructure around the world. These would take whatever tactical form militants could muster—actions against pipelines, power lines, tankers, and refineries, perhaps using electromagnetic pulses (EMPs) to do damage. Unlike in the previous scenario, no attempt would be made to keep pace with aboveground activists. The attacks would be as persistent as the militants could manage. Fossil fuel energy availability would decline by 90 percent. Greenhouse gas emissions would plummet.

The industrial economy would come apart. Manufacturing and transportation would halt because of frequent blackouts and tremendously high prices for fossil fuels. Some, perhaps most, governments would institute martial law and rationing. Governments that took an authoritarian route would be especially targeted by militant resisters. Other states would simply fail and fall apart.

In theory, with a 90 percent reduction in fossil fuel availability, there would still be enough to aid basic survival activities like growing food, heating, and cooking. Governments and civil institutions could still attempt a rapid shift to subsistence activities for their populations, but instead, militaries and the very wealthy would attempt to suck up virtually all remaining supplies of energy. In some places, they would succeed in doing so and widespread hunger would result. In others, people would refuse the authority of those in power. Most existing large-scale institutions would simply collapse, and it would be up to local people to either make a stand for human rights and a better way of life or give in to authoritarian power. The death rate would increase, but as we have seen in examples from Cuba and Russia, civic order can still hold despite the hardships.

What happens next would depend on a number of factors. If the attacks could persist and oil extraction were kept minimal for a prolonged period, industrial civilization would be unlikely to reorganize itself. Well-guarded industrial enclaves would remain, escorting fuel and resources under arms. If martial law succeeded in stopping attacks after the first few waves (something it has been unable to do in, for example, Nigeria), the effects would be uncertain. In the twentieth century, industrial societies have recovered from disasters, as Europe did after World War II. But this would be a different situation. For most areas, there would be no outside aid. Populations would no longer be able to outrun the overshoot currently concealed by fossil fuels. That does not mean the effects would be the same everywhere; rural and traditional populations would be better placed to cope.

In most areas, reorganizing an energy-intense industrial civilization would be impossible. Even where existing political organizations persist, consumption would drop. Those in power would be unable to project force over long distances, and would have to mostly limit their activities to nearby areas. This means that, for example, tropical biofuel plantations would not be feasible. The same goes for tar sands and mountain-top removal coal mining. The construction of new large-scale infrastructure would simply not be possible.

Though the human population would decline, things would look good for virtually every other species. The oceans would begin to recover rapidly. The same goes for damaged wilderness areas. Because greenhouse emissions would have been reduced to a tiny fraction of their previous levels, runaway global warming would likely be averted. In fact, returning forests and grasslands would sequester carbon, helping to maintain a livable climate.

Nuclear war would be unlikely. Diminished populations and industrial activities would reduce competition between remaining states. Resource limitations would be largely logistical in nature, so escalating resource wars over supplies and resource-rich areas would be pointless.

This scenario, too, has its implementation and plausibility caveats. It guarantees a future for both the planet and the human species. This scenario would save trillions upon trillions upon trillions of living creatures. Yes, it would create hardship for the urban wealthy and poor, though most others would be better off immediately. It would be an understatement to call such a concept unpopular (although the militants in this scenario would argue that fewer people will die than in the case of runaway global warming or business as usual).

There is also the question of plausibility. Could enough ecologically motivated militants mobilize to enact this scenario? No doubt for many people the second, more moderate scenario seems both more appealing and more likely.

There is of course an infinitude of possible futures we could describe. We will describe one more possible future, a combination of the previous two, in which a resistance movement embarks on a strategy of Decisive Ecological Warfare.

DECISIVE ECOLOGICAL WARFARE STRATEGY

Goals

The ultimate goal of the primary resistance movement in this scenario is simply a living planet—a planet not just living, but in recovery, growing more alive and more diverse year after year. A planet on which humans live in equitable and sustainable communities without exploiting the planet or each other.

Given our current state of emergency, this translates into a more immediate goal, which is at the heart of this movement's grand strategy:

Goal 1: To disrupt and dismantle industrial civilization; to thereby remove the ability of the powerful to exploit the marginalized and destroy the planet.

This movement's second goal both depends on and assists the first:

Goal 2: To defend and rebuild just, sustainable, and autonomous human communities, and, as part of that, to assist in the recovery of the land.

To accomplish these goals requires several broad strategies involving large numbers of people in many different organizations, both aboveground and underground. The primary strategies needed in this theoretical scenario include the following:

Strategy A: Engage in direct militant actions against industrial infrastructure, especially energy infrastructure.

Strategy B: Aid and participate in ongoing social and ecological justice struggles; promote equality and undermine exploitation by those in power.

Strategy C: Defend the land and prevent the expansion of industrial logging, mining, construction, and so on, such that more intact land and species will remain when civilization does collapse.

Strategy D: Build and mobilize resistance organizations that will support the above activities, including decentralized training, recruitment, logistical support, and so on.

Strategy E: Rebuild a sustainable subsistence base for human

societies (including perennial polycultures for food) and localized, democratic communities that uphold human rights.

In describing this alternate future scenario, we should be clear about
some shorthand phrases like "actions against industrial infrastructure."
Not all infrastructure is created equal, and not all actions against infrastructure are of equal priority, efficacy, or moral acceptability to the
resistance movements in this scenario. As Derrick wrote in *Endgame*,
you can't make a moral argument for blowing up a children's hospital.
On the other hand, you can't make a moral argument against taking
out cell phone towers. Some infrastructure is easy, some is hard, and
some is harder.

On the same theme, there are many different mechanisms driving
collapse, and they are not all equal or equally desirable. In the Decisive
Ecological Warfare scenario, some of the mechanisms are intention-
ally accelerated and encouraged, while others are slowed or reduced.
For example, energy decline by decreasing consumption of fossil fuels
is a mechanism of collapse highly beneficial to the planet and (especially in the medium to long term) humans, and that mechanism is
encouraged. On the other hand, ecological collapse through habitat
destruction and biodiversity crash is also a mechanism of collapse
(albeit one that takes longer to affect humans), and that kind of collapse
is slowed or stopped whenever and wherever possible.

Collapse, in the most general terms, is a rapid loss of complexity.[16] It
is a shift toward smaller and more decentralized structures—social,
political, economic—with less social stratification, regulation, behavioral control and regimentation, and so on.[17] Major mechanisms of
collapse include (in no particular order):

- *Energy decline* as fossil fuel extraction peaks, and a growing,
  industrializing population drives down per capita availability.
- *Industrial collapse* as global economies of scale are ruined by
  increasing transport and manufacturing costs, and by economic decline.
- *Economic collapse* as global corporate capitalism is unable to
  maintain growth and basic operations.

- *Climate change* causing ecological collapse, agricultural failure, hunger, refugees, disease, and so on.
- *Ecological collapse* of many different kinds driven by resource extraction, destruction of habitat, crashing biodiversity, and climate change.
- *Disease*, including epidemics and pandemics, caused by crowded living conditions and poverty, along with bacteria diseases increasingly resistant to antibiotics.
- *Food crises* caused by the displacement of subsistence farmers and destruction of local food systems, competition for grains by factory farms and biofuels, poverty, and physical limits to food production because of drawdown.
- *Drawdown* as the accelerating consumption of finite supplies of water, soil, and oil leads to rapid exhaustion of accessible supplies.
- *Political collapse* as large political entities break into smaller groups, secessionists break away from larger states, and some states go bankrupt or simply fail.
- *Social collapse* as resource shortages and political upheaval break large, artificial group identities into smaller ones (sometimes based along class, ethnic, or regional affinities), often with competition between those groups.
- *War and armed conflict*, especially resource wars over remaining supplies of finite resources and internal conflicts between warlords and rival factions.
- *Crime and exploitation* caused by poverty and inequality, especially in crowded urban areas.
- *Refugee displacement* resulting from spontaneous disasters like earthquakes and hurricanes, but worsened by climate change, food shortages, and so on.

In this scenario, each negative aspect of the collapse of civilization has a reciprocal trend that the resistance movement encourages. The collapse of large authoritarian political structures has a countertrend of emerging small-scale participatory political structures. The collapse of global industrial capitalism has a countertrend of local systems of exchange, cooperation, and mutual aid. And so on. Generally speaking, in this alternate future, a small number of underground people bring down the big bad structures, and a large number of aboveground people cultivate the little good structures.

In his book *The Collapse of Complex Societies*, Joseph Tainter argues that a major mechanism for collapse has to do with societal complexity. Complexity is a general term that includes the number of different jobs or roles in society (e.g., not just healers but epidemiologists, trauma surgeons, gerontologists, etc.), the size and complexity of political structures (e.g., not just popular assemblies but vast sprawling bureaucracies), the number and complexity of manufactured items and technology (e.g., not just spears, but many different calibers and types of bullets), and so on. Civilizations tend to try to use complexity to address problems, and as a result their complexity increases over time.

But complexity has a cost. The decline of a civilization begins when the costs of complexity begin to exceed the benefits—in other words, when increased complexity begins to offer declining returns. At that point, individual people, families, communities, and political and social subunits have a disincentive to participate in that civilization. The complexity keeps increasing, yes, but it keeps getting more expensive. Eventually the ballooning costs force that civilization to collapse, and people fall back on smaller and more local political organizations and social groups.

Part of the job of the resistance movement is to increase the cost and decrease the returns of empire-scale complexity. This doesn't require instantaneous collapse or global dramatic actions. Even small actions can increase the cost of complexity and accelerate the good parts of collapse while tempering the bad.

Part of Tainter's argument is that modern society won't collapse in the same way as old societies, because complexity (through, for example, large-scale agriculture and fossil fuel extraction) has become the physical underpinning of human life rather than a side benefit. Many historical societies collapsed when people returned to villages and less complex traditional life. They chose to do this. Modern people won't do that, at least not on a large scale, in part because the villages are gone, and traditional ways of life are no longer directly accessible to them. This means that people in modern civilization are in a bind, and many will continue to struggle for industrial civilization even when continuing it is obviously counterproductive. Under a Decisive Ecological Warfare scenario, aboveground activists facilitate this aspect of collapse by developing alternatives that will ease the pressure and encourage people to leave industrial capitalism by choice.

◪ ◪ ◪

There's something admirable about the concept of protracted popular warfare that was used in China and Vietnam. It's an elegant idea, if war can ever be described in such terms; the core idea is adaptable and applicable even in the face of major setbacks and twists of fate.

But protracted popular warfare as such doesn't apply to the particular future we are discussing. The people in that scenario will never have the numbers that protracted popular warfare requires. But they will also face a different kind of adversary, for which different tactics are applicable. So they will take the essential idea of protracted popular warfare and apply it to their own situation—that of needing to save their planet, to bring down industrial civilization and keep it down. And they will devise a new grand strategy based on a simple continuum of steps that flow logically one after the other.

In this alternate future scenario, Decisive Ecological Warfare has four phases that progress from the near future through the fall of industrial civilization. The first phase is *Networking & Mobilization*. The second phase is *Sabotage & Asymmetric Action*. The third phase is *Systems Disruption*. And the fourth and final phase is *Decisive Dismantling of Infrastructure*.

Each phase has its own objectives, operational approaches, and organizational requirements. There's no distinct dividing line between the phases, and different regions progress through the phases at different times. These phases emphasize the role of militant resistance networks. The aboveground building of alternatives and revitalization of human communities happen at the same time. But this does not require the same strategic rigor; rebuilding healthy human communities with a subsistence base must simply happen as fast as possible, everywhere, with timetables and methods suited to the region. This scenario's militant resisters, on the other hand, need to share some grand strategy to succeed.

PHASE I: NETWORKING & MOBILIZATION

Preamble: In phase one, resisters focus on organizing themselves into networks and building cultures of resistance to sustain those networks. Many sympathizers or potential recruits are unfamiliar with serious resistance strategy and action, so efforts are taken to spread that information. But key in this phase is actually forming the above- and underground organizations (or at least nuclei) that will carry out organizational recruitment and decisive action. Security culture and resistance culture are not very well developed at this point, so extra efforts are made to avoid sloppy mistakes that would lead to arrests, and to dissuade informers from gathering or passing on information.

Training of activists is key in this phase, especially through low-risk (but effective) actions. New recruits will become the combatants, cadres, and leaders of later phases. New activists are enculturated into the resistance ethos, and existing activists drop bad or counterproductive habits. This is a time when the resistance movement gets organized and gets serious. People are putting their individual needs and conflicts aside in order to form a movement that can fight to win.

In this phase, isolated people come together to form a vision and strategy for the future, and to establish the nuclei of future organizations. Of course, networking occurs with resistance-oriented organizations that already exist, but most mainstream organizations are not willing to adopt positions of militancy or intransigence with regard to those in power or the crises they face. If possible, they should be encouraged to take positions more in line with the scale of the problems at hand.

This phase is already underway, but a great deal of work remains to be done.

Objectives:
- To build a culture of resistance, with all that entails.
- To build aboveground and underground resistance networks, and to ensure the survival of those networks.

Operations:

- Operations are generally lower-risk actions, so that people can be trained and screened, and support networks put in place. These will fall primarily into the sustaining and shaping categories.
- Maximal recruitment and training is very important at this point. The earlier people are recruited, the more likely they are to be trustworthy and the longer time is available to screen them for their competency for more serious action.
- Communications and propaganda operations are also required for outreach and to spread information about useful tactics and strategies, and on the necessity for organized action.

Organization:

- Most resistance organizations in this scenario are still diffuse networks, but they begin to extend and coalesce. This phase aims to build organization.

PHASE II: SABOTAGE & ASYMMETRIC ACTION

Preamble: In this phase, the resisters might attempt to disrupt or disable particular targets on an opportunistic basis. For the most part, the required underground networks and skills do not yet exist to take on multiple larger targets. Resisters may go after particularly egregious targets—coal-fired power plants or exploitative banks. At this phase, the resistance focus is on practice, probing enemy networks and security, and increasing support while building organizational networks. In this possible future, underground cells do not attempt to provoke overwhelming repression beyond the ability of what their nascent networks can cope with. Furthermore, when serious repression and setbacks do occur, they retreat toward the earlier phase with its emphasis on organization and survival. Indeed, major setbacks probably do happen at this phase, indicating a lack of basic rules and structure and signaling the need to fall back on some of the priorities of the first phase.

The resistance movement in this scenario understands the importance of decisive action. Their emphasis in the first two phases has not been on direct action, but not because they are holding back. It's because they are working as well as they damned well can, but doing so while putting one foot in front of the other. They know that the planet (and the future) need their action, but understand that it won't benefit from foolish and hasty action, or from creating problems for which they are not yet prepared. That only leads to a morale whiplash and disappointment. So their movement acts as seriously and swiftly and decisively as it can, but makes sure that it lays the foundation it needs to be truly effective.

The more people join that movement, the harder they work, and the more driven they are, the faster they can progress from one phase to the next.

In this alternate future, aboveground activists in particular take on several important tasks. They push for acceptance and normalization of more militant and radical tactics where appropriate. They vocally support sabotage when it occurs. More moderate advocacy groups use the occurrence of sabotage to criticize those in power for failing to take action on critical issues like climate change (rather than criticizing the saboteurs). They argue that sabotage would not be necessary if civil society would make a reasonable response to social and ecological problems, and use the opportunity and publicity to push solutions to the problems. They do not side with those in power against the saboteurs, but argue that the situation is serious enough to make such action legitimate, even though they have personally chosen a different course.

At this point in the scenario, more radical and grassroots groups continue to establish a community of resistance, but also establish discrete organizations and parallel institutions. These institutions establish themselves and their legitimacy, make community connections, and particularly take steps to found relationships outside of the traditional "activist bubble." These institutions also focus on emergency and disaster preparedness, and helping people cope with impending collapse.

Simultaneously, aboveground activists organize people for civil disobedience, mass confrontation, and other forms of direct action where appropriate.

Something else begins to happen: aboveground organizations establish coalitions, confederations, and regional networks, knowing that there will be greater obstacles to these later on. These confederations maximize the potential of aboveground organizing by sharing materials, knowledge, skills, learning curricula, and so on. They also plan strategically themselves, engaging in persistent planned campaigns instead of reactive or crisis-to-crisis organizing.

Objectives:

- Identify and engage high-priority individual targets. These targets are chosen by these resisters because they are especially attainable or for other reasons of target selection.
- Give training and real-world experience to cadres necessary to take on bigger targets and systems. Even decisive actions are limited in scope and impact at this phase, although good target selection and timing allows for significant gains.
- These operations also expose weak points in the system, demonstrate the feasibility of material resistance, and inspire other resisters.
- Publically establish the rationale for material resistance and confrontation with power.
- Establish concrete aboveground organizations and parallel institutions.

Operations:

- Limited but increasing decisive operations, combined with growing sustaining operations (to support larger and more logistically demanding organizations) and continued shaping operations.
- In decisive and supporting operations, these hypothetical resisters are cautious and smart. New and unseasoned cadres have a tendency to be overconfident, so to compensate they pick only operations with certain outcomes; they know that in this stage they are still building toward the bigger actions that are yet to come.

Organization:

- Requires underground cells, but benefits from larger underground networks. There is still an emphasis on recruitment at this point. Aboveground networks and movements are proliferating as much as they can, especially since the work to come requires significant lead time for developing skills, communities, and so on.

PHASE III: SYSTEMS DISRUPTION

Preamble: In this phase resisters step up from individual targets to address entire industrial, political, and economic systems. Industrial systems disruption requires underground networks organized in a hierarchal or paramilitary fashion. These larger networks emerge out of the previous phases with the ability to carry out multiple simultaneous actions.

Systems disruption is aimed at identifying key points and bottlenecks in the adversary's systems (electrical, transport, financial, and so on) and engaging them to collapse those systems or reduce their functionality. This is not a one-shot deal. Industrial systems are big and can be fragile, but they are sprawling rather than monolithic. Repairs are attempted. The resistance members understand that. Effective systems disruption requires planning for continued and coordinated actions over time.

In this scenario, the aboveground doesn't truly gain traction as long as there is business as usual. On the other hand, as global industrial and economic systems are increasingly disrupted (because of capitalist-induced economic collapse, global climate disasters, peak oil, peak soil, peak water, or for other reasons) support for resilient local communities increases. Failures in the delivery of electricity and manufactured goods increases interest in local food, energy, and the like. These disruptions also make it easier for people to cope with full collapse in the long term—short-term loss, long-term gain, even where humans are concerned.

Dimitry Orlov, a major analyst of the Soviet collapse, explains that the dysfunctional nature of the Soviet system prepared people for its eventual disintegration. In contrast, a smoothly functioning industrial economy causes a false sense of security so that people are unprepared, worsening the impact. "After collapse, you regret not having an unreliable retail segment, with shortages and long bread lines, because then people would have been forced to learn to shift for themselves instead of standing around waiting for somebody to come and feed them."[18]

Aboveground organizations and institutions are well-established by this phase of this alternate scenario. They continue to push for reforms, focusing on the urgent need for justice, relocalization, and resilient communities, given that the dominant system is unfair, unreliable, and unstable.

Of course, in this scenario the militant actions that impact daily life provoke a backlash, sometimes from parts of the public, but especially from authoritarians on every level. The aboveground activists are the frontline fighters against authoritarianism. They are the only ones who can mobilize the popular groundswell needed to prevent fascism.

Furthermore, aboveground activists use the disrupted systems as an opportunity to strengthen local communities and parallel institutions. Mainstream people are encouraged to swing their support to participatory local alternatives in the economic, political, and social spheres. When economic turmoil causes unemployment and hyperinflation, people are employed locally for the benefit of their community and the land. In this scenario, as national governments around the world increasingly struggle with crises (like peak oil, food shortages, climate chaos, and so on) and increasingly fail to provide for people, local and directly democratic councils begin to take over administration of basic and emergency services, and people redirect their taxes to those local entities (perhaps as part of a campaign of general noncooperation against those in power). This happens in conjunction with the community emergency response and disaster preparedness measures already undertaken.

In this scenario, whenever those in power try to increase exploitation or authoritarianism, aboveground resisters call for people to withdraw support from those in power, and divert it to local, democratic political bodies. Those parallel institutions can do a better job than those in power. The cross demographic relationships established in previous phases help to keep those local political structures accountable, and to rally support from many communities.

Throughout this phase, strategic efforts are made to augment existing stresses on economic and industrial systems caused by peak oil, financial instability, and related factors. The resisters think of themselves as pushing on a rickety building that's already starting to lean. Indeed, in this scenario many systems disruptions come from within the system itself, rather than from resisters.

This phase accomplishes significant and decisive gains. Even if the main industrial and economic systems have not completely collapsed, prolonged disruption means a reduction in ecological impact; great news for the planet, and for humanity's future survival. Even a 50 percent decrease in industrial consumption or greenhouse gas emissions is a massive victory (especially considering that emissions have continued to rise in the face of all environmental activism so far), and that buys resisters—and everyone else—some time.

In the most optimistic parts of this hypothetical scenario, effective resistance induces those in power to negotiate or offer concessions. Once the resistance movement demonstrates the ability to use real strategy and force, it can't be ignored. Those in power begin to knock down the doors of mainstream activists, begging to negotiate changes that would co-opt the resistance movements' cause and reduce further actions.

In this version of the future, however, resistance groups truly begin to take the initiative. They understand that for most of the history of civilization, those in power have retained the initiative, forcing resistance groups or colonized people to stay on the defensive, to respond to attacks, to be constantly kept off balance. However, peak oil and systems disruption has caused a series of emergencies for those in power; some caused by organized resistance groups, some caused by civil unrest and shortages, and some caused by the social and ecological consequences of centuries—millennia—of exploitation. For perhaps the first time in history, those in power are globally off balance and occupied by worsening crisis after crisis. This provides a key opportunity for resistance groups, and autonomous cultures and communities, to seize and retain the initiative.

Objectives:

- Target key points of specific industrial and economic systems to disrupt and disable them.
- Effect a measurable decrease in industrial activity and industrial consumption.
- Enable concessions, negotiations, or social changes if applicable.
- Induce the collapse of particular companies, industries, or economic systems.

Operations:

- Mostly decisive and sustaining, but shaping where necessary for systems disruption. Cadres and combatants should be increasingly seasoned at this point, but the onset of decisive and serious action will mean a high attrition rate for resisters. There's no point in being vague; the members of the resistance in this alternate future who are committed to militant resistance go in expecting that they will either end up dead or in jail. They know that anything better than that was a gift to be won through skill and luck.

Organization:

- Heavy use of underground networks required; operational coordination is a prerequisite for effective systems disruption.
- Recruitment is ongoing at this point; especially to recruit auxiliaries and to cope with losses due to attrition. However, during this phase there are multiple serious attempts at infiltration. The infiltrations are not as successful as they might have been, because underground networks have recruited heavily in previous stages (before large-scale action) to ensure the presence of a trusted group of leaders and cadres who form the backbone of the networks.
- Aboveground organizations are able to mobilize extensively because of various social, political, and material crises.
- At this point, militant resisters become concerned about backlash from people who *should* be on their side, such as

many liberals, especially as those in power put pressure on aboveground activists.

PHASE IV: DECISIVE DISMANTLING OF INFRASTRUCTURE

Preamble: Decisive dismantling of infrastructure goes a step beyond systems disruption. The intent is to permanently dismantle as much of the fossil fuel–based industrial infrastructure as possible. This phase is the last resort; in the most optimistic projection, it would not be necessary: converging crises and infrastructure disruption would combine with vigorous aboveground movements to force those in power to accept social, political, and economic change; reductions in consumption would combine with a genuine and sincere attempt to transition to a sustainable culture.

But this optimistic projection is not probable. It is more likely that those in power (and many everyday people) will cling more to civilization even as it collapses. And likely, they will support authoritarianism if they think it will maintain their privilege and their entitlement.

The key issue—which we've come back to again and again—is time. We will soon reach (if we haven't already reached) the trigger point of irreversible runaway global warming. The systems disruption phase of this hypothetical scenario offers selectivity. Disruptions in this scenario are engineered in a way that shifts the impact toward industry and attempts to minimize impacts on civilians. But industrial systems are heavily integrated with civilian infrastructure. If selective disruption doesn't work soon enough, some resisters may conclude that all-out disruption is required to stop the planet from burning to a cinder.

The difference between phases III and IV of this scenario may appear subtle, since they both involve, on an operational level, coordinated actions to disrupt industrial systems on a large scale. But phase III requires some time to work—to weaken the system, to mobilize people and organizations, to build on a series of disruptive actions. Phase III also gives "fair warning" for regular people to prepare. Furthermore, phase III gives time for the resistance to develop itself logistically and organizationally, which is required to proceed to phase IV. The difference between the two phases is capacity and restraint. For resisters in this scenario to proceed from phase III to phase IV, they need two things: the organizational capacity to take on the scope of action required under phase IV, and the certainty that there is no longer any point in waiting for societal reforms to succeed on their own timetable.

In this scenario, both of those phases save lives, human and non-human alike. But if large-scale aboveground mobilization does not happen once collapse is underway, phase IV becomes the most effective way to save lives.

Imagine that you are riding in a streetcar through a city crowded with pedestrians. Inside the streetcar are the civilized humans, and outside is all the nonhuman life on the planet, and the humans who are not civilized, or who do not benefit from civilization, or who have yet to be born. Needless to say, those outside far outnumber the few of you inside the streetcar. But the driver of the streetcar is in a hurry, and is accelerating as fast as he can, plowing through the crowds, maiming and killing pedestrians en masse. Most of your fellow passengers don't seem to particularly care; they've got somewhere to go, and they're glad to be making progress regardless of the cost.

Some of the passengers seem upset by the situation. If the driver keeps accelerating, they observe, it's possible that the streetcar will crash and the passengers will be injured. Not to worry, one man tells them. His calculations show that the bodies piling up in front of the streetcar will eventually slow the vehicle and cause it to safely come to a halt. Any intervention by the passengers would be reckless, and would surely provoke a reprimand from the driver. Worse, a troublesome passenger might be kicked off the streetcar and later run over by it.

You, unlike most passengers, are more concerned by the constant carnage outside than by the future safety of the streetcar passengers. And you know you have to do something. You could try to jump out the window and escape, but then the streetcar would plow on through the crowd, and you would lose any chance to intervene. So you decide to try to sabotage the streetcar from the inside, to cut the electrical wires, or pull up the flooring and activate the brakes by hand, or derail it, or do whatever you can.

As soon as the other passengers realize what you are doing, they'll try to stop you, and maybe kill you. You have to decide whether you are going to stop the streetcar slowly or speedily. The streetcar is racing along so quickly now that if you stop it suddenly, it may fling the passengers against the seats in front of them or down the aisle. It may kill some of them. But if you stop it slowly, who knows how many innocent people will be struck by the streetcar while it is decelerating? And if you just slow it down, the driver may be able to repair the damage and get the streetcar going again.

So what do you do? If you choose to stop the streetcar as quickly as possible, then you have made the same choice as those who would implement phase IV. You've made the decision that stopping the destruction as rapidly as possible is more important than any particular program of reform. Of course, even in stopping the destruction as rapidly as possible, you can still take measures to reduce casualties on board the streetcar. You can tell people to sit down or buckle up or brace themselves for impact. Whether they will listen to you is another story, but that's their responsibility, not yours.

It's important to not misinterpret the point of phase IV of this alternate future scenario. The point is *not* to cause human casualties. The point is to stop the destruction of the planet. The enemy is not the civilian population—or any population at all—but a sociopathological sociopolitical and economic *system*. Ecological destruction on this planet is primarily caused by industry and capitalism; the issue of population is tertiary at best. The point of collapsing industrial infrastructure in this scenario is not to harm humans any more than the point of stopping the streetcar is to harm the passengers. The point is to reduce the damage as quickly as possible, and in doing so to account for the harm the dominant culture is doing to *all* living creatures, past and future.

This is not an easy phase for the abovegrounders. Part of their job in this scenario is also to help demolish infrastructure, but they are mostly demolishing exploitative political and economic infrastructure, not physical infrastructure. In general, they continue to do what they did in the previous phase, but on a larger scale and for the long term. Public support is directed to local, democratic, and just political and economic systems. Efforts are undertaken to deal with emergencies and cope with the nastier parts of collapse.

Objectives:

- Dismantle the critical physical infrastructure required for industrial civilization to function.
- Induce widespread industrial collapse, beyond any economic or political systems.
- Use continuing and coordinated actions to hamper repairs and replacement.

Operations:

- Focus almost exclusively on decisive and sustaining operations.

Organization:

- Requires well-developed militant underground networks.

IMPLEMENTING DECISIVE ECOLOGICAL WARFARE

It's important to note that, as in the case of protracted popular warfare, Decisive Ecological Warfare is not necessarily a linear progression. In this scenario resisters fall back on previous phases as necessary. After major setbacks, resistance organizations focus on survival and networking as they regroup and prepare for more serious action. Also, resistance movements progress through each of the phases, and then recede in reverse order. That is, if global industrial infrastructure has been successfully disrupted or fragmented (phase IV) resisters return to systems disruption on a local or regional scale (phase III). And if that is successful, resisters move back down to phase II, focusing their efforts on the worst remaining targets.

And provided that humans don't go extinct, even this scenario will require some people to stay at phase I indefinitely, maintaining a culture of resistance and passing on the basic knowledge and skills necessary to fight back for centuries and millennia.

The progression of Decisive Ecological Warfare could be compared to ecological succession. A few months ago I visited an abandoned quarry, where the topsoil and several layers of bedrock had been stripped and blasted away, leaving a cubic cavity several stories deep in the limestone. But a little bit of gravel or dust had piled up in one corner, and some mosses had taken hold. The mosses were small, but they required very little in the way of water or nutrients (like many of the shoestring affinity groups I've worked with). Once the mosses had grown for a few seasons, they retained enough soil for grasses to grow.

Quick to establish, hardy grasses are often among the first species to reinhabit any disturbed land. In much the same way, early resistance organizations are generalists, not specialists. They are robust and rapidly spread and reproduce, either spreading their seeds aboveground or creating underground networks of rhizomes.

The grasses at the quarry built the soil quickly, and soon there was soil for wildflowers and more complex organisms. In much the same way, large numbers of simple resistance organizations help to establish communities of resistance, cultures of resistance, that can give rise to more complex and more effective resistance organizations.

◙ ◙ ◙

The hypothetical actionists who put this strategy into place are able to intelligently move from one phase to the next: identifying when the correct elements are in place, when resistance networks are sufficiently mobilized and trained, and when external pressures dictate change. In the US Army's field manual on operations, General Eric Shinseki argues that the rules of strategy "require commanders to master transitions, to be adaptive. Transitions—deployments, the interval between initial operation and sequels, consolidation on the objective, forward passage of lines—sap operational momentum. Mastering transitions is the key to maintaining momentum and winning decisively."

This is particularly difficult to do when resistance does not have a central command. In this scenario, there is no central means of dispersing operational or tactical orders, or effectively gathering precise information about resistance forces and allies. Shinseki continues: "This places a high premium on readiness—well trained Soldiers; adaptive leaders who understand our doctrine; and versatile, agile, and lethal formations." People resisting civilization in this scenario are not concerned with "lethality" so much as effectiveness, but the general point stands.

Resistance to civilization is inherently decentralized. That goes double for underground groups which have minimal contact with others. To compensate for the lack of command structure, a general grand strategy in this scenario becomes widely known and accepted. Furthermore, loosely allied groups are ready to take action whenever the strategic situation called for it. These groups are prepared to take advantage of crises like economic collapses.

Under this alternate scenario, underground organizing in small cells has major implications for applying the principles of war. The ideal entity for taking on industrial civilization would have been a large, hierarchal paramilitary network. Such a network could have engaged in the training, discipline, and coordinated action required to implement decisive militant action on a continental scale. However, for practical reasons, a single such network never arises. Similar arrangements in the history of resistance struggle, such as the IRA or various territory-controlling insurgent groups, happened in the absence of the modern surveillance state and in the presence of a well-developed culture of resistance and extensive opposition to the occupier.

Although underground cells can still form out of trusted peers along kinship lines, larger paramilitary networks are more difficult to form in a contemporary anticivilization context. First of all, the proportion of potential recruits in the general population is smaller than in any anticolonial or antioccupation resistance movements in history. So it takes longer and is more difficult to expand existing underground networks. The option used by some resistance groups in Occupied France was to ally and connect existing cells. But this is inherently difficult and dangerous. Any underground group with proper cover would be invisible to another group looking for allies (there are plenty of stories from the end of the war of resisters living across the hall from each other without having realized each other's affiliation). And in a panopticon, exposing yourself to unproven allies is a risky undertaking.

A more plausible underground arrangement in this scenario is for there to have been a composite of organizations of different sizes, a few larger networks with a number of smaller autonomous cells that aren't directly connected through command lines. There are indirect connections or communications via cutouts, but those methods are rarely consistent or reliable enough to permit coordinated simultaneous actions on short notice.

Individual cells rarely have the numbers or logistics to engage in multiple simultaneous actions at different locations. That job falls to the paramilitary groups, with cells in multiple locations, who have the command structure and the discipline to properly carry out network disruption. However, autonomous cells maintain readiness to engage in opportunistic action by identifying in advance a selection of appropriate local targets and tactics. Then once a larger simultaneous action happened (causing, say, a blackout), autonomous cells take advantage of the opportunity to undertake their own actions, within a few hours. In this way unrelated cells engage in something close to simultaneous attacks, maximizing their effectiveness. Of course, if decentralized groups frequently stage attacks in the wake of larger "trigger actions," the corporate media may stop broadcasting news of attacks to avoid triggering more. So, such an approach has its limits, although large-scale effects like national blackouts can't be suppressed in the news (and in systems disruption, it doesn't really matter what caused a blackout in the first place, because it's still an opportunity for further action).

◩ ◩ ◩

When we look at some struggle or war in history, we have the benefit of hindsight to identify flaws and successes. This is how we judge strategic decisions made in World War II, for example, or any of those who have tried (or not) to intervene in historical holocausts. Perhaps it would be beneficial to imagine some historians in the distant future— assuming humanity survives—looking back on the alternate future just described. Assuming it was generally successful, how might they analyze its strengths and weaknesses?

For these historians, phase IV is controversial, and they know it had been controversial among resisters at the time. Even resisters who agreed with militant actions against industrial infrastructure hesitated when contemplating actions with possible civilian consequences. That comes as no surprise, because members of this resistance were driven by a deep respect and care for all life. The problem is, of course, that members of this group knew that if they failed to stop this culture from killing the planet, there would be far more gruesome civilian consequences.

A related moral conundrum confronted the Allies early in World War II, as discussed by Eric Markusen and David Kopf in their book *The Holocaust and Strategic Bombing: Genocide and Total War in the Twentieth Century*. Markusen and Kopf write that: "At the beginning of World War II, British bombing policy was rigorously discriminating—even to the point of putting British aircrews at great risk. Only obvious military targets removed from population centers were attacked, and bomber crews were instructed to jettison their bombs over water when weather conditions made target identification questionable. Several factors were cited to explain this policy, including a desire to avoid provoking Germany into retaliating against non-military targets in Britain with its then numerically superior air force."[19]

Other factors included concerns about public support, moral considerations in avoiding civilian casualties, the practice of the "Phoney War" (a declared war on Germany with little real combat), and a small air force which required time to build up. The parallels between the actions of the British bombers and the actions of leftist militants from the Weather Underground to the ELF are obvious.

The problem with this British policy was that it simply didn't work. Germany showed no such moral restraint, and British bombing crews were taking greater risks to attack less valuable targets. By February of 1942, bombing policy changed substantially. In fact, Bomber Command began to deliberately target enemy civilians and civilian morale—particularly that of industrial workers—especially by destroying homes around target factories in order to "dehouse" workers. British strategists believed that in doing so they could sap Germany's will to fight. In fact, some of the attacks on civilians were intended to "punish" the German populace for supporting Hitler, and some strategists believed that, after sufficient punishment, the population would rise up and depose Hitler to save themselves. Of course, this did not work; it almost never does.

So, this was one of the dilemmas faced by resistance members in this alternate future scenario: while the resistance abhorred the notion of actions affecting civilians—even more than the British did in early World War II—it was clear to them that in an industrial nation the "civilians" and the state are so deeply enmeshed that any impact on one will have some impact on the other.

Historians now believe that Allied reluctance to attack early in the war may have cost many millions of civilian lives. By failing to stop Germany early, they made a prolonged and bloody conflict inevitable. General Alfred Jodl, the German Chief of the Operations Staff of the Armed Forces High Command, said as much during his war crimes trial at Nuremburg: "[I]f we did not collapse already in the year 1939 that was due only to the fact that during the Polish campaign, the approximately 110 French and British divisions in the West were held completely inactive against the 23 German divisions."[20]

Many military strategists have warned against piecemeal or half measures when only total war will do the job. In his book *Grand Strategy: Principles and Practices,* John M. Collins argues that timid attacks may strengthen the resolve of the enemy, because they constitute a provocation but don't significantly damage the physical capability or morale of the occupier. "Destroying the enemy's resolution to resist is far more important than crippling his material capabilities . . . studies of cause and effect tend to confirm that violence short of total devastation may *amplify* rather than erode a people's determination."[21] Consider, though, that in this 1973 book Collins may underestimate the importance of technological infrastructure and decisive strikes on them. (He advises elsewhere in the book that computers "are of limited utility."[22])

Other strategists have prioritized the material destruction over the adversary's "will to fight." Robert Anthony Pape discusses the issue in *Bombing to Win,* in which he analyzes the effectiveness of strategic bombing in various wars. We can wonder in this alternate future scenario if the resistors attended to Pape's analysis as they weighed the benefits of phase III (selective actions against particular networks and systems) vs. phase IV (attempting to destroy as much of the industrial infrastructure as possible).

Specifically, Pape argues that targeting an entire economy may be more effective than simply going after individual factories or facilities:

Strategic interdiction can undermine attrition strategies, either by attacking weapons plants or by smashing the industrial base as a whole, which in turn reduces military production. Of the two, attacking weapons plants is the less effective. Given the substitution capacities of modern industrial economies, "war"production is highly fungible over a period of months. Production can be maintained in the short term by running down stockpiles and in the medium term by conservation and substitution of alternative materials or processes. In addition to economic adjustment, states can often make doctrinal adjustments.[23]

This analysis is poignant, but it also demonstrates a way in which the goals of this alternate scenario's strategy differed from the goals of strategic bombing in historical conflicts. In the Allied bombing campaign (and in other wars where strategic bombing was used), the strategic bombing coincided with conventional ground, air, and naval battles. Bombing strategists were most concerned with choking off enemy supplies to the battlefield. Strategic bombing alone was not meant to win the war; it was meant to support conventional forces in battle. In contrast, in this alternate future, a significant decrease in industrial production would itself be a great success.

The hypothetical future historians perhaps ask, "Why not simply go after the worst factories, the worst industries, and leave the rest of the economy alone?" Earlier stages of Decisive Ecological Warfare did involve targeting particular factories or industries. However, the resistors knew that the modern industrial economy was so thoroughly integrated that anything short of general economic distruption was unlikely to have lasting effect.

This, too, is shown by historical attempts to disrupt economies. Pape continues, "Even when production of an important weapon system is seriously undermined, tactical and operational adjustments may allow other weapon systems to substitute for it. . . . As a result, efforts to remove the critical component in war production generally fail." For example, Pape explains, the Allies carried out a bombing campaign on German aircraft engine plants. But this was not a decisive factor in the struggle for air superiority. Mostly, the Allies defeated the *Luftwaffe* because they shot down and killed so many of Germany's best pilots.

Another example of compensation is the Allied bombing of German ball bearing plants. The Allies were able to reduce the German production of ball bearings by about 70 percent. But this did not force a corresponding decrease in German tank forces. The Germans were able to compensate in part by designing equipment that required fewer bearings. They also increased their production of infantry antitank weapons. Early in the war, Germany was able to compensate for the destruction of factories in part because many factories were running only one shift. They were not using their existing industrial capacity to its fullest. By switching to double or triple shifts, they were able to (temporarily) maintain production.

Hence, Pape argues that war economies have no particular point of collapse when faced with increasing attacks, but can adjust incrementally to decreasing supplies. "Modern war economies are not brittle. Although individual plants can be destroyed, the opponent can reduce the effects by dispersing production of important items and stockpiling key raw materials and machinery. Attackers never anticipate all the adjustments and work-arounds defenders can devise, partly because they often rely on analysis of peacetime economies and partly because intelligence of the detailed structure of the target economy is always incomplete."[24] This is a valid caution against overconfidence, but the resisters in this scenario recognized that his argument was not fully applicable to their situation, in part for the reasons we discussed earlier, and in part because of reasons that follow.

Military strategists studying economic and industrial disruption are usually concerned specifically with the production of war materiel and its distribution to enemy armed forces. Modern war economies are economies of *total war* in which all parts of society are mobilized and engaged in supporting war. So, of course, military leaders can compensate for significant disruption; they can divert materiel or rations from civilian use or enlist civilians and civilian infrastructure for military purposes as they please. This does not mean that overall production is unaffected (far from it), simply that military production does not decline as much as one might expect under a given onslaught.

Resisters in this scenario had a different perspective on compensation measures than military strategists. To understand the contrast, pretend that a military strategist and a militant ecological strategist both want to blow up a fuel pipeline that services a major industrial area. Let's say the pipeline is destroyed and the fuel supply to industry is drastically cut. Let's say that the industrial area undertakes a variety of typical measures to compensate—conservation, recycling, efficiency measures, and so on. Let's say they are able to keep on producing insulation or refrigerators or clothing or whatever it is they make, in diminished numbers and using less fuel. They also extend the lifespan of their existing refrigerators or clothing by repairing them. From the point of view of the military strategist, this attack has been a failure— it has a negligible effect on materiel availability for the military. But from the perspective of the militant ecologist, this is a victory. Ecological damage is reduced, and with very few negative effects on civilians. (Indeed, some effects would be directly beneficial.)

And modern economies in general *are* brittle. Military economies mobilize resources and production by any means necessary, whether that means printing money or commandeering factories. They are economies of crude necessity. Industrial economies, in contrast, are economies of luxury. They mostly produce things that people don't need. Industrial capitalism thrives on manufacturing desire as much as on manufacturing products, on selling people disposable plastic garbage, extra cars, and junk food. When capitalist economies hit hard times, as they did in the Great Depression, or as they did in Argentina a decade ago, or as they have in many places in many times, people fall back on necessities, and often on barter systems and webs of mutual aid. They fall back on community and household economies, economies of necessity that are far more resilient than industrial capitalism, and even more robust than war economies.

Nonetheless, Pape makes an important point when he argues, "Strategic interdiction is most effective when attacks are against the economy as a whole. The most effective plan is to destroy the transportation network that brings raw materials and primary goods to manufacturing centers and often redistributes subcomponents among various industries. Attacking national electric power grids is not effective because industrial facilities commonly have their own backup power generation. Attacking national oil refineries to reduce backup power generators typically ignores the ability of states to reduce consumption through conservation and rationing." Pape's analysis is insightful, but again it's important to understand the differences between his premises and goals, and the premises and goals of Decisive Ecological Warfare.

The resisters in the DEW scenario had the goals of reducing consumption and reducing industrial activity, so it didn't matter to them that some industrial facilities had backup generators or that states engaged in conservation and rationing. They believed it was a profound ecological victory to cause factories to run on reduced power or for nationwide oil conservation to have taken place. They remembered that in the whole of its history, the mainstream environmental movement was never even able to stop the *growth* of fossil fuel consumption. To actually reduce it was unprecedented.[25]

No matter whether we are talking about some completely hypothetical future situation or the real world right now, the progress of peak oil will also have an effect on the relative importance of different transportation networks. In some areas, the importance of shipping imports will increase because of factors like the local exhaustion of oil. In others, declining international trade and reduced economic activity will make shipping less important. Highway systems may have reduced usage because of increasing fuel costs and decreasing trade. This reduced traffic will leave more spare capacity and make highways less vulnerable to disruption. Rail traffic—a very energy-efficient form of transport—is likely to increase in importance. Furthermore, in many areas, railroads have been removed over a period of several decades, so that remaining lines are even now very crowded and close to maximum capacity.

Back to the alternative future scenario: In most cases, transportation networks were not the best targets. Road transportation (by far the most important form in most countries) is highly redundant. Even rural parts of well-populated areas are crisscrossed by grids of county roads, which are slower than highways, but allow for detours.

In contrast, targeting energy networks was a higher priority to them because the effect of disrupting them was greater. Many electrical grids were already operating near capacity, and were expensive to expand. They became more important as highly portable forms of energy like fossil fuels were partially replaced by less portable forms of energy, specifically electricity generated from coal-burning and nuclear plants, and to a lesser extent by wind and solar energy. This meant that electrical grids carried as much or more energy as they do now, and certainly a larger percentage of all energy consumed. Furthermore, they recognized that energy networks often depend on a few major continent-spanning trunks, which were very vulnerable to disruption.

◫ ◫ ◫

There is one final argument that resisters in this scenario made for actions against the economy as a whole, rather than engaging in piecemeal or tentative actions: the element of surprise. They recognized that sporadic sabotage would sacrifice the element of surprise and allow their enemy to regroup and develop ways of coping with future actions. They recognized that sometimes those methods of coping would be desirable for the resistance (for example, a shift toward less intensive local supplies of energy) and sometimes they would be undesirable (for example, deployment of rapid repair teams, aerial monitoring by remotely piloted drones, martial law, etc.). Resisters recognized that they could compensate for exposing some of their tactics by carrying out a series of decisive surprise operations within a larger progressive struggle.

On the other hand, in this scenario resisters understand that DEW depended on relatively simple "appropriate technology" tactics (both aboveground and underground). It depended on small groups and was relatively simple rather than complex. There was not a lot of secret tactical information to give away. In fact, escalating actions with straightforward tactics were beneficial to their resistance movement. Analyst John Robb has discussed this point while studying insurgencies in countries like Iraq. Most insurgent tactics are not very complex, but resistance groups can continually learn from the examples, successes, and failures of other groups in the "bazaar" of insurgency.

Decentralized cells are able to see the successes of cells they have no direct communication with, and because the tactics are relatively simple, they can quickly mimic successful tactics and adapt them to their own resources and circumstances. In this way, successful tactics rapidly proliferate to new groups even with minimal underground communication.

Hypothetical historians looking back might note another potential shortcoming of DEW: that it required perhaps too many people involved in risky tactics, and that resistance organizations lacked the numbers and logistical persistence required for prolonged struggle. That was a valid concern, and was dealt with proactively by developing effective support networks early on. Of course, other suggested strategies—such as a mass movement of any kind—required far more people and far larger support networks engaging in resistance. Many underground networks operated on a small budget, and although they required more specialized equipment, they generally required far fewer resources than mass movements.

◨ ◨ ◨

Continuing this scenario a bit furthert, historians asked: how well did Decisive Ecological Warfare rate on the checklist of strategic criteria we provided at the end of the Introduction to Strategy (Chapter 12, page 385).

Objective: This strategy had a clear, well-defined, and attainable objective.

Feasibility: This strategy had a clear A to B path from the then-current context to the desired objective, as well as contingencies to deal with setbacks and upsets. Many believed it was a more coherent and feasible strategy than any other they'd seen proposed to deal with these problems.

Resource Limitations: How many people are required for a serious and successful resistance movement? Can we get a ballpark number from historical resistance movements and insurgencies of all kinds?

- *The French Resistance. Success indeterminate.* As we noted in the "The Psychology of Resistance" chapter: The French Resistance at most comprised perhaps 1 percent of the adult population, or about 200,000 people.[26] The postwar French government officially recognized 220,000 people[27] (though one historian estimates that the number of active resisters could have been as many as 400,000[28]). In addition to active resisters, there were perhaps another 300,000 with substantial involvement.[29] If you include all of those people who were willing to take the risk of reading the underground newspapers, the pool of sympathizers grows to about 10 percent of the adult population, or two million people.[30] The total population of France in 1940 was about forty-two million, so recognized resisters made up one out of every 200 people.

- *The Irish Republican Army. Successful.* At the peak of Irish resistance to British rule, the Irish War of Independence (which built on 700 years of resistance culture), the IRA had about 100,000 members (or just over 2 percent of the population of 4.5 million), about 15,000 of whom participated in the guerrilla war, and 3,000 of whom were fighters at any one time. Some of the most critical and decisive militants were in the "Twelve Disciples," a tiny number of people who swung the course of the war. The population of occupying England at the time was about twenty-five million, with another 7.5 million in Scotland and Wales. So the IRA membership comprised one out of every forty Irish people, and one out of every 365 people in the UK. Collins's Twelve Disciples were one out of 300,000 in the Irish population.[31]

- *The antioccupation Iraqi insurgency. Indeterminate success.* How many insurgents are operating in Iraq? Estimates vary widely and are often politically motivated, either to make the occupation seem successful or to justify further military crackdowns, and so on. US military estimates circa 2006 claim 8,000–20,000 people.[31] Iraqi intelligence estimates are higher. The total population is thirty-one million, with a land area about 438,000 square kilometers. If there are 20,000 insurgents, then that is one insurgent for every 1,550 people.

- *The African National Congress. Successful.* How many ANC members were there? Circa 1979, the "formal political underground" consisted of 300 to 500 individuals, mostly in larger urban centers.[33] The South African population was about twenty-eight million at the time, but census data for the period is notoriously unreliable due to noncooperation. That means the number of formal underground ANC members in 1979 was one out of every 56,000.

- *The Weather Underground. Unsuccessful.* Several hundred initially, gradually dwindling over time. In 1970 the US population was 179 million, so they were literally one in a million.

- *The Black Panthers. Indeterminate success.* Peak membership was in late 1960s with over 2,000 members in multiple cities.[34] That's about one in 100,000.

- *North Vietnamese Communist alliance during Second Indochina War. Successful.* Strength of about half a million in 1968, versus 1.2 million anti-Communist soldiers. One figure puts the size of the Vietcong army in 1964 at 1 million.[35] It's difficult to get a clear figure for total of combatants and noncombatants because of the widespread logistical support in many areas. Population in late 1960s was around forty million (both North and South), so in 1968, about one of every eighty Vietnamese people was fighting for the Communists.

- *Spanish Revolutionaries in the Spanish Civil War. Both successful and unsuccessful.* The National Confederation of Labor (CNT) in Spain had a membership of about three million at its height. A major driving force within the CNT was the anarchist FAI, a loose alliance of militant affinity groups. The Iberian Anarchist Federation (FAI) had a membership of perhaps 5,000 to 30,000 just prior to revolution, a number which increased significantly with the onset of war. The CNT and FAI were successful in bringing about a revolution in part of Spain, but were later defeated on a national scale by the Fascists. The Spanish population was about 26 million. So about one in nine Spaniards were CNT members, and

(assuming the higher figure) about one in 870 Spaniards was
FAI members.

- *Poll tax resistance against Margaret Thatcher circa 1990. Successful.* About fourteen million people were mobilized. In a population of about fifty-seven million, that's about one in four (although most of those people participated mostly by refusing to pay a new tax).
- *British suffragists. Successful.* It's hard to find absolute numbers for all suffragists. However, there were about 600 nonmilitant women's suffrage societies. There were also militants, of whom over a thousand went to jail. The militants made all suffrage groups—even the nonmilitant ones—swell in numbers. Based on the British population at the time, the militants were perhaps one in 15,000 women, and there was a nonmilitant suffrage society for every 25,000 women.[36]
- *Sobibór uprising. Successful.* Less than a dozen core organizers and conspirators. Majority of people broke out of the camp and the camp was shut down. Up to that point perhaps a quarter of a million people had been killed at the camp. The core organizers made up perhaps one in sixty of the Jewish occupants of the camp at the time, and perhaps one in 25,000 of those who had passed through the camp on the way to their deaths.

It's clear that a small group of intelligent, dedicated, and daring people can be extremely effective, even if they only number one in 1,000, or one in 10,000, or even one in 100,000. But they are effective in large part through an ability to mobilize larger forces, whether those forces are social movements (perhaps through noncooperation campaigns like the poll tax) or industrial bottlenecks.

Furthermore, it's clear that if that core group can be maintained, it's possible for it to eventually enlarge itself and become victorious.

All that said, future historians discussing this scenario will comment that DEW was designed to make maximum use of small numbers, rather than assuming that large numbers of people would materialize for timely action. If more people had been available, the strategy would have become even more effective. Furthermore, they might comment that this strategy attempted to mobilize people from a wide variety of backgrounds in ways that were feasible for them; it didn't rely solely on militancy (which would have excluded large numbers of people) or on symbolic approaches (which would have provoked cynicism through failure).

Tactics: The tactics required for DEW were relatively simple and accessible, and many of them were low risk. They were appropriate to the scale and seriousness of the objective and the problem. Before the beginnings of DEW, the required tactics were not being implemented because of a lack of overall strategy and of organizational development both above- and underground. However, that strategy and organization were not technically difficult to develop—the main obstacles were ideological.

Risk: In evaluating risk, members of the resistance and future historians considered both the risks of acting and the risks of not acting: the risks of implementing a given strategy and the risks of not implementing it. In their case, the failure to carry out an effective strategy would have resulted in a destroyed planet and the loss of centuries of social justice efforts. The failure to carry out an effective strategy (or a failure to act at all) would have killed billions of humans and countless nonhumans. There were substantial risks for taking decisive action, risks that caused most people to stick to safer symbolic forms of action. But the risks of inaction were far greater and more permanent.

Timeliness: Properly implemented, Decisive Ecological Warfare was able to accomplish its objective within a suitable time frame, and in a reasonable sequence. Under DEW, decisive action was scaled up as rapidly as it could be based on the underlying support infrastructure. The exact point of no return for catastrophic climate change was unclear, but if there are historians or anyone else alive in the future, DEW and other measures were able to head off that level of climate change. Most other proposed measures in the beginning weren't even trying to do so.

Simplicity and Consistency: Although a fair amount of context and knowledge was required to carry out this strategy, at its core it was very simple and consistent. It was robust enough to deal with unexpected events, and it could be explained in a simple and clear manner without jargon. The strategy was adaptable enough to be employed in many different local contexts.

Consequences: Action and inaction both have serious consequences.

A serious collapse—which could involve large-scale human suffering—was frightening to many. Resisters in this alternate future believed first and foremost that a terrible outcome was not inevitable, and that they could make real changes to the way the future unfolded.

☒ ☒ ☒

Q: How can I be sure my actions won't hasten or cause the extinction of the very species I'm trying to save? How can I be sure my actions won't result in hungry people killing every last wild animal in the area for food or cutting down every last tree for fuel?

Derrick Jensen: We can't be absolutely certain of anything. The only thing we can be certain of is that if civilization continues, it will kill every last being on earth. But let's take a reasonable worst-case scenario for a cataclysmic event. Chernobyl was a horrible disaster. Yet it has had a spectacularly positive ecological outcome: humans have been kept out of the area and wildlife is returning. Do you know what that means? The day-to-day workings of civilization are worse than a nuclear catastrophe. It would be hard to do worse than Chernobyl.

Yes, be smart and attend to those questions. But if we fail to act there will be nothing left. What the world needs is to be left alone. What the world needs is to have this culture—that is continuously cutting it, torturing it, murdering it—stopped.

PART IV: **THE FUTURE**

Chapter 15

Our Best Hope
by Lierre Keith

Fairy Tales are more than true; not because they tell us that dragons exist, but because they tell us that dragons can be beaten.

—G. K. Chesterton

The IRA had Sinn Fein. The abolition movement had the Underground Railroad, Nat Turner and John Brown, and Bloody Kansas. The suffragists had organizations that lobbied and educated, and then the militant WSPU that burned down train stations and blew up golf courses. The original American patriots had printers and farmers and weavers of homespun domestic cloth, and also Sons of Liberty who were willing to bodily shut down the court system. The civil rights movement had the redefinition of blackness in the Harlem Renaissance and the stability, dignity, and community spirit of the Pullman porters, and then four college students willing to sit down at a lunch counter and face the angry mob. The examples are everywhere across history. A radical movement grows from a culture of resistance, like a seed from soil. And just like soils must have the cradling roots and protective cover of plants, without the actual resistance, no community will win justice or human rights against an oppressive system.

Our best hope will never lie in individual survivalism. Nor does it lie in small groups doing their best to prepare for the worst. Our best and only hope is a resistance movement that is willing to face the scale of the horrors, gather our forces, and fight like hell for all we hold dear. These, then, are the principles of a Deep Green Resistance movement.

1. Deep Green Resistance recognizes that this culture is insane.
A DGR movement understands that power is sociopathic and hence there will not be a voluntary transformation to a sustainable way of life. Providing "examples" of sustainability may be helpful for specific projects geared toward people who are anxious about their survival, but they are not a broad solution to a culture addicted to power and domination.

Since this culture went viral out of the Tigris-Euphrates River Valley, it has encountered untold numbers of sustainable societies, some of them profoundly peaceful and egalitarian, and its response has been to wipe them out with a sadism that is incomprehensible. As one example among millions, Christopher Columbus's officers preferred their rape victims between the ages of ten and twelve.

The pattern repeats itself across time and culture wherever civilization has risen. Civilization requires empire, colonies to dominate and gut. Domination requires a steady supply of hierarchy, objectification, and violence. As Ellen Gabriel, one of the participants in the Oka uprising, said, "We were fighting something without a spirit. . . . [t]hey were like robots."[1] The result is torture, rape, and genocide. And the deep heart of this hell is the authoritarian personality structured around masculinity with its entitlement and violation imperative. Lundy Bancroft, writing about the mentality of abusive men, writes, "The roots [of abuse] are ownership, the trunk is entitlement, and the branches are control."[2] You could not find a clearer description of civilization's 10,000 year reign of terror.

2. Deep Green Resistance embraces the necessity of political struggle.

DGR is not a liberal movement. Oppression is not a mistake, and changing individual hearts and minds is not a viable strategy. Political struggle must happen on every level and in every arena if we're to avert the worst ecological disasters and create a culture worth the name. By political struggle, I mean specifically institutional change, whether by reform or replacement or both. It's institutions that shape those hearts and minds. A project of individual change would take lifetimes, if it worked at all. The individual has never been the target of any liberation movement for the simple reason that it's not a feasible strategy, as our previous chapters have explained.

Fighting injustice is never easy. History tells us that the weight of power will come down on any potential resistance, a weight of violence and sadism designed to crush the courageous and anyone who might consider joining them. This is what abusive men do when women in their control fight back. It's what slave owners do to slaves. It's what imperial armies do to the colonized, and what the civilized do to the indigenous. The fact that there will be retaliation is no reason to give up before we begin. It is a reality to be recognized so that we can prepare for it.

The necessity of political struggle especially means confronting and contradicting those on the left who say that resistance is futile. Such people have no place in a movement for justice. For actionists who choose to work aboveground, this confrontation with detractors—and some of these detractors reject the idea of resistance of any kind—is one of the small, constant actions you can take. Defend the possibility of resistance, insist on a moral imperative of fighting for this planet, and argue for direct action against perpetrators. Despite what much of the left has now embraced, we are *not* all equally responsible. There are a few corporations that have turned the planet into a dead commodity for their private wealth, destroying human cultures along with it.

As we have said, their infrastructures—political, economic, physical—are, in fact, immensely vulnerable. Perhaps the gold standard of resistance against industrial civilization is MEND, the Movement for the Emancipation of the Niger Delta. The oil industry has earned literally hundreds of billions of dollars from taking Nigeria's oil. The country currently takes in $3 billion a month from oil, which accounts for 40 percent of its GDP.[3] The Niger Delta is the world's largest wetland, but it could more readily be called a sludgeland now. The indigenous people used to be able to support themselves by fishing and farming. No more. They're knee-deep in oil industry waste. The fish population has been "decimated" and the people are now sick and starving.[4] The original resistance, MOSOP, was led by poet-activist Ken Saro-Wiwa. Theirs was a nonviolent campaign against Royal Dutch/Shell and the military regime. Saro-Wiwa and eight others were executed by the military government, despite international outcry and despite their nonviolence.

MEND is the second generation of the resistance. They conduct direct attacks against workers, bridges, office sites, storage facilities, rigs and pipelines, and support vessels. They have reduced Nigeria's oil output by a dramatic one-third.[5] In one single attack, they were able to stop 10 percent of the country's production. And on December 22, 2010, MEND temporarily shut down three of the country's four oil refineries by damaging pipelines to the facilities.[6] Their main tactic is the use of speedboats in surprise attacks against simultaneous targets toward the goal of disrupting the entire system of production.

According to Nnamdi K. Obasi, West Africa senior analyst at the International Crisis Group, "MEND seems to be led by more enlightened and sophisticated men than most of the groups in the past."[7] They have university educations and have studied other militant movements. Their training in combat is so good that they have fought and won in skirmishes against both Shell's private military and Nigeria's elite fighting units. They've also won "broad sympathy among the Niger Delta community."[8] This sympathy has helped them maintain security and safety for their combatants as the local population has not turned them in. These are not armed thugs, but a true resistance. And they number just a few hundred.

Understand: a few hundred people, well-trained and organized, have reduced the oil output of Nigeria by one-third. MEND has said, "It must be clear that the Nigerian government cannot protect your workers or assets. Leave our land while you can or die in it. . . . Our aim is to totally destroy the capacity of the Nigerian government to export oil."[9] I can guarantee that 98 percent of the people who are reading this book have more resources individually than all of MEND put together when they started. Resistance is not just theoretically possible. It is happening now. The only question is, will we join them?

3. Deep Green Resistance must be multilevel.

There is work to be done—desperately important work—aboveground and underground, in the legal sphere and the economic realm, locally and internationally. We must not be divided by a diversionary split between radicalism and reformism. One more time: the most militant strategy is not always the most radical or the most effective. The divide between militance and nonviolence does not have to destroy the possibility of joint action. People of conscience can disagree. They can also respectfully choose to work in different arenas requiring different tactics. I can think of no scenario in which a program to provide school cafeterias with food straight from local, grass-based farms would be advanced by explosives. In contrast, a project to save the salmon would do well to consider such an option.

Every movement for justice struggles with the subject. Violence, including property destruction, should not be undertaken without serious reflection and ethical, even spiritual, investigation. Better to accept that as individuals we will arrive at different answers—but that we have to build a successful movement despite those differences.

That shouldn't be hard, considering that this entire culture has to be replaced. We need every level of action and every passion brought to bear. The Spanish Anarchists stand as a great example of a broad and deep effort to transform an entire society. Writes Murray Bookchin, "The great majority of these [affinity] groups were not engaged in terrorist actions. Their activities were limited mainly to general propaganda and to the painstaking but indispensable job of winning over individual converts."[10] The café was where all that discussion and proselytizing happened, just as it happened in pubs for the Irish and the IRA. The anarchists in Barcelona took over railroads, factories, public utilities, retail and wholesale businesses, and ran them by workers' committees. They also created their own police force to patrol their neighborhoods, and revolutionary tribunals to mete out justice. Before the Fascist victory, the anarchists in Andalusia created communal land tenure arrangements, stopped using money for internal exchange of goods, and established directly democratic popular assemblies for their governance. They also started over fifty alternative schools across the country. Their educational ideas spread through Europe, landing in England where they were taken up by A. S. Neil at his famous Summerhill. From there, the concept of free schools migrated to the US. If you are involved with any student-directed, alternative education, you are a direct descendent of the Spanish Anarchists.

Every institution across this culture must be reworked or replaced by people whose loyalty to the planet and to justice is absolute. A DGR movement understands the necessity of both aboveground and underground work, of confronting unjust institutions as well as building alternative institutions, of every effort to transform the economic, political, and social spheres of this culture. Whatever you are called to do, it needs to be done.

It is unlikely that a political candidate on the national or even state level will have a chance of winning on a platform of truth telling, at least not in the United States. The industrial world needs to reduce its energy consumption to that of Brazil or Sri Lanka. That this reduction is inevitable doesn't make it any more palatable to the average American, who will likely only give up his entitlement when it's pried from his cold, dead fingers.[11] At the local level, the political process may be more amenable to radical truth telling, especially in progressive enclaves. For those with the skills and interest, running for local office could yield results worth the effort. It could also scale up to other communities. What kinds of institutional change could be affected at the local level is a question worth asking.

From outside, a vast amount of pressure is needed to stop fossil fuel and other industrial extractions. Legislative initiatives, boycotts, direct action, and civil disobedience must be priorities. We need to form groups like Climate Camp, which started in 2006 with a teach-in and protest at a coal-fired plant in West Yorkshire, England. They've blockaded the European Carbon Exchange in London, protested runway expansion at Heathrow Airport, and are taking action against BP for its participation in the tar sands. In their own words,

> The climate crisis cannot be solved by relying on governments and big businesses with their "techno-fixes" and other market-driven approaches. . . . We must therefore take responsibility for averting climate change, taking individual and collective action against its root causes and to develop our own truly sustainable and socially just solutions. We must act together and in solidarity with all affected communities—workers, farmers, indigenous peoples and many others—in Britain and throughout the world.[12]

If the referendums and court decisions and market solutions fail, if the civil disobedience and blockades aren't enough, a Deep Green Resistance is willing to take the next step to stop the perpetrators.

In the UK, someone is feeling the urgency. On April 12, 2010, the machinery at Mainshill coal site was sabotaged, machinery that was Mordorian in its destructive power: "a 170 tonne face scrapping earth mover." The coal was slated for the Drax Power Station, "recognised as the most polluting in the UK."[13]

According to their communiqué:

> Sabotage against the coal industry will continue until its expansion is halted.

This is a simple vow, an "I do" to every living creature. Deep Green Resistance remembers that love is a verb, a verb that must call us to action.

4. Deep Green Resistance requires repair of the planet.

This principle has the built-in prerequisite, of course, of stopping the destruction. Burning fossil fuels has to stop. Likewise, industrial logging, fishing, and agriculture have to stop. Denmark and New Zealand, for instance, have outlawed coal plants—there's no reason the rest of the world can't follow.

Stopping the destruction requires an honest look at the culture that a true solar economy can support. We need a new story, but we don't need fairy tales, and the bread crumbs of windfarms and biofuels will not lead us home.

To actively repair the planet requires understanding the damage. The necessary repair—the return of forests, prairies, and wetlands—could happen over a reasonable fifty to one hundred years if we were to voluntarily reduce our numbers. This is not a technical problem: we actually do know where babies come from and there are a multitude of ways to keep them from coming. As discussed in Part I, overshoot is a social problem caused by the intersections of patriarchy, civilization, and capitalism.

People are still missing the correct information. Right now, the grocery stores are full here. In poor areas, the so-called food deserts may be filled with cheap carbohydrates and vegetable oil, but they are still full. But how many people could any given local foodshed actually support, and support sustainably, indefinitely? Whatever that number is, it needs to be emblazoned like an icon across every public space and taken up as the baseline of the replacement culture. Our new story has to end, "And they lived happily ever after at 20,000 humans from here to the foothills."

This is a job for the Transitioners and the permaculture wing, and so far, they're getting it wrong. The Peak Oil Task Force in Bloomington, Indiana, for instance, put out a report entitled *Redefining Prosperity: Energy Descent and Community Resilience.* The report recognizes that the area does not have enough agricultural land to feed the population. They claim, however, that there is enough land within the city using labor-intensive cultivation methods to feed everyone on a "basic, albeit mostly vegetarian diet."[14] The real clue is that "vegetarian diet." What they don't understand is that soil is not just dirt. It is not an inert medium that needs nothing in order to keep producing food for humans. Soil is alive. It is kept alive by perennial polycultures—forests and prairies. The permanent cover protects it from sun, rain, and wind; the constant application of dead grass and leaves adds carbon and nutrients; and the root systems are crucial for soil's survival, providing habitat for the microfauna that make land life possible.

Perennials, both trees and grasses, are deeply rooted. Annuals are not. Those deep roots reach into the rock that forms the substrate of our planet and pull up minerals, minerals which are necessary for the entire web of life. Without that action, the living world would eventually run out of minerals. Annuals, on the other hand, literally mine the soil, pulling out minerals with no ability to replace them. Every load of vegetables off the farm or out of the garden is a transfer of minerals that must be replaced. This is a crucial point that many sustainability writers do not understand: organic matter, nitrogen, and minerals all have to be replaced, since annual crops use them up.

John Jeavons, for instance, claims to be able to grow vast quantities of food crops with only vegetable compost as an input on his Common Ground demonstration site.[15] But as one observer writes,

> Sustainable Laytonville visited Common Ground. The gardens could only supply one meal a day *because they didn't have enough compost.* The fallacy with Biointensive/Biodynamic and Permaculture is that they all require outside inputs whether it's rock phosphate or rock dusts, etc. There is no way to have perpetual fertility and take a crop off and replace lost nutrients with the "leftovers" from the area under culti-

vation . . . even if the person's urine, poop and bones were added back.[16]

I have built beautiful garden soil, dark as chocolate and with a scent as deep, using leaves, spoiled hay, compost, and chickens. But I eventually was forced to realize the basic arithmetic in the math left a negative number. *I was shifting fertility, not building it.* The leaves and hay may have been throwaways to the lawn fetishists and the farmers, but they were also nutrients needed by the land from which they were taken. The suburban backyard that produced those leaves needed them. If I was using the leaves, the house owner was using packaged fertilizer instead. The addition of animal products—manure, bloodmeal, bonemeal—is essential for nitrogen and mineral replacement, and they are glaringly absent in most calculations I've seen for food self-sufficiency. Most people, no matter how well-intentioned, have no idea that both soil and plants need to eat.

Annual crops use up the organic matter in the soil, whereas perennials build it. Processes like tilling and double digging not only mechanically destroy soil, they add oxygen, which causes more biological activity. That activity is the decay of organic matter. This releases both carbon and methane. One article in *Science* showed that *all tillage systems are contributors to global warming*, with wheat and soy as the worst.[17] This is why, historically, agriculture marks the beginning of global warming. In contrast, because perennials build organic matter, they sequester both carbon and methane, at about 1,000 pounds per acre.[18] And, of course, living forests and prairies will not stay alive without their animal cohorts, without the full complement of their community.

So be very wary of claims of how many people can be supported per acre in urban landscapes. It is about much more than just acreage. If you decide to undertake such calculations, consider that the soil in garden beds needs permanent cover. Where will that mulch come from? The soil needs to eat; where will the organic material and minerals come from? And people need to eat. We cannot live on the thin calories of vegetables, no matter how organic, to which 50,000 nerve-damaged Cubans can attest. So far, the Transitioners, even though many of them have a permaculture background, seem unaware of the biological constraints of soil and plants, which are, after all, living creatures with physical needs. In the end, the only closed loops that are actually closed are the perennial polycultures that this planet naturally organizes—the communities that agriculture has destroyed.

But as we have said, people's backyard gardens are of little concern to the fate of our planet. Vegetables take up maybe 4 percent of agricultural land. What is of concern are the annual monocrops that provide the staple foods for the global population. Agriculture is the process that undergirds civilization. That is the destruction that must be repaired. Acre by acre, the living communities of forests, grasslands, and wetlands must be allowed to come home. We must love them enough to miss them and miss them enough to restore them.

The best hope for our planet lies in their restoration. Perennials build soil, and carbon is their building block. A 0.5 percent increase in organic matter—which even an anemic patch of grass can manage—distributed over 75 percent of the earth's rangelands (11.25 billion acres) would equal 150 billion tons of carbon removed from the atmosphere. The current carbon concentrations are at 390 ppm. The prairies' repair would drop that to 330 ppm.[19] Peter Bane's calculations show that restoring grasslands east of the Dakotas would instantaneously render the United States a carbon-sequestering nation.[20] Ranchers Doniga Markegard and Susan Osofsky put it elegantly: "As a species, we need to shift from carbon-releasing agriculture to carbon-sequestering agriculture."[21]

That repair should be the main goal of the environmental movement. Unlike the Neverland of the Tilters' solutions, we have the technology for prairie and forest restoration, and we know how to use it. And the grasses will be happy to do most of the work for us.

The food culture across the environmental movement is ideologically attached to a plant-based diet. That attachment is seriously obstructing our ability to name the problem and start working on the obvious solutions. Transition Town originator Rob Hopkins writes, "Reducing the amount of livestock will also be inevitable, as large-scale meat production is an absurd and unsustainable waste of resources."[22] Raising animals in factory farms—concentrated animal feeding operations

(CAFOs)—and stuffing them with corn is absurd and cruel. But animals are necessary participants in biotic communities, helping to create the only sustainable food systems that have ever worked: they're called forests, prairies, wetlands. In the aggregate, a living planet.

That same ideological attachment is the only excuse for the blindness to Cuban suffering and for the comments that 30 percent of Cubans are "still obese." That figure is supposed to reassure us: see, nobody starves in this regime. What such comments betray is a frank ignorance about human biology. Eating a diet high in carbohydrates will make a large percentage of the population gain weight. Eating any sugar provokes a surge of insulin, to control the glucose levels in the bloodstream. The brain can only function within a narrow range of glucose levels. Insulin is an emergency response, sweeping sugar out of the blood and into the cells for storage. Insulin has been dubbed "the fat storage hormone" because this is one of its main functions. Its corresponding hormone, glucagon, is what unlocks that stored energy. But in the presence of insulin, glucagon can't get to that energy. This is why poor people the world over tend to be fat: all they have to eat is cheap carbohydrate, which trigger fat storage. If the plant diet defenders knew the basics of human biology, *that weight gain would be an obvious symptom of nutritional deficiencies, not evidence of their absence.* Fat people are probably the most exhausted humans on the planet, as minute to minute their bodies cannot access the energy they need to function. Instead of understanding, they are faced with moral judgment and social disapproval across the political spectrum.

I don't want any part of a culture that inflicts that kind of cruelty and humiliation on anyone. Shaun Chamberlin writes, "The perception of heavy meat eaters could be set to change in much the same way that the perception of [SUV] drivers has done."[23] Even if he was right that meat is inherently a problem, this attitude of shaming people for their simple animal hunger is repugnant. Half the population—the female half—already feels self-loathing over every mouthful, no matter what, and how little, is on their plates. Food is not an appropriate arena for that kind of negative social pressure, especially not in an image-based culture saturated in misogyny. Food should be a nourishing and nurturing part of our culture, including our culture of resistance. If

Chamberlin wants an appropriate target for social shaming, he can start with men who rape and batter, and then move on to men who refuse to get vasectomies—that would be a better use of his moral approbation.

Getting past that ideological attachment would also bring clarity to the bewildered attitude that underlies many of these "radical" writers' observations about dietary behavior. Accepting that humans have a biological need for nutrient-dense food, it's no longer a surprise that when poor people get more money, they will buy more meat. They're not actually satisfied on the nutritional wonders of a plant-based diet. Ideology is a thin gruel and imposing it on people who are chronically malnourished is not only morally suspect, it won't work. The human animal will be fed. And if we had stuck to our original food, we would not have devoured the planet.

Restoring agricultural land to grasslands with appropriate ruminants has multiple benefits beyond carbon sequestration. It spells the end of feedlots and factory farming. It's healthier for humans. It would eliminate essentially all fertilizer and pesticides, which would eliminate the dead zones at the mouths of rivers around globe. The one in the Gulf of Mexico, for instance, is the size of New Jersey. It would stop the catastrophic flooding that results from annual monocrops, flooding being the obvious outcome of destroying wetlands.

It also scales up instantly. Farmers can turn a profit the first year of grass-based farming. This is in dramatic contrast to growing corn, soy, and wheat, in which they can never make a profit. Right now six corporations, including Monsanto and Cargill, control the world food supply. Because of their monopoly, they can drive prices down below the cost of production. The only reason farmers stay in business is because the federal government—that would be the US taxpayers—make up the difference, which comes to billions of dollars a year. The farmers are essentially serfs to the grain cartels, and dependent on handouts from the federal government. But grass-fed beef and bison can liberate them in one year. We don't even need government policy to get started on the most basic repair of our planet. We just need to create the demand and set up the infrastructure one town, one region at a time.

Land with appropriate rainfall can grow two steers per acre. But those steers can be raised in two ways. You can destroy the grasses, plant corn, and feed that corn to CAFO steers, making them and their human consumers sick in the process. Or you can skip the fossil fuels and the torture, the habitat destruction, the dead zones that used to be bays and oceans, and let those steer eat grass. Either method produces the same amount of food for humans, but one destroys the cycle of life while the other participates in it. I can tell you with certainty which food the red-legged frogs and the black-footed ferrets are voting for: let them eat grass.

Repairing those grasslands will also profoundly restore wildlife habitat to the animals that need a home. Even if the rest of the above reasons weren't true, that repair would still be necessary. The acronym HANPP stands for "human appropriation of net primary production." It's a measure of how much of the biomass produced annually on earth is used by humans. Right now, 83 percent of the terrestrial biosphere is under direct human influence, and 36 percent of the earth's bioproductive surface is completely dominated by humans.[24] By any measure, that is vastly more than our share. Humans have no right to destroy everyone else's home, 200 species at a time. It is our responsibility not just to stop it, but to fix it. Civilizations are, in the end, cultures of human entitlement, and they've taken all there is to take.

5. Deep Green Resistance means repair of human cultures.

That repair must, in the words of Andrea Dworkin, be based on "one absolute standard of human dignity."[25] That starts in a fierce loyalty to everyone's physical boundaries and sexual integrity. It continues with food, shelter, and health care, and the firm knowledge that our basic needs are secure. And it opens out into a democracy where all people get an equal say in the decisions that affect them. That includes economic as well as political decisions. There's no point in civic democracy if the economy is hierarchical and the rich can rule through wealth. People need a say in their material culture and their basic sustenance.

For most of our time on this planet, we had that. Even after the rise of civilization, there were many social, legal, and religious strictures that protected people and society from the accumulation of wealth.

There exists an abundance of ideas on how to transform our communities away from domination and accumulation and toward justice and human rights. We don't lack analysis or plans; the only thing missing is the decision to see them through.

We also need that new story that so many of the Transitioners prioritize. It's important to recognize first that not everyone has lost their original story. There are indigenous peoples still holding on to theirs. According to Barbara Alice Mann,

> The contrast between western patriarchal and Iroquoian matriarchal thought could not be more clear. . . . I do not think it is possible to examine the real impetus behind mother-right unless we walk boldly up to the spiritual underpinnings of its systems. By the same token, we cannot free ourselves of the serious damage of patriarchy, unless we appreciate where matriarchy's spiritual allegiances lie.
>
> The Iroquois are unapologetic about the fact that spirit informs and undergirds all our social, economic, and governmental structures. Every council of any honor begins with thanksgiving, that is, an energy-out broadcast, to make way for the energy in-gathering required by the One Good Mind of Consensus. When a council fails, people just assume that the faithkeeper who opened it did a poor job in the thanksgiving department. In a thanksgiving address, all the spirits of Blood and Breath (or Earth and Sky) are properly gathered and acknowledged, with the ultimate acknowledgment being that the One Good Mind of Consensus requires the active participation of not just an elite but everyone in the community. This is a foundational insight of all matriarchies.[26]

She describes a culture where "things happen by consultation, not by fiat," based on a spiritual understanding of everyone's participation in the cosmos rather than the "paranoid isolation" brought on by the temper tantrums of a sociopathic God. This is the difference between cultures of matriarchy and patriarchy, egalitarianism and domination, participation and power.

Such stories need to be told, but more, they need to be instituted. All the stories in the world will do no good if they end with the telling.

One institution that deserves serious consideration is a true people's militia. Right now in the United States only the right wing is organizing itself into an armed force. In 2009, antigovernment militias, described by the Southern Poverty Law Center as "the paramilitary arm of the Patriot movement," grew threefold, from forty-two to 127.[27] We should be putting weapons in the hands of people who believe in human rights and who are sworn to protect them, not in those of people who feel threatened because we have a black president. If jackbooted, racist, and increasingly paranoid thugs coalesce into an organized movement with its eye on political power, we don't need to relive Germany in 1936 to know where it may end, especially as energy descent and economic decline continue. Contemporaneous with a people's militia would be training in both the theory and practice of mass civil disobedience to reject illegitimate government or a coup if that comes to pass. Gene Sharp's *Civilian-Based Defense* explains how this technique works with successful examples from history. His book is a curriculum that should be added to Transition Towns and other descent preparation initiatives.

But if the people with the worst values are the ones with the guns and the training, we may be very sorry. This is a dilemma with which progressives and radicals should be grappling. A large and honorable proportion of the left believes in nonviolence, a belief that for many reaches a spiritual calling. But societies through history and currently around the globe have degenerated into petty tyrannies with competing atrocities. Personal faith in the innate goodness of human beings is not enough of a deterrent or shield for me.

A true people's militia would be sworn to uphold human rights, including women's rights. The horrors of history include male sexual sadism on a mass scale. Women are afraid of men with guns for good reason. But rape is not inevitable. It's a behavior that springs from specific social norms, norms that a culture of resistance can and must confront and counteract, whether or not we have a people's militia. We need a zero tolerance policy for abuse, especially sexual abuse.

Military organizations, like any other culture, can promote rape or stop it. Throughout history, soldiers, especially mercenary soldiers, have often been granted the "right" to rape and plunder as part of their payment. Other militaries have taken strong stands against rape. Writes Jean Bethke Elshtain, "The Israeli army . . . are scrupulous in prohibiting their soldiers to rape. The British and United States armies, as well, have not been armies to whom rape was routinely 'permitted,' with officers looking the other way, although British and American soldiers have committed 'opportunistic' rapes. Even in the Vietnam War, where incidents of rape, torture and massacre emerged, raping was sporadic and opportunistic rather than routine."[28] The history of military atrocities against civilians is a history; it's not universal, and it's not inevitable. Elshtain continues, "War is not a freeform unleashing of violence; rather, fighting is constrained by considerations of war aims, strategies and permissible tactics. Were war simply an unbridled release of violence, wars would be even more destructive than they are." Western nations, over hundreds of years, assembled an unwritten code of conduct for militaries, known as the "customary law of war," which tried to limit the suffering of soldiers and to safeguard civilians. This was eventually codified into the Hague Convention Number IV of 1907 and the Geneva Convention of 1949. These attempted to limit looting and property destruction and to protect noncombatants. The Uniform Code of Military Justice is very clear that rape is unacceptable, and even gets the finer points of how "consent" with an armed assailant is a pretty meaningless concept. Elshtain also notes that "[the] maximum punishment for rape is death. Thus, interestingly, rape is a capital offense under the Uniform Code of Military Justice, by contrast to most civilian legal codes."

Getting the command structure to take rape and human rights abuses seriously is, of course, the next step. As Elshtain points out, "It is difficult to bring offenders to trial unless the leaders of the military forces are themselves determined to ferret out and punish tormentors of civilian populations. Needless to say, if the strategy is itself one of tormenting civilians, rapists are not going to be called before a bar of justice."

It will be up to the founders and the officers of new communities to set the norms and to make those norms feminist from the beginning. The following would go a long way toward helping create a true people's militia, and not just another organization of armed thugs to "trample the grass"—the women and girls who so often suffer when men fight for power.

1. Female officers. Women must be in positions of authority from the beginning, and their authority needs absolute respect from male officers.

2. Training curriculum that includes feminism, rape awareness, and abuse dynamics, and a code of conduct that emphasizes honorable character in protecting and defending human rights.

3. Zero tolerance for misogynist slurs, sexual harassment, and assault amongst all members.

4. Clear policies for reporting infringements and clear consequences.

5. Background checks to exclude batterers and sex offenders from the militia.

6. Severe consequences for any abuse of civilians.

A people's militia could garner widespread support by following a model of community engagement, much as the Black Panthers grew through their free breakfast program. Besides basic activities like weapons training and military maneuvers, the militia could help the surrounding community with the kind of services that are always appreciated: delivering firewood to the elderly or fixing the roof of the grammar school. The idea of a militia will make some people uneasy, and respectful personal and community relationships would help overcome their reticence.

6. Deep Green Resistance recognizes the necessity of militant action.

If we had enough people for a mass movement or enough time to build one, we could shut down the activities that are destroying our planet using only determined human bodies. People armed with nothing but courage ended segregation in the American South and others pulled down the Berlin wall. Enough people could shut down the oil refineries, the coal plants, the relentless horror of the tar sands and strip mines and clear-cuts. In the fall of 2010, French workers went on strike to protest a proposed raise in the retirement age. Protestors used trucks, burning tires, and human chains to blockade fuel depots and close all twelve of France's oil refineries. The major oil terminal was offline for three weeks, stranding thirty oil tankers. When the government tried to open the country's emergency reserves, protestors blockaded twenty more terminals. In a few weeks, the whole economy was slowing toward a halt for lack of fuel. Even after fuel trucks were able to access the terminals, it took a few days for the affected gas stations to resume regular business "since transportation and other strikes have tangled each step of the distribution process."[29] As Jean-Louis Schilansky, president of the French Oil Association said, "We have considerable bottlenecks."

The French strikers did what every military and every insurgency does: interrupted key nodes of infrastructure. They were well on their way to completely shutting down the economy, and they did it using nonviolence. I would vastly prefer to wage our struggle nonviolently. As the French strike has shown, it could be done. If we had enough people, we could shut this party down by midnight using human blockades. But my longing will not produce the necessary numbers. And it's a little late in the day for millenarianism.

◙ ◙ ◙

This is the question on which the world entire may depend: Are you willing to accept the only strategy left to us? Are you willing to set aside your last, fierce dream of that brave uprising of millions strong? I know what I am asking. The human heart needs hope as it needs air. But the existence of those brave millions is the empty hope of the desperate, and they're not coming to our rescue.

But a few hundred exist, answering to the name of MEND. They are the direct descendants of Ken Saro-Wiwa and eight others who chose to fight with their lives as their only weapons. Those lives ended in a tight noose of vicious power, tied by the military on behalf of the globally wealthy. But their struggle didn't end. On that grim day, power didn't win. Because now MEND is willing to say to the oil industry, "Leave our land or you will die in it."

Are we willing to do the same? To say: Leave our mountains, our wetlands, our last, ancient trees. Leave our kin of feathers and fur, who every second are slipping away from the world and into memory. Are we willing to fight for this planet?

The Black Panthers wanted to have a national gathering but there was no safe place for them to hold it. The Quakers of the Philadelphia Meeting, despite serious and profound differences on the issue of nonviolence, offered their largest meetinghouse. And more: they circled the building to keep the Black Panthers safe from the police in an act of extraordinary solidarity, putting their lives between power and the resistance.[30]

Will you offer your meetinghouse? There are polar bears and black-footed ferrets, bison and coho salmon who need a safe place. Where will you put your life?

Will you offer your meetinghouse? The resistance needs a place, too, a place to gather its forces, find its courage, and launch the final battle.

The carbon is swelling; the heat is rising; the rivers are fading and somewhere a black tern is giving up in exhaustion. The same noose that took Ken Saro-Wiwa is tightening, and there is only time for one last breath. Will you close your eyes and let the earth fall, with a sickening snap of species and forests and rivers? Or will you fight?

Whatever you love, it is under assault. Love is a verb. So take that final breath and fight.

A STORY

All of this will come to nothing without direct action against infrastructure, without an actual resistance. Dissidence has never yet brought down a system of power, and it never will. We know what will happen without resistance: it's happening now as mountains fall to monstrous greed, and this day dawned across oceans turned empty and acid.

And so, a story. It doesn't begin with once upon a time, because it didn't happen long ago. This is a story of a future, a future as fragile as the first line of dawn. Our protagonists aren't built of blood and bone. They're made of words gathered from whispers and dreams. And the rules of our grammar will rise as despair refuses its own eschatology. Because life wants to live: sturdy fact and fledgling miracle both. Life wants to live: our communion and our battle cry.

Our story begins on a day like today. Somewhere, the question happens: will you join me? The question is a risk, but weighed against the aching despair of another 200 species and twenty-nine degrees of heat, the risk is taken. The test of intellectual agreement and the trust of character will already have been established. All that's left is for someone to ask it: will you join me?

The question is asked again, and then again, six times, ten times. The first meeting is held, tactics discussed, tasks disbursed. Someone's job is to keep asking that one question, to find others, to multiply outward until there are enough.

Enough doesn't mean just numbers. Enough means trained bodies, disciplined habits, dependable behavior, an unshakable moral core. Our lack of purpose is not our strength, it is our profound downfall, and this revolution is not for the hell of it. Enough means courage and an acceptance of the sacrifices that may well lie ahead. And enough also means thoroughly understanding the strategy we are proposing.

We are not the Weather Underground or the Red Army Faction. Jeremy Varon, in his history of both, writes, "1960s radicals were driven by an apocalyptic impulse resting on a chain of assumptions: that the existing order was thoroughly corrupt and had to be destroyed; that its destruction would give birth to something radically new and better; and that the transcendent nature of this leap rendered the future a largely blank or unrepresentable utopia."[31] DGR is not secular millennialism. The revolution is not nigh, general chaos is not going to bring it on, and essentially symbolic attacks on people or on property are of no strategic use. Cross those attacks—all of them—off the list now. This is true whether the targets are large or even monumental, or small and accessible. DGR is not a desperate call to act on whatever targets are at hand. If this is a struggle against an opponent with a scorched earth policy, then our strategy has to aim a wee bit higher than the windows of your local corporate outpost. So put away your bricks and spray paint: those are not weapons for the serious. From this point forward, we aim to be effective.

And because our detractors will be determined to misunderstand: DGR is also not a call for an armed insurgency. The idea is absurd. A few radical environmentalists could not possibly take on the US military. A DGR strategy does not include pitched battles, ever, or Theodore Roszak's "fistfight with the nearest cop." Our goal is not to bring down the US government or any government. The realm of broad and transformative political change is best left to the aboveground groups working on scales from the local to the international. Such campaigns demand mass movements and, in all probability, nonviolent tactics, and there are examples from around the world for study.

DGR is a fight against a singular enemy: industrial civilization. This makes us different from every other struggle in history. It has some similarities with the original Luddites (news flash: they were right). It also has overlap with indigenous peoples trying to forestall displacement—extinction—due to dams and mining. But those indigenous are mostly having to fight while rooted in place, protecting their land and their survival. They cannot win in pitched battles against the might of armies in the service of capitalist profits. MEND offers a more successful model for DGR, using the flexibility and surprise that are the strengths of a guerrilla strategy.

But because the enemy is not a military, we are left with wrenching ethical decisions. If there are people between us and our targets, they are not soldiers. We can say that civilization is a war against the living world, but that does not answer the moral dilemma of putting living beings at risk. I have no answer, only an emergency the size of land, sea, and sky. I never asked to be in this position. Insisting that there is still somewhere a win-win solution that leads the civilized to that voluntary transformation is a stance of willful denial against all the facts, and I cannot turn away from my moral agency to ease my moral agony. I am forced to weigh the fact that actions have consequences, sometimes dramatic and unpredictable ones, against Oxfam's words on climate change: "The prospects are very bleak for hundreds of millions of people, most of them amongst the world's poorest."[32] This is not a dilemma that I created. It is the dilemma I am trying with all I have to stop. I can't put my longings for emotional ease and innocence above the emergency—biotic and human—to which history has abandoned us. All I can do is beg the people who might read this book: please do everything you can to spare all sentient life.

To those of you shaking your heads in horror: do you have enough bodies to shut down one-third of the oil industry and drive BP from this land? How about the whole industry and mountain-top removal along with it?

And I know that loosening our basic moral precepts also has consequences, especially when mixed with youth, fanaticism, and masculinity. The Remembrance Day bombing is an incident that stands in for too many. On November 8, 1987, the IRA exploded a bomb in Enniskillen (County Fermanagh, Northern Ireland) during a ceremony to honor veterans who fought in the British military in World War I. Eleven people died, most of them elderly women, and sixty-three people were injured. Danny Morrison, Sinn Fein's spokesperson, said, "I was shattered, as I think many of our supporters were, to find the IRA was involved."[33] A second bomb, four times larger, mercifully failed to explode at another ceremony, this one led by Boys' and Girls' Brigades. The IRA's strategy, as stated in their *Green Book* training manual, includes "a war of attrition against enemy personnel." That would be the British army, not the Girls' Brigade. There was a "universal wave of anger and disgust" along with statements of outrage from politicians in the Republic of Ireland.[34] It would take fourteen years for Sinn Fein's electoral support to recover. The IRA leadership disavowed the bombing and the entire Fermanagh Brigade was stood down. Three IRA units were behind the bombs, and as many as thirty IRA volunteers were involved in setting up the explosives. Out of all those men, not one thought to talk sense into the others?

The communiqué from the IRA claimed that British army troops and Royal Ulster Constabulary were the intended targets. No one accepted their rationale. This action didn't go wrong—it was wrong. Neither a willful targeting of civilians nor such callous disregard for collateral damage are morally acceptable. To refer back to the diagram on page 407, this action was neither just nor strategic.

There is no reason for a resistance to make these mistakes. But even in writing this, I am reminded of a commentator who, in regards to the Enniskillen bombing, spoke of "the grimly metaphysical tone of the debates about death . . . that tries to distinguish between justified and unjustified political killings."[35] I know all the ways this can go wrong, how easily extremism unmoors its own moral compass. And I also know that my planet is dying, with the most vulnerable always first in line.

No one who does not feel the burden of the moral risks of serious action should be making these decisions. Extremism has its own addictive thrills; violence feeds masculinity too easily, and the human heart is quite capable of justifying atrocity. And I also know that decisions have to be made, life-and-death decisions, the decisions of the desperate. For those without the stamina, better work awaits. But for the desperate, pick leaders with more character than charisma, and never forget the goal, the strategy, and the real target. A forester once said to me, "The day you stop being afraid of a chainsaw is the day you stop using one." Remember explosives directed at the Girls' Brigade: the day you go numb to morality is the day to stop making moral decisions.

DGR has a very different goal from anticolonial struggles. The *Green Book*, for instance, puts the goal of those struggles clearly: "To make the six counties . . . ungovernable except by colonial military rule." This has no parallel in the Decisive Ecological Warfare strategy we are describing: none. DEW has only that one goal at the heart of its strategy: to disrupt and dismantle industrial civilization and to thereby remove the ability of the powerful to exploit the powerless and to destroy the planet.

Our actionists are not trying to change consciousness. They're not trying to get press. They're not after a new government or a seat at a political table. They are trying to stop the burning of fossil fuels and industrial-scale destruction of the life-support systems of their planet. That is the goal of DGR, and DEW is their strategy.

The infrastructure of industrial civilization is both vulnerable and accessible, but the environmental movement is not used to thinking in terms of infrastructure. This is the language of war, not petitions. But it is long past time for this war to have two sides. To date, environmentalists have not suggested the level of engagement that we're discussing here, though surely in the long hours between midnight and dawn others have longed for it.

And surely there are some that don't just long, but believe. If those few start thinking like a resistance, it might be possible. And that possibility holds the whole future.

The underground cells that form are unlikely to connect into a single network, given the realities of surveillance technologies and the atomized nature of modern life. But a few networks the size of MEND could easily be built, backed up by stand-alone affinity groups. Those cells and networks would take their training, their leaders, and their security seriously. Resistance is not a game for hopeful children or overwrought adolescents of whatever age. Above all, these groups would understand the grand strategy of DEW.

The resistance's organization is likely to be fractured. The unconnected networks will not be able to coordinate but they can still act in concert and multiply each other's efforts. They can only do this, though, if they understand the overall strategy of DEW, which is why strategy is the beating heart of our actionists' training.

The infrastructure of fossil fuels would be their highest priority, and the nodes with the densest criticality are their best targets. Those targets have two factors that have got to be weighed. The first is access: can it be reached? The second is moral: should it be done? Are there risks to sentient life that are unacceptable? And what counts as unacceptable?

The deepest wish of my weary soul is that that question could answer itself with the simple relief of *never*, that no action that included such risks should be considered. I understand people who need to answer with that, and I respect that our ultimate decisions both ache and diverge. I also know that the strategy that eases that ache will not work in the time our planet—200 species today—has left.

Ten minutes on the Internet will tell anyone where the oil comes from, where the tankers dock, where the refineries blister in clumps along the coasts. All of this information is easy, and public, and obvious.

And right now, as I write this, the BP Deepwater Horizon accident is still pouring oil into the ocean, unleashing a smothering hell. The first seabird, a brown pelican, recently taken off the endangered species list, was found covered in oil, the first leatherback turtle found gasping for air. By the time this book waits in your hands, who knows how many casualties there will be, the collateral damage of everyday life in this culture.

And the part no one is talking about, the facts that mean nothing because the environmental movement doesn't think like a resistance: a year after Katrina, 12 percent of the oil and 9.5 percent of the natural gas production had stopped for good.[36] The facilities haven't been rebuilt because the reserves left aren't worth the costs of construction. But you and your fictional friends understand: actions against infrastructure will get less desultory every year.

In our story, both the laws of physics and the reality of economics hold true. A few more minutes of research will yield maps, gas and oil pipelines, the rail lines that carry coal. A few more, the addresses of corporate headquarters. Another search reveals a tiny handful of factories that make the monstrous equipment for mountaintop removal. Once more: those sites could be shut down using civil disobedience, but unless you have a nonviolent army of thousands, all that you and your twenty friends will accomplish is a morning of symbolic action. But thinking like a resistance, you and your twenty friends could stop mountaintop removal.

Will they build it back? A lot of it. And the resistance will bring it down again because that's what resistance movements do. Will someone get caught? Probably, but there are others ready to take their place. Will there be consequences and fallout that no one foresaw? Yes. Releasing mink from European fur farms has devastated ground-dwelling birds and native mink populations. Someone more knowledgeable could have predicted that, but the liberators didn't know and didn't ask. DEW requires cadres, not just combatants, people who will research, study, and think. But in the end, all the planning in the world will not save DGR actionists from the moral grief and adult sorrow that our responsibilities hold.

What would be the unwelcome consequences of industrial sabotage? Flammable materials do not always play well with others, so utmost care must be taken if oil and gas infrastructure are targeted. Environmentally, the risks could be quite minimal, as pipelines have control houses as well as shut-off valves. Washington State, as one example, requires leak detection systems that can locate a leak of 8 percent of maximum flow within fifteen minutes or less. In some places, the figure is 2 percent.

Targeting the Internet would take more specialized skill, so its accessibility is limited, but it's a target that involves no risks to sentient beings. To be clear: the Internet does not exist so people can tag each other on Facebook. It was originally created for the military, and was quickly adopted by corporations. It's what makes the vast and instant transfer of capital possible; without it, there would be no globalization. And the electric grid is 300,000 critical kilometers of accessibility. Even intermittent disruption could bring industrialization to a near halt.

And every day of that halt is that much less carbon in the sky, that much more breathing room for bison and black terns, that much more of a chance for the poor the world over whose lives and lands are being gutted by weapons made of power and wealth. Poor people are not hungry for lack of American imports. They're being driven off their land and into starvation by the dumping of cheap agricultural commodities. Six corporations essentially control the world food supply, and they've wrecked self-sufficient subsistence economies the world over. The sooner the imports of the grain cartels are stopped, the more likely it will be that the impoverished can reclaim their land and their lifeways. Remember that the poor are impoverished because the rich are stealing from them. On our planet right now, the wealthiest 20 percent (that would include you and me) account for 76.6 percent of all private consumption. The poorest fifth get just 1.5 percent.[37] The authors of this book have been accused of suggesting genocide: meanwhile, the genocide is happening now. Anything that stops the rich can only ease the burdens on the poor, including the burden of starvation.

And every disruption in daily life in the rich countries helps break through the denial that this way of life is stable and permanent. Remember, the end is inevitable: anything that encourages people to start preparing will ease our collective way into energy descent with less suffering. Nothing that our actionists do is going to bring industrial civilization crashing down in twenty-four hours. DEW will not result in sudden mass starvation, here or abroad. It will result in disruptions, and if it works, those disruptions will become more or less permanent over a few years' time.

The disruptions of DEW will give the global rich an opportunity to realize the vulnerability inherent in their dependence on industrial civilization, and start rebuilding the resilient communities that are the core project of the Transition Town movement. The need for those local economies and local democracies is urgent from the impending reality of peak oil and catastrophic climate change. The faster we can make the industrially cushioned feel that urgency, the more time they will have to prepare. It takes time to learn to grow food, to accumulate skills, and build the required infrastructure. It takes even more skills and infrastructure to create a functioning democracy.

And never forget there are other people being hurt right now, people who have no choice about oil or coal or iPods, starting with a brown pelican and a leatherback turtle. They have a right to not be choking on sludge, and they have a right to a future for their children as well. They have no choice about denial or preparation, and no possible transition to a way of life on a planet too many degrees hotter than anything their ancestors knew.

In our story, the first direct hit to industrial infrastructure is likely to be something more pragmatic and less daring, like the electric grid. Our actionists have planned well. Remember the four criteria for target selection: the grid is accessible, vulnerable, and critical, and while it is recuperable, the abundance of the first three criteria could potentially make that recuperability more theoretical than practical.

The underground networks can hit a few nodes at once, and the unconnected affinity groups, well versed in DEW and the DGR grand strategy, can follow up on the vulnerable targets to which they have access. The first DGR blackout could last days or even weeks.

An instructive event to consider from recent history is the Northeast Blackout of 2003. On August 14, a huge power surge caused a rolling blackout over a large section of northeastern US and Canada, affecting fifty-five million people. This event brought home how very delicate power grids are. Because electricity can't really be stored, it has to be consumed within a second of being produced or else dumped. Supply and demand have to be matched very precisely or costly infrastructure can be seriously damaged by either too much or too little power. The grid has built-in protective relays to guard against flashovers, which disconnect any line that has a sudden surge in power. But with such tight correspondences, it's amazing that any of us have reliable electricity.

August 14 saw a cascading failure that started with electric arcs between a few overhead lines and some trees in northeast Ohio. By the time the grid had finished responding, power plants all across the Northeast had gone offline and a full-fledged blackout was on. A total of 256 electric power plants shut down, and electricity generation dropped by 80 percent.

But the phrase "cascading failure" applies to a lot more than the grid. Oil refineries couldn't operate and neither could the nine nuclear power plants in the region. Gas stations couldn't pump gas. Air, rail, and even car traffic halted. The financial centers of Chicago and Manhattan were immobilized, and Wall Street was completely shut down. The Internet only worked for dial-up users, and then only as long as their batteries lasted.[38] Most industries had to stop, and many weren't running again until August 22. That last includes the auto industry. The major television and cable networks had disruptions in their broadcasts. In New York City, both restaurants and neighbors cooked up everything on hand and gave it away for free as the perishables were just going to have to be thrown out. Meanwhile, the Indigo Girls concert went on as planned in Central Park. And the New Jersey Turnpike stopped collecting tolls.

I don't know about you, but I'm not seeing any drawbacks here. The cascade was broad and deep, if short. Fossil fuel use was seriously decreased; nuclear power plants rendered useless; oil went unrefined in northern New Jersey, my child's eyes' vision of Mordor in that last whisper of wetlands; the rich were kept from draining the poor; and the flood of lies and vicious media images stopped drowning our hearts, our children, and our culture for a brief night. And there were parties with neighbors and music on top of that.

The DEW activists will be soundly condemned, and not just by the mainstream, but by Big Eco, and by many grassroots activists. This is to be expected. Our actionists need to prepare for it emotionally, socially, organizationally. It can't be helped. Remember the goal: to disrupt and dismantle industrial civilization. Judged by that goal, our actionists' first attack on the electric grid has been a raging success.

And nothing breeds success like success. More groups form, more cells divide in the network. Maybe a whole arm is dedicated to the grid while others go on to other targets. Like the tar sands. The pipelines carrying tar sands oil from Alberta to the coast are 800 miles long; sabotage is too easy. Meanwhile, the equipment necessary for the massive scale of the tar sands extraction is almost inconceivable: twenty stories high and counting. Some of it has to be carried on trucks with ninety tires on twenty-four axles, weighing a total of 917,000 pounds, which is so heavy that two auxiliary trucks are needed to help push.[39] These trucks need special permits and are only allowed on the highway during daylight hours.

Our story is accelerating. A victory for the Tar Sands Brigade comes on the night the draglines are torched, and a few of the factories that make them are incinerated. Does Suncor get more? Yes. And those are burned as well, somewhere on their vulnerable route between their arrival point in Bellingham, Washington, and their departure point in Fort McMurray, Alberta.

Again, Big Oil, Big Coal, and Big Eco all condemn the activists. The public overwhelmingly hates them. But in the Athabasca River, the northern pike and the tundra swans love them. More equipment is purchased. Our actionists respond by sinking the replacements on the boats before they even touch shore and, for added emphasis, a midnight demolition of a corporate headquarters or two. Native Athabasca Chipewyan and Mikisew Cree elders and more than a few Clan Mothers are smiling all week. The warriors, meanwhile, ask some questions, starting with: kakipewîcîhwin cî? Will you come and join me? It's up to them to decide whether to move from protecting their community to offensive action. The young, of course, are all "Yes." When the next DGR blackout rolls through the middle of the continent, a sudden blast blazes across the night as a key bridge comes down on Provincial Route 63. Try getting that million-pound equipment across the river now.

Only a few hundred people are involved at this point. There are three networks, one in the northeast US, one in the Pacific Northwest, and a smaller one in the upper Midwest. There are also affinity groups in Vancouver, Asheville, Burlington, Austin, Guelph, Montreal, and some of the First Nations' warrior societies are now involved.

And in this story, there are people who want to join, but can't. They make the decisions they have to make, and do what they can instead. They translate a scaled-down version of this book—the marrow, the soul—into Hindi and Spanish and Mandarin and Sámi. *Deep Green Resistance* becomes *Résistance Verte Profonde* and then *Molaskaskwi Aodwagan*, slipping south into *Resistencia Verde Radical*, crossing oceans into *Djúpur Grænn Mótspyrna, Dunkelgrüner Widerstand, Mörktgrönt Motstånd, Paglaban Malalim Berde*. The question only changes its sound, never its heart: K'widzawidzi nia? Ti unirai a me? Kayo ay sumali sa akin? The question is asked and asked and asked, whispered like a prayer in that moment the heart shifts from petition to thanksgiving: will you join me? Until "me" becomes "us," because finally a resistance has quickened.

The resistance never loses sight of the targets, though it may lose combatants over it. Better to have a reliable few then an unstable more, especially when potentially dangerous activities are involved. The targets hold steady: fossil fuels, industrial logging, industrial fishing, industrial agriculture, and industrial capitalism.

Industrial logging is ripping the lungs from the earth, and the people from their homes. The Amazon rain forest once sheltered ten million indigenous, reduced now to under 200,000. If you want to talk about genocide, there is a trail of tears still wet with blood leading to the actual perpetrators: Mitsubishi, Georgia-Pacific, Unocal (now Chevron). Unocal, for instance, was sued by Burmese villagers for complicity in rape, torture, forced labor, and murder, abuses inflicted on them when Unocal put in its pipeline. They were also forcibly relocated, the happily ever after of this story every single time it is told.

DGR requires a trail of solidarity, a trail that is build up into a protective barrier, an unbreachable line of determination against industrial assault. Our actionists draw that line around every rainforest and every last stand of old growth, and they build that barrier with transfers of funds and training and materiel. They also build it with risks and courage, as corporate infrastructure is within reach of people in the

United States and Canada, especially the white, native English speakers who can dress the part.

Industrial logging requires a chain of command, a flow of capital, specialized equipment, transportation routes, and end points in manufacturing centers. Every item in that list reads like a command to a general officer. Our actionists, steeped in strategy, understand what needs to be done, and some of the elves come out of the trees to join them, picking up the weapons of this war.

Industrial fishing is made possible by gigantic trawlers three stories high, with steel rollers on the bottom. The rollers crush everything, starting with the oceans' forests, coral. Coral reefs are the oldest living communities on earth, some of them over fifty million years old. Read that again: fifty million years. They are home to one-quarter of ocean life.

Industrial fishing is the murder of the oceans along with the people who once subsisted on them. That murder—the vicious lines, the voracious nets, the silent drain of life—is an emergency that displaces metaphors. We use the oceans as a stand-in for anything vast, ineffable, eternal. But the vast is being emptied, the ineffable priced for pennies, and the eternal—fifty million years of it—is being crushed to dust.

And that murder has infrastructure, just like logging and oil and coal. It has a small handful of command centers, a few weapons manufacturers, some perpetrators, and some supply lines. Every DEW strike against fossil fuels and the electric grid will slow the industry down indirectly; direct attacks will bring it down faster. Remember this: there used to be whales in the Mediterranean. Will our children learn that there used to be fish in the ocean? Remember this as well: two out of three animal breaths are made possible by plankton, by the oxygen they produce. We owe everything to those tiny creatures, creatures whose home of water is acidifying with every hour that industrialization burns on. If the oceans go down, we go down with them.

Industrial agriculture may present fewer targets, but those targets are crucial to fish and forests and the last scraps of prairie. They're also crucial for food security and cultural survival in the majority world. Fish are at risk because agriculture requires water, especially those Green Revolution crops, and that water is either pumped from aquifers or drained from dammed rivers. A dammed river is a dead river, and what dies first are the fish. Next to die are the trees that need the nutrients the fish bring. Trees also need the ground water that has now sunk a mile below the surface, drained out for cotton and rice. An engineered river is the exact opposite of a wetland, which were once the most species-dense habitats on the planet. Without the wetlands, the birds are gone. Rivers are essentially the blood of the world, pumped by a heart of seasonal floods and spring thaws, and their veins have been emptied for cheap agricultural commodities that leverage too well into power and wealth.

Many dams score high for industrial criticality. The Mississippi has been tamed into concrete not just for agricultural draws, but so that the waterway can be used for transport. Huge barges carrying grain travel downstream, out the Gulf, and around the world, while oil traces the opposite path. Dam removal is also critical for biotic survival, and the demolition of dams would be a cascading success for birds and fish, for wetlands and forests, for the disappearing deltas and the slim hope of prairies.

In our story, there are houses on those once and future floodplains. Our actionists warn people and warn them well, because DEW has to mean it. These are not symbolic attacks meant for media coverage. These are the last chances for that long, slow pulse of life now bleeding out around the globe.

◻ ◻ ◻

The end of industrial agriculture could be an opening where the culture of resistance gets serious. Somebody has to start repairing the prairie. That industrial carbon has got to be sequestered, and the bison brought home to help. A political migration to Kansas happened once; there's no reason it can't happen again. People felt the emergency of slavery and knew that the entire west could fall if Kansas didn't hold as a free state. Thousands of abolitionists moved to the middle of nowhere—the cultural edge of the universe for Boston urbanites—to stop slavery, and they succeeded. If environmentalists would only understand that the prairie is desperate to return and do its part, that all it needs is people willing to help it, then acre by acre hope could take root. The young and idealistic have been willing to fight fascism in Spain, to harvest sugar in Cuba, to pick coffee in Nicaragua. They're needed now to plant prairies, only no one is calling them. Let this be the first call: repair, restore, rejoin. Repair the broken rivers, the exhausted soil. Restore the grasses and their animal cohorts. Rejoin as participants, never again to dominate. Stop buying barely edible industrial waste products manufactured from soybeans, and start dreaming of prairies. The land itself is cheap. Understand that corporations don't own the land. They are very clear that if they owned the land, they'd have to pay farmers as employees. Now, they can command prices below production costs, and the federal government makes up the difference. Farming, according to the US Department of Labor, is a statistically insignificant occupation. There are ghost towns across the Midwest without enough children to fill a baseball team, let alone a high school. And just like MEND is financially self-sufficient, grass-based farmers can make money in the first year. So gather your friends and your deep green vision and go. Thousands of people did it in 1854. Another 100,000 followed Helen and Scott Nearing to Vermont. Follow the Pasque flower, the first one to open in the tall grass prairie; follow the bison. Their beauty and sturdy grace alone could call a generation. The Dakota Indians sing when the Pasque flower blooms, to encourage the rest to follow. Let your acre of prairie be that first flower, and sing for all you're worth.

The last two generations have seen a mass migration from rural life to urban, both in the US and around the globe. Those dislocations, caused by economic pressures ultimately based on the application of fossil fuels to civilization, are a billion tears in the weave of human cultures and human hearts. As the oil age shudders and dims, those migrations will naturally reverse if they can. People go where the hope is, especially the hope of basic survival. I say "if they can," though, because the land they have left behind has in many places been reduced to salt flats and sand. Nothing is beyond repair—life wants to live—but their repair may take resources and time that starving people don't have, as well as democratic decision making in areas ruled by corruption. How this will play out is anyone's guess. It's a horrifying race between the forces for life and justice and the accumulated power of the entitled. Kenya's Green Belt Movement is forty-five million trees and one Nobel Peace Prize strong, and is as rooted in democracy and feminism as it is in the regreening slopes of their mountains. "Failure to act now will be catastrophic. This means that we are the only generation of humans ever who are able to effectively respond to this challenge," said the Prize recipient, founder Wangari Maathai.[40] Dust storms from China, meanwhile, have affected air quality in Colorado, 94 percent of Iran's agricultural land is degraded, and one-third of Pakistan is under risk of desertification.[41] All of this shows how absolutely necessary the aboveground and the militants are to each other. DEW alone cannot stop processes of desertification, while all the committed efforts of human rights and democracy activists will not produce the essential changes needed in the time left to our planet.

The crumbling of the global economy could easily mean that in the majority world, where the impoverished majority live, the rural poor get to stay home and the urban poor can return home. For the minority world, where the rich and powerful minority live, Europe is in a very different situation from the United States and Canada, because Europe's built environment was in place long before the age of the automobile, and it was designed to human scale. They have also done a much better job at protecting the farmland outside towns and cities. Sweden, for instance, outlawed shopping malls. Anyone in the US who suggested that would be either tried for treason or burned as a heretic. The average bite of food in the US travels 2,000 miles, in part because it has to: the land around towns and cities has been devoured by asphalt, the sacrifice demanded by the God of Gasoline. As the inimitable James Howard Kunstler puts it, the suburbs are "a living arrangement with no future." That future is almost here, and urbanites in the US and Canada need to face it now, before the laws of physics enforce their own facts. This is true whether or not DGR actionists get serious.

The coming of energy descent and biotic collapse, in whatever proportions, do not have to mean mass starvation. To be very blunt, it is up to us whether we starve or eat. Will the energy left to society go to more useless crap for the wealthy or will it go to transport basic sustenance while local economies struggle into existence? Are we willing to tell the wealthy that they can't have a personal mountain of electronic junk, not while we lack for food? And 90¢ of every food dollar in the US goes to *processed* food. Right now, subsidies to the grain cartels make agricultural commodities the cheapest calories on the market. The food supply is structured for corporate profits. So unstructure it. It is our mutual fault if we starve, our failure to take back our power. Our denial is, in the words of Kunstler, "wholly incompatible with anything describable as our collective responsibility to the future."[42]

That responsibility includes the final target of industrial capital. Fourteen hundred people control the world economy. This one is simple: they have our wealth and we aim to take it back. Once more, this will necessitate the combined dedication of the aboveground and the militants. The destruction of the physical infrastructure of capitalism is only a stopgap so long as law structures organized theft, and that theft is backed by force. But the activism and initiatives to redirect our economies to human needs are not winning, not anywhere on the globe. Those initiatives need help. Targeting the infrastructure of global capitalism involves little threat to human life. There are twenty major stock exchanges. All of them are profoundly dependent on electricity. All of them close at night. All of them are in large cities that require transportation for millions of commuters. And once more: without the Internet, globalization would not be possible. Believing that the poor are dependent on the rich is just an updated version of the White Man's Burden. They don't need America's grain, GMOs, technology, or corporations. They definitely don't need the rich to transform their "resource base"—their land, trees, fish, oil, sunlight, or labor—into wealth and then loan it back to them. The more our actionists can disrupt the flow of capital, the more breathing room there is for fragile radicles of justice.

◙ ◙ ◙

Here is the emotional tension in our story. Our actionists have a very fine balance to walk. There is tremendous strain between the need for action and its necessary preparation. The pull to decisive targets is set against the ethical weight of possible casualties. The desperate need for

serious impact leads to a moral dilemma inherent in uncertain consequences. And there will be so many more dilemmas, some requiring
decisions and offering no time except for regrets.

DEW will require sacrifices, some of them harsh and permanent.
Our actionists may have to choose this work over love, family, friends.
They will have to take recruitment seriously and security breaches even
more seriously. They may have to go to prison for half a lifetime rather
than turn on their comrades. They may have to risk their lives, and
what's often harder, the lives of others. There will be no heroes' welcome, not for the nonindigenous. There will be secrecy and trauma and
betrayal, and it will wear them to the bone.

But because this is our best hope, there's also the possibility of victory. The strikes will be decisive, but the victory will be more like the
slow search of roots through soil. From above, today looks no different
than yesterday, but the roots don't give up, not today, not the next day,
or the next. Until finally the fragile filaments find water, and then all
things are possible. You will find water when the answer is yes. You
will find more water when six yeses meet to draw a map of the possible,
a list of the tasks, an arrow aimed at the heart of hell. Strength is only
half the pull. Steady your hands as you take aim. It will take a few
months to let it loose. But that first arrow will be fletched with the
feathers of passenger pigeons and great auks, and every flying thing
will wish it home.

In six months, you've scored a few and lost a few, but you're ready
for more. More means your success has parlayed into recruitment and
a small network is almost in place. In nine months, they're trained. In
ten, the need-to-know order ripples through. Two days later, the grid
goes dead, the bridge comes down, the equipment sinks or burns up
in the night. You have bought life on this planet—from the tiny green
constancy of plankton to the patient grace of bison—a few more hours,
maybe a day.

And the joy you weren't expecting: across the continent or halfway
around the world, someone else answers in kind, a "Yes" in the clear
language of resistance. People you will never meet darken the sky above
Berlin or Bangkok, light up the night in Fort McMurray, kill computers
in the Bombay Stock Exchange. The war is on.

In a year you've crashed the grid twice; the Forest Brigade has taken out equipment and roads, two factories, and a nice chunk of Plum Creek's corporate headquarters. And Fish Defense got the Swan's Falls and Minidoka dams. Twelve are dead, three have been captured. And the response by those in power has been swift, severe, and indiscriminate. Two hundred people have disappeared, taken by the police or by corporate goons. Some may be actionists, some may be aboveground activists. Some may have nothing to do with the resistance at all. Those three who are captured don't talk, and the message comes that they won't. All you can do is mourn in the minutes between sleep and waking. Some day you can break and let tears come. But not now. Now all focus is forward.

It takes a few more months, but one morning the news is everywhere: in the night, three draglines in West Virginia were melted to scrap. "Leave our mountains or you will die in them" is the single communiqué. You don't know these actionists, but you know the rhythm of their hearts. The Oil Brigade has left for Louisiana, committed to taking down the rigs, a toxic mimic of a forest rising above the sea, a sea that has been slick with oil for twenty years. The dams on the Mississippi are attacked, one by one by one. Then a whole cell is caught in the Midwest, eight of them rounded up. Paranoia spreads like a plague, the rumors, the purges. Your network holds because you built it to do that. Only the serious were asked, and they were trained. They also had to swear on everything they held sacred to hold to discipline and act with honor.

Aboveground, Big Eco and the public intellectuals of the environmental movement have nothing good to say about you and your unmet comrades. It doesn't matter. Under the surface, people are talking, and the young, ever fearless, want to join. And it doesn't matter because what does matter is the goal, the strategy, and the targets. Convincing the readers of *Yes! Magazine* was never the plan.

By the end of the second year, the grid is no longer dependable. The economy is stuttering, and the American public is ready to drink your blood. But somewhere a black tern is feeding her young, and when they fledge, they will carry the hope of their entire species on their small wings. In Burma and the Amazon, a few elders still speak their native languages, dense with words for plants and rain and spirit. Just outside Boulder and Lincoln and Des Moines, there are bison again, and a few brave acres of perennial grasses. The I-70 underpass into Lawrence is emblazoned "You are in Free Lawrence! Deep Green or die!" A thin stream of repairers has made its way here, from Baltimore and Seattle and Oakland, rewilding not their psyches, but the world. The first ones teach the next how to plant, how to keyline for precious water while the grass takes root, how to keep respectful watch on a bison heifer expecting her first. And they teach evolution, birth control, and democracy in their alternative school.

In year three, oil hits $200 a barrel, then $210, $220. A little higher and the system will start to crash in upon itself. Carbon is at 400 ppm and still climbing. The network in the east sends successful shiploads of homemade materiel south. The people have more than spears to fight with now. The Belo Monte dam is stopped forever: 20,000 people and the forest get to stay home.

The People's Militia in rural Wisconsin and Maine set up firewood deliveries to the elderly in the winter. Vermont votes for independence by a slim 2 percent. Cascadia starts talking. Farmers in India stage a Bengal Tea Party, dumping cheap commodities from the US into the bay, then blocking the roads from the ports. Nonviolent activists are able to completely shut down the G-20 meeting that year. An amendment to the US Constitution to strike corporate personhood is making its way through the states. People who bring soy products to permaculture potlucks start getting funny looks. Las Vegas goes dark and even those who hate you have to smile. But there is no air conditioning in New York City or Washington or Atlanta, the trees are long gone, and the summers are hotter than anything this planet has known. The frail and the elderly are hit hard. And there are widening gaps on the supermarket shelves.

But urban chickens have eased the way for the return of goats and pigs. Lawns give way to browse; people learn to calculate the carbon sequestration number—affectionately called "seek"—of their small patches of perennials. The Transitioners write a new platform, a third generation Transition Town manifesto, based on direct democracy, human rights, feminism, steady state economies. Some run for local office; a few win. In Eugene and Madison and Pittsburgh, there are monumental efforts on behalf of civic literacy and then participatory democracy. In Berkeley, corporations are declared illegal. Gulabi Gangs start in Boston, Northampton, and Ithaca, then in London and Amsterdam and Mexico City. The Gangs send books to girls' schools in Pakistan and Sudan, and emergency contraception disguised as baby aspirin. Rumor has it they also send guns.

The rewilders, eyes gleaming, pledge to buy up the flood plain of the Mississippi River, acre by acre. Since the current inhabitants can't get insurance any more, some of them sell. Others are intrigued, tour the restored wetlands, look over the accounting books, and sign up for some summer interns. In minor league baseball, the Peoria Chiefs become the Peoria Prairie Dogs. There are bison visible now along the Trans-Canada Highway and I-90.

Your action group has gotten good at speedboats and the geography of oil rigs, the landscape of pipelines, and you are fluent in the language of megawatts. There are tracts of old-growth forest now fertilized by the blood of your friends, but the trees still stand. There are sixteen-year-olds in Lima and Chennai and suburban Minneapolis desperate to say yes. Your numbers keep rising, but so does the carbon. It's a grim race to the end.

And from here the story is uncertain. I can't finish it. Only you can. Whatever work you are called to do, the world can wait no longer. Power in all its versions—the arrogant, the sadistic, the stupid—is poised to kill every last living being. If we falter, it will win. Gather your heart and all its courage; fletch love into an arrow that will not bend; and take aim.

Will you join me? The clock starts now, the moment you put down this book and think as hard as you've ever thought: who can I ask to join me? Our clock doesn't tick off seconds; it advances by species and carbon. How many and how much since you started this book? Will you join me?

In the time it takes to write that question, another amphibian has dropped into the abyss of extinction, another flower will never stretch and bloom, another native elder slips with her language from the world. And in the time it takes to say yes, there's still time to make the possible real. There is still time for amphibians as a class, still time for justice to win against power and its rancid pleasures of domination. Will you join me?

Pass that question not from mouth to ear, but from heart to heart. It will have to be whispered, but it can still blaze. K'widzawidzi nia? Te joindras-tu à moi? Ndicele undincede? Apni ki amar sathey jog deben? Let it circle the globe until it comes all the way back: will you join me?

Yes is still possible. But yes, like love, needs to be a verb, our best and only hope. Let yes guide your aim.

Then let it loose.

Getting Started

The secret of our success is that we never, never give up.
—Wilma Mankiller, Cherokee chief

As the saying goes, a journey of 1,000 miles begins with a single step. Sometimes the hardest part of doing something is just getting started. The more ambitious and challenging the journey, the more daunting the first steps may seem, and bringing down civilization is definitely an ambitious undertaking.

Personal productivity writer David Allen (of *Getting Things Done*) often asks people working on a project to write down two things: the ultimate goal of the project, and the next step they need to take in order to make progress on that project. We've defined the goal. For most people, the obstacle is that next step. Thousands of people write to us or ask us at talks and conferences: "But what do I do now?" Hopefully having read this far, you have a pretty good idea. But it never hurts to have a list to choose from.

To that end, here is a not-even-remotely-exhaustive list of some (low-risk) entry points to the grand strategy, a few ways of getting started or expanding your resistance activity, out of the thousands or hundreds of thousands of options available. We've broken it down into above-ground and underground actions, but the lists are not mutually exclusive.

INITIAL ACTIONS AND ENTRY POINTS

Aboveground:
- Read or watch inspiring and informative media about resistance. Organize a movie night or series of movie nights to watch films with others, and to discuss them and how they apply to action in the here and now.
- Make a list of the skills you want to learn. Once you have a

list, make a plan for how to learn the skills and where you can get them from. Prepare a schedule and set aside time each day or at least each week to learn and practice.

- Engage in prisoner support and general solidarity work. Writing to political prisoners is a good way of getting started, and there's certainly no shortage of them. General solidarity work with various struggles is also a good way of getting experience, building alliances, and seeing different perspectives and methods of struggle.
- Be a distributor of propaganda. Pass on your favorite political books, movies, and other media to receptive friends and acquaintances.
- Start or join a radical community sufficiency group in your area.
- Start or join an affinity group for political action and mutual support. Meet on a regular basis to assess political activism in your area, to identify actions you can undertake that would improve that situation, and to develop long-term goals and strategies. An affinity group can help you keep focused and accountable. (A word of caution from experience: it's not necessarily a good idea to live with and/or polyamorously date everyone in your affinity group.)
- Practice being interrogated. Take turns playing "police" and "activist" in an arrest situation. Remember that the police threaten, manipulate, and lie.
- Role play breaches of security culture. Pretend someone in your social circle has bad security behaviors like asking if Jane is an agent or if Jorge is involved in the underground. How do you confront and educate this person?
- Go back to the lists of aboveground tactics in the previous chapter. Pick out something you want to do, plan it, and do it.
- Get your household prepared for when the grid crashes.
- Get to know your landbase and the other creatures who live on it. One of the easiest ways to do this is to pack a field guide on edible plants, a pair of binoculars, and a water

bottle, and just go and spend some time in a relatively wild area near you.

- Build community sufficiency in your area.
- Mobilize people to undertake civil disobedience or related tactics for current struggles in your area. This will help build aboveground movements and train people in how to fight power. You will want as many allies as possible in your area for collapse.

Underground:
- Read the histories of successful and unsuccessful underground groups from the past century, including the Underground Railroad, the ANC, escape lines in Occupied Europe, dissident groups in the Baltics and other Soviet-occupied countries, and the student movement that led to Tiananmen Square. Think about why they succeeded or failed, and what can be learned from them.
- As part of the above, study the challenging realities of life as part of an underground resistance cell (which can often be tedious, anxiety-inducing, or dangerous) and consider whether it is something that you have the deep commitment and constitution to undertake.
- Practice keeping a low profile, and take measures to make yourself an inconspicuous candidate for underground activity. This also means disguising your social networks, not using Facebook, Myspace, etc.
- Read over section on recruitment and screening in the first part of the book. Consider ways to screen others and engage in mutual recruitment in your life. Consider (without writing it down) who in your life would be candidates for forming an underground group and how you might approach them. If you don't have enough candidates, figure out where you would meet them.
- Form a "precautionary group," a group of trusted friends without a name or a mission statement who meet on a discreet basis to discuss, in general terms, the pros and cons of

potential underground action. This group would *not* exist to undertake action, but exists to provide a safe space for discussion; however, it should engage in basic security culture and have explicit limits on what (if anything) about the group can be mentioned to nonmembers.

- Study skills that would be relevant to underground groups but that are perfectly normal and legal to learn in general society. This might include computer encryption and codes, mechanics, first aid, and firearms safety.
- Practice being interrogated. Take turns playing "police" and "activist" in an arrest situation. Remember that the police threaten, manipulate, and lie.
- Role play breaches of security culture. Pretend someone in your social circle has bad security behaviors like asking if Jane is an agent or if Jorge is involved in the underground. How do you confront and educate this person?
- Practice self-discipline in general. Underground cadres and combatants need very high levels of self-discipline and self-control. Establishing regular routines for general training and exercise can be part of this—even a regular jogging routine. Abstaining from drugs and alcohol would be another means of practicing self-discipline. Other acts of omission are also candidates.
- Learn basic survival skills and learn how to cope and improvise under difficult circumstances. Here's an example: plan a weekend camping trip, and pack two bags that contain everything you need and nothing more. Then, just before you leave, flip a coin and leave one of them behind.

Notes

Chapter 1: The Problem

1. Mongabay.com, "Two-thirds of polar bears at risk."
2. Butler, "Climate Change."
3. Wilson, *The Future of Life*, p. 74. See also Olson, "Species Extinction Rate."
4. Ravilious, "Only Zero Emissions."
5. Aitkenhead, "Enjoy Life."
6. Jensen and McMillan, *As the World Burns*, p. 15.
7. Jensen and McBay, *What We Leave Behind*.
8. Jensen, *Endgame*, p. 36.
9. Leber, "Trash Course," p. 21.

Chapter 2: Civilization and Other Hazards

1. See for example Derrick Jensen's *Endgame* or Eric Fromm's *The Anatomy of Human Destructiveness*.
2. See Sanderson et al., "The Human Footprint and the Last of the Wild."
3. In terms of food production, the population spike is mostly the result of the use of synthetic fertilizers, especially the energy-intensive Haber-Bosch process for fixing nitrogen, and the substitution of fossil fuel energy for animal energy (including humans and draft animals). Industrial agriculture uses roughly ten fossil fuel calories for every calorie of food energy produced, a net loss of 900 percent.
4. This dependency is exhaustively covered in the peak oil literature. See the introductory chapter of Aric's *Peak Oil Survival: Preparation for Life after Gridcrash* as well as the suggested readings on energy at the end of this book.
5. *Homo habilis* was a tool-using human species that appeared about 2.5 million years ago. If we assume human civilization is 10,000 years old, then civilization has only existed for 0.004 percent of human history. If we compressed the entirety of human history to one year—365 days—then civilization would appear at around 8:00 pm on December 31.
6. The word itself comes from the word for "city"; see etymological discussion in *Endgame*, vol. 1.
7. Lewis Mumford's monumental work *The City in History: Its Origins, Its Transformations, and Its Prospects* is perhaps the best book ever written about the history of the city as an entity. It discusses many effects of city development in great detail.
8. Mumford, *The City in History*, p. 53.
9. Diamond, *In Search of the Primitive: A Critique of Civilization*, p. 1.
10. Gibbon, *Decline and Fall*, pp. 173–174.
11. It's ironic and deeply sad that for a species to be plentiful chiefly means that it takes a long time for the dominant culture to wipe them out.
12. Hansen et al., "Target Atmospheric CO_2: Where Should Humanity Aim?" Of course, even 350 ppm does not offer a guarantee.

13. This and subsequent figures are from the Intergovernmental Panel on Climate Change's *Climate Change 2007*. Between writing and editing this, estimates have gone up even more. Virtually every scientific estimate seems to be deemed overly conservative within a few years of its publication, not a pattern I would like to bet the planet on.

14. *Climate Change 2007*, p. 17.

15. See, for example, Ward, *Under a Green Sky: Global Warming, the Mass Extinctions of the Past, and What They Can Tell Us about Our Future.*

16. Officially called the Permian-Triassic extinction event.

17. Mrasek, "Melting Methane." Also Hanley, "Climate Trouble May be Bubbling Up in Far North."

18. Blakemore, "NASA: Danger Point Closer Than Though From Warming."

19. See Sample, "Warming Hits 'Tipping point.'" Also Blegley, "Climate-Change Calculus."

20. *Science Daily*, "Pollution Causes 40 Percent Of Deaths Worldwide, Study Finds."

21. Toy, "China Covers Up Pollution Deaths."

22. Bullard et al., "Toxic Wastes and Race at Twenty."

23. Marufu et al., "The 2003 North American Electrical Blackout."

24. Derrick and Aric wrote in much more detail about garbage, waste, and pollution in their book *What We Leave Behind.*

25. Sourced from endangered-languages.com and ethnologue.com, especially Gordon, "Ethnologue."

26. Milanovic, "True World Income Distribution, 1988 and 1993."

27. Robbins, *Global Problems and the Culture of Capitalism*, p. 354.

28. According to David Pimentel, professor of ecology and agricultural sciences at Cornell University. Also *Science Daily*, "Pollution Causes 40 Percent Of Deaths Worldwide, Study Finds."

29. See Shah, "Structural Adjustment."

30. Jubilee Debt Campaign, "Debt and Public Services."

31. The estimated 2007 US Defense Budget was $504 billion dollars, divided by the 8760 hours in a year. See http://www.whitehouse.gov/omb/budget/fy2007/defense.html.

32. Shah, "World Military Spending."

33. Hellman, "The Runaway Military Budget."

34. Duncan, "The Peak of World Oil Production and the Road to the Olduvai Gorge."

35. It is true, although capitalism and science are capable of employing rough substitutes for many particular metals, something that's not true for energy. Some minerals do represent sources of energy. Fissionable materials like uranium may be in short supply, especially after peak oil dramatically increases demand for non-petroleum-based energy supplies.

36. Gordon et al., "Metal Stocks and Sustainability."

37. Salonius, "Intensive Crop Culture for High Population is Unsustainable."

38. Pearce, "Asian Farmers Sucking the Continent Dry."

39. Brown, "Emerging Water Shortages."

40. Ibid.

41. Paulson, "The Lowdown on Topsoil."

42. Mongabay.com, "Africa May Be Able to Feed Only 25% of Its Population by 2025."

43. Facilitation Committee, "Interlinkages Between Drought, Desertification and Water."

44. Millenium Ecosystem Assessment, 2005 (Synthesis.)

45. Ibid.

46. Sanderson, "The Human Footprint and the Last of the Wild."

47. Salonius, "A 10,000 Year Misunderstanding."

48. Worldwatch Institute, *Vital Signs 2007–2008.*

49. Myers et al., "Rapid Worldwide Depletion of Predatory Fish Communities."

50. Black, "Only 50 Years Left for Sea Fish."
51. EurekAlert, "Bottom Trawling Impacts, Clearly Visible from Space."
52. Australian Marine Conservation Society, *Orange Roughy: Down and Out.*
53. Culum Brown, "Animal Minds: Clever Fish."
54. BBC News, "Scientists Highlight Fish 'Intelligence.'"
55. McGinn, "Promoting Sustainable Fisheries."
56. CBC News, "Pine Beetle Outbreak Adds to Greenhouse Gas Woes."
57. Heaney, "Phillipine Deforestation"; ReforestHaiti, "International Conference on Reforestation and Environmental Regeneration of Haiti"; Mongabay.com, "Nigeria: Environmental Profile."
58. Wilson, *The Future of Life.*
59. Laurance, "Reflections on the Tropical Deforestation Crisis."
60. There are a few mathematical assumptions here. I'm assuming that your eyes are just shy of 5.5 feet off the ground, which is taller than average globally, but which makes the width of our swath work out to almost 10 km wide. And I'm assuming that you can hike 30 km a day every day without breaks, a very good clip for a backpacker.
61. Worldwatch Institute, *Vital Signs 2007–2008.*
62. Butler, "Soybeans May Worsen Drought in the Amazon Rainforest."
63. Butler, "Deforestation in the Amazon."
64. Lean, "Dying Forest: One Year to Save the Amazon."
65. Rosenthal, "An Amazon Culture Withers as Food Dries Up."
66. Lean, "Amazon Rainforest Could Become a Desert."
67. For numerous examples of this, see a blog that specializes on the subject: *Shifting Baselines: The Cure for Planetary Amnesia* at http://www.scienceblogs.com/shiftingbaselines/.
68. Name since changed to protect the guilty.
69. See Hubbert, "Nuclear Energy and the Fossil Fuels."
70. The discovery of the greenhouse effect is credited to Joseph Fourier. Arrhenius mistakenly predicted that global warming would take about 3,000 years to develop because he failed to anticipate the dramatic increases in fossil fuel consumption that were about to begin. His predictions of the amount of increase (about 5°C for a doubling of atmospheric CO_2) are consonant with current estimates, but a lack of knowledge about ecology and meteorology led him to suggest that the change would be largely beneficial.
71. Of course, even the unsustainable corporations are doing a pretty good job of bankrupting themselves these days. Obviously there is some advantage to taking a "green" approach to business, since many people are willing to pay more for a green premium. But show me an example of a company that is anything more than "slightly greener" than its close competitors. It would arguably be nice in theory if companies would race each other to become less destructive, but the "green" trend in business seems to function mostly to devalue and debase the idea of genuine sustainability to gain a slight competitive edge.
72. See Rainforest Action Network, "About the Campaign" and Reuters, "Jane Goodall Says Biofuel Crops Hurt Rain Forests."
73. Censky, "GE: 7,000 Tax Returns, $0 U.S. Tax Bill."
74. Aric and Derrick took on this very idea at length in *What We Leave Behind.*
75. See Derrick's extensive discussion of this in *Endgame.*

Chapter 3: Liberals and Radicals

1. Mrasek, "Melting Methane."
2. MacKinnon, *Towards a Feminist Theory of the State,* p. 47.
3. Smedley, *Race in North America,* p. 23.

4. Kennedy, "Institutionalized Oppression," p. 492.

5. Frye, *The Politics of Reality*, p 33.

6. Dworkin, *Letters from a War Zone*, p. 266.

7. Ibid.

8. Ibid.

9. Ibid.

10. Ibid., p. 267.

11. Roszak, *The Making*, p. 61.

12. Barber, "Leading the Vanguard," p. 110. Barber argues that the reason SDS fell was because it failed to address racism, sexism, and empire.

13. Talk at Bluestockings Bookstore, New York, New York, October 4, 2007.

14. Varon, *Bringing the War Home*, p. 58.

15. Ibid., pp. 59–60.

16. Ibid., p. 60.

17. Ibid.

18. Mather, "A History," p. 109.

19. Cleaver, *Soul on Ice*.

20. Pearson, *The Shadow*, p. 292.

21. Fasulo, "The Revolution," p. 23. See also Ogbar, "Rainbow Radicalism."

22. Ibid.

23. Fasulo, "The Revolution," p. 24. See also Ogbar. "Rainbow Radicalism."

24. See also Abramson, *Palante*.

25. Fasulo, "The Revolution," p. 23.

26. Allen, "Reactionary Black Nationalism." See also Neal, *New Black Man*, and Collins, *Black Feminist Thought*.

27. This is William Harper's 1837 defense of slavery: "President Dew [another speaker at the conference where he first delivered this message] has shown that the institution of Slavery is a principal cause of civilization. Perhaps nothing can be more evident than that it is the sole cause. If anything can be predicated as universally true of uncultivated man, it is that he will not labour beyond what is absolutely necessary to maintain his existence. Labour is pain to those who are unaccustomed to it, and the nature of man is averse to pain. Even with all the training, the helps and motives of civilization, we find that this aversion cannot be overcome in many individuals of the most cultivated societies. The coercion of Slavery alone is adequate to form man to habits of labour. Without it, there can be no accumulation of property, no providence for the future, no taste for comforts or elegancies, which are the characteristics and essentials of civilization. He who has obtained the command of another's labour, first begins to accumulate and provide for the future, and the foundations of civilization are laid. . . . Since the existence of man upon the earth, with no exception whatever, either of ancient or modern times, every society which has attained civilization has advanced to it through this process."

28. Fanon, *The Wretched of the Earth*, p. 270.

29. Morgan, *The Demon Lover*, p. 49.

30. Kennedy, "Institutionalized Oppression."

31. Maier, *From Resistance*, p. 16.

32. Ibid.

33. Bushman, "Massachusetts Farmers," p. 79.

34. Maier, *From Resistance*, p. 54.

35. Conser et al., p. 5.

36. Ibid., p. 251.

37. Ibid., p. 8.

38. Raphael, *The First*.

39. Ibid, p. 65.
40. Ibid.
41. Ibid., p. 67.
42. Ibid., p. 78.
43. Ibid., p. 2.
44. Ibid., p. 3.
45. Conser et al.
46. Quinn, *Beyond Civilization*, p. 95.
47. Ibid., p. 6.
48. Ishmael.org.
49. Ibid.
50. Heinberg, *Powerdown*, p. 139.
51. Heinberg, *Peak Everything*, p. 45.
52. *Philosophy Dictionary*.
53. The Xhosa Cattle-Killing Movement.
54. Harrison, *Visions of Glory*, p. 16.
55. Watters, "When Prophecies Fail."
56. Festinger, *When Prophecy Fails*.
57. Sharp, *The Politics*, p. 64.
58. Ibid., p. 67.
59. Quoted in Jensen, *Endgame*, Vol. 2, p. 515.
60. For an example of "eco-friendly wedding" advice, see "Green Guru," p. 46.

Chapter 4: Culture of Resistance

1. There are untold numbers of cultures around the globe, and each one will resist in ways that are unique to it. That means they have their own specific histories and strengths, their own blind spots and failings. In the contemporary US, leftists across the progressive to radical spectrum come from diverse backgrounds. Sometimes these groups have been able to build working coalitions with functioning community norms. What I am critiquing here is the culture created by largely white, more or less privileged young men. To the extent that the values and norms of that group became hegemonic across many movements, this critique needs to happen, as those values will create neither a successful resistance nor the coalitions necessary for social transformation. But my analysis may not have much bearing on other cultures of resistance, nor would it be appropriate for me to criticize cultures of which I am not a member.
2. *Catholic Encyclopedia*, "Adamites."
3. Wikipedia contributors, "Diggers."
4. Thanks to Annemarie Monahan.
5. I need to point out that this man, who is famous for his treatise on child-rearing and education, abandoned as many as five of his own infant children to foundling homes, where most likely they died.
6. This is not to say that nonhumans do not create cultures. It is merely to say that all humans do.
7. Just and Lust, *Return to Nature*, p. 4.
8. Kennedy and Ryan, *Hippie Roots*.
9. Ibid.
10. Ibid.
11. Eley, *From Unification*, p. 172.
12. Blackborn, *History of Germany*, p. 302.
13. Wikipeida contributors, "Talk: Völkisch movement."

14. Wikipeida contributors, "Johann Gottfried Herder."
15. Lunn, *Prophet of Community*, p. 101.
16. Ibid., p. 5.
17. Ibid.
18. Ibid., p. 111.
19. Weindling, *Health, Race, and German Politics*, p. 78.
20. Blackborn, *History of Germany*, p. 373.
21. Biehl and Staudenmaier. *Ecofascism.*
22. Biehl and Staudenmaier, *Ecofascism.* Barbaric is precisely the wrong word. *Barbarian* is what the civilized Greeks called the indigenous of Greece, making fun of their language. Whatever atrocities those indigenous may have committed, they are not of the same magnitude nor motivation of those committed by mechanized societies like modern Germany.
23. Ibid., p. 33.
24. Ibid., pp. 48–49.
25. Crossman, "Introduction," *Young Germany*, p. xviii. And the next wave of the German youth movement in the 1920s was, of course, far less ambiguous.
26. Laqueur, *Young Germany*, p. 49.
27. Dworkin, p. 30.
28. Wilson, *Bohemians*, p. 2.
29. Ibid.
30. Spender, *World Within World*, p. 134.
31. Wilson, *Bohemians*, p. 3.
32. Ibid.
33. Rousseau, *Émile*, p. 169.
34. Melville, *Communes*, p. 101.
35. Melville, p. 64.
36. Blackborn, p. 265.
37. Walsh, *Why Do They*, p. 16.
38. Ibid., p. 31.
39. Ibid., p. 44.
40. Ibid., p. 149.
41. Ibid., p. 37.
42. Ibid., p. 37.
43. Ibid., p. 46.
44. Roszak, *The Making*, p. 39.
45. Melville, *Communes*, p. 12.
46. Ibid., 55.
47. Ibid., pp. 55–56
48. Hoffman, *Revolution*, p. 36.
49. Ibid., p. 27.
50. EST (Erhard Seminars Training) was a seminar for "personal transformation" that started in 1971, kicking off the whole workshop culture.
51. Though it has since been taken up by right-wing ideologues as an excuse for their racist, misogynist, and entitled attitudes.
52. Kennedy, "Institutionalized Oppression."
53. Thompson, *A Discussion.*
54. See Fendrich, "Radicals Revisited," p. 161, and Nassil and Abramowitz, "Transition or Transformation," p. 21.
55. Tye, *Rising from the Rails*, p. 252.
56. Ibid., p. 253.

57. Mackenzie, *Shoulder to Shoulder*, p. 205.
58. Obomsawin, *Kanehsatake*.
59. Mackenzie, *Shoulder to Shoulder*, p. 2.
60. Ibid., p. 19.
61. Ibid., p. 20.
62. Ibid., p. 199.
63. Queenan, *Malcontents*, p. 519.
64. Jensen, "Pornography."
65. Dworkin, *Right-Wing Women*, p. 226.
66. Dines, *Pornland*.
67. Haddow, "Pornocalypse."
68. Jeffreys, *The Industrial Vagina*.
69. Dworkin, *Pornography*, p. 70.
70. Camus, *The Rebel*, p. 47.
71. Jeffreys, *The Lesbian Heresy*.
72. *Science Daily*, "Empathy."
73. Rich, *Dream*, p. 30.
74. Ibid., p. 5.
75. Ibid., p. 14.
76. Murphy, *Plan C*, p. 102.
77. Congressional Public Health Summit, "Joint Statement."
78. American Academy of Pediatrics, "Media Violence," pp. 1222–1226.
79. Jackson, *Distracted*, p. 12.
80. American Psychological Association, "Report of the APA Task Force."
81. Murphy, *Plan C*, p. 248.
82. Bookchin, "Introductory Essay," p. xix.
83. The war on drugs had direct and deleterious effects on my life. I've had a degenerative disease for twenty-six years, and for the first seventeen I was refused any pain relief. There are 16,000 people who die every year from lack of pain medications, because we're forced to take unsafe doses of over-the-counter analgesics. On more than one occasion I have taken thirty-two ibuprofen: that's 6400 mg in a four-hour period. You have to be desperate to do that. I know the exact thought that ran through my head: I'm either going to die or feel better, and either is fine. It's also degrading to be treated like a medication-seeking addict by doctors when you're dealing with life-altering pain and need help.
84. Melville, *Communes*, p. 183.
85. Ross, *The Mood Cure*.
86. Ibid., pp. 105–6.
87. Kerouac, *On the Road*, pp. 179–180.
88. James, *Varieties of Religious Experience*, p. 419.
89. Ibid., p. 425.
90. I said a few bland sentences about looking forward to my garden.
91. Dworkin, *Right-Wing Women*, p. 218.
92. Rich, *Dream*.
93. Sharp, *Social Power*, p. 6.
94. Ibid.
95. Hobsbawm, *Revolutionaries*, p. 76.
96. MacKenzie, *Shoulder to Shoulder*, p. 54.
97. Ibid., p. 332.
98. Ibid., p. 318.
99. Ibid., p. 228.
100. Ibid., p. 230.

101. Ibid., p. 231.
102. Ibid., p. 248.
103. Ibid.
104. Ibid., p. 64.
105. Berger, *Outlaws*, p. 158.
106. Tye, *Rising From the Rails*, p. 222.
107. Bookchin, "Introductory Essay," p. xvii.
108. Wittig, *Les Guérillères*, p. 89.
109. de Longbhuel, "The Gaelic Revival's."
110. Dworkin, *The Making of a Radical Feminist*.
111. Graham, *Loving to Survive*.
112. Kent, *From Slogans*, p. 23.
113. Zablocki, "Forward," p. xii.
114. Kent, *From Slogans*, p. 42.

Chapter 5: Other Plans

1. Hopwood and Cohen, "Greenhouses Gases."
2. It is interesting to note that small engines were invented in ancient Greece and China, but without a cheap source of dense energy they amounted to nothing but toys to amuse the wealthy.
3. Trainer, *Towards a Sustainable Economy*, pp. 82–83.
4. Capra, *The Turning Point*, pp. 201–202.
5. Trainer, *Towards a Sustainable Economy*, p. 6.
6. Trainer, *Renewable Energy*, pp. 7–8.
7. Brown, *Plan B 4.0*.
8. Ibid., p. 24.
9. Ibid., p. 25.
10. Ibid., p. 15.
11. Ibid., p. 17.
12. Williams and Zabel, "An Open Letter."
13. Ibid.
14. Brown, *Plan B 4.0*, p. 113.
15. Trainer, *Renewable Energy*, p. 11.
16. Ibid., p. 40.
17. Ibid., p. 21.
18. Ibid., p. 33.
19. Brown, *Plan B 4.0*, p. 122.
20. Trainer, *Rewewable Energy*, p. 45.
21. Ibid., p. 61.
22. Ibid., p. 57.
23. Ibid., p. 66.
24. Ibid., p. 62.
25. Ibid., p.68.
26. Brown, *Plan B 4.0*, p. 131.
27. Ibid., p. 131.
28. Trainer, *Renewable Energy*, p. 73.
29. Brown, *Plan B 4.0*, p. 131.
30. Newman, quoted in Trainer, *Renewable Energy*, p. 117.
31. Allport, *The Primal Feast*.
32. Brown, *Plan B 4.0*, p. 24.

33. Ibid., p. 210.
34. Ibid., p. 237.
35. Ibid., p. 40.
36. Greer, *The Long Descent*, p. 26.
37. Ibid., p. 27.
38. Ibid., p. 116.
39. Trainer, *Renewable Energy*, p. 154.
40. Ibid., p. 133.
41. Quinn, *Beyond Civilization*, p. 102.
42. Heinberg, *Powerdown*, p. 45.
43. Murphy, *Plan C*, p. 114.
44. Ibid., p. 118.
45. Ibid., p. 129.
46. Ibid.
47. Fitz, "Energy, Environment," p. 21.
48. Ibid., p. 23.
49. Ibid., p. 141.
50. Heinberg, *Powerdown*, p. 15.
51. Transition Network.
52. Chamberlin, *The Transition Timeline*, p. 33.
53. Office of Global Analysis, "Cuba's Food & Agriculture."
54. Díaz-Briquets and Pérez-López, "The Special Period," p. 9.
55. Murphy, *Plan C*, p. 220.
56. Cuban Health Statistics Bureau, "Annual Health Statistics."
57. Johnson, "Crisis Cubana," pp. 1A–5A.
58. Ibid.
59. Omestad, "Cuba Plans."
60. Orlov, *Reinventing Collapse*, p. 8.
61. Ibid., p. 9.
62. Ibid., p. 10.
63. Ibid., p. 11.
64. Eberstadt, "Russia."
65. Rosenberg. "Negative Population."
66. Eberstadt, "Russia."
67. Ibid.
68. Ibid.
69. Ibid.
70. Ibid.
71. Grady, "Chernobyl's Voles."
72. Connor, "Chernobyl: Lost World."
73. Population Institute, "Iran's Family Program."
74. Brown, p. 182.
75. Ibid.
76. Stearns, *The Hole*, pp. 156–157.
77. Växjö kommun, "Sustainable Development."
78. Ibid.
79. C40 Cities Climate Leadership Group, "Renewables."
80. Leahy, "Biodiversity."
81. Caldicott, "Nuclear Power."
82. All statistics from Mongabay.com.
83. Ericsson et al., "A Forest."

84. Ekberg, "The Swedish Law."
85. Wielgolaski et al., *Plant Ecology*, p. 332.
86. Farnsworth."Sweden's Biggest."
87. Swedish Institute, "The Sami People."
88. Ibid.
89. Ibid.
90. Baldwin, "Be There," p. 25.
91. Ibid., p. 2.
92. Bryan, "Vermont's Genetic Code," p. 4.
93. Naylor, "Rebel Against."
94. Commonwealth of Vermont Working Group, "A 21st Century Statement."
95. Sale, "Dispersions," p. 27.
96. Bryan, "Vermont's Genetic Code," p. 4.
97. Ibid.
98. Ibid., p. 5.
99. Flomenhoft, "Vermont's Common," p. 23.
100. Bryan, "Vermont's Genetic Code," p. 4.

Chapter 6: A Taxonomy of Action

1. And there's no need to say "if." Full-Spectrum Dominance is appallingly, yet unsurprisingly, a stated goal of the US government, through military and other means.
2. Hart, *Strategy*, 2nd ed., p. 335.
3. See the *Oxford English Dictionary*, 2nd ed., "strike" as a verb. For Egypt, see Daumas, *Ägyptische Kultur im Zeitalter der Pharaonen*.
4. Foster, *Modern Ireland*.
5. Bureau of Labor Statistics, "Consumer Expenditures 2008."
6. Blackfriars Communications, Inc., "U.S. Marketing Spending to Exceed $1 Trillion in 2005."
7. After his 1904 book of the same name.
8. Margolies, "North Kansas City Company Settles Charge Related to Boycott of Israel."
9. Bureau of Industry and Security, "Antiboycott Compliance."
10. BBC, "The Story of Africa: Tax Wars." Also see Pakenham, *The Scramble*, pp. 497–498.
11. Marx, "No Tax Payments!"
12. Hodge, "Number of Americans Paying Zero Federal Income Tax Grows to 43.4 Million."
13. Civil Rights Movement Veterans, "Birmingham Segregation Laws."
14. *Time*, "Dogs, Kids and Clubs."
15. Fairclough, *To Redeem the Soul*, p. 113.
16. Nunnelley, *Bull Connor*, p. 132.
17. Garrow, *Bearing the Cross*, pp. 176–177.
18. Bass, *Blessed Are the Peacemakers*, p. 105.
19. Carpenter et al., "A Call for Unity."
20. Inflation calculated using "The Inflation Calculator" at http://www.westegg.com/inflation which uses the Consumer Price Index.
21. *Newsweek*, "Birmingham USA: Look at Them Run."
22. Eskew, *But for Birmingham*, p. 264.
23. Cotman, *Birmingham, JFK, and the Civil Rights Act*, p. 45.
24. These bombings took place at various times during the larger campaign. See, for example, Shuttlesworth, "Interview."
25. Gado, "Bombingham."

26. General references for this entire section: United States Holocaust Memorial Museum, "Sobibór." Holocaust Education & Archive Research Team, "The Sobibór Death Camp." *Sobibór: The Forgotten Revolt* (website and books) by Sobibór survivor Thomas Blatt at http://www.sobibor.info/.

    The events of the escape were dramatized in the 1987 movie *Escape from Sobibór*. The film is remarkably true to actual events, having been written in part by survivors of the camp. I would recommend the film, which is currently available for download at the Internet Archive at archive.net.

27. Blatt, "The Hero."

28. Prisoner numbers from Blatt, "Dragnet."

29. If that sounds hyperbolic, remember Gandhi's letter to Hitler explaining why killing Jews was wrong, and would Hitler please stop doing it.

30. Bakunin, "Letter to a Frenchman on the Present Crisis."

31. Obomsawin, *Kanehsatake: 270 Years of Resistance*.

32. Ibid.

33. York and Pindera, *People of the Pines*, p. 321.

34. Ibid., p. 390.

35. Carter, "The MST and Democracy in Brazil."

36. See, in particular, Stalin and Kamo: Radzinsky, *Stalin*, pp. 256–259.

37. Widely known, but see for example: Stephen Kinzer, *All the Shah's Men. An American Coup and the Roots of Middle East Terror*.

38. OSS, *Simple Sabotage Field Manual*, p. 6.

39. Ibid., p. 7.

40. Davis, *The Civil War, Strange and Fascinating Facts*.

41. Neely, "Was the Civil War a Total War?"

42. Catton, *The Civil War*, p.3.

43. Ibid., *The Civil War*, p. 239.

44. Sherman, "Field Order 120."

45. Catton, *The Civil War*, p. 224.

46. Eicher, *The Longest Night*, p. 768.

47. Malcolm X spoke and wrote many times on this very idea: "There is nothing in our book, the *Koran*, that teaches us to suffer peacefully. Our religion teaches us to be intelligent. Be peaceful, be courteous, obey the law, respect everyone; but if someone puts his hand on you, send him to the cemetery. That's a good religion." Malcolm X, "Message to the Grass Roots."

48. Council on Foreign Relations, "Hamas: Background Q&A".

49. United States Army, *Guerrilla War and Special Forces Operations*.

50. And there is also the issue of "kill ratios." If a small revolutionary group were to depend primarily on killing those in positions of power, it would have to kill large numbers of those people for every revolutionary lost in action. The combined deaths caused by RAF action were negligible as a percentage of those in power, but each casualty in action diminished the fighting power of the RAF by several percent.

51. Kuhn, *The Structure of Scientific Revolutions*.

52. Students of orbital mechanics also have access to a set of data that is exhaustive, unambiguous, and mostly numerical.

53. Kuhn, *The Structure of Scientific Revolutions*, p. 151.

54. The Pew Research Center, "A Deeper Partisan Divide over Global Warming."

Chapter 7: The Psychology of Resistance

1. Asch, "Effects of Group Pressure upon the Modification and Distortion of Judgments."

2. They also found that that conformity dropped only slightly even when the experiment was modified so that participants were highly certain and anonymous. Deutsch and Gerard, "A Study of Normative and Informational Social Influences upon Individual Judgment."

3. Milgram, "Behavioral Study of Obedience."

4. See Sheridan and King, "Obedience to Authority with an Authentic Victim."

5. Seligman and Maier, "Failure to Escape Traumatic Shock."

6. Peterson, "Learned Helplessness and Explanatory Style."

7. Rasenberger, "Nightmare on Austin Street."

8. Darley, "Bystander Intervention in Emergencies: Diffusion of Responsibility."

9. Latané, "Group Inhibition of Bystander Intervention in Emergencies."

10. Darley, "This Week's Classic Citation."

11. Wolfenstein, *Disaster*, p. 11.

12. Ibid., p. 13.

13. Ibid., p. 5.

14. Ibid., p. 7.

15. Quoted in Jackson, *France*, p. 403. Jackson sources Claude Bourdet, *L'Aventure incertaine: De la Résistance à la Restauration*, Paris: Stock, 1975, p. 26–27.

16. Jackson, *France*, p. 403. Jackson sources H. R. Kedward, *Resistance in Vichy France: A Study of Ideas and Motivations in the Southern Zone*, London: Oxford University Press, 1978: 76–77.

17. Ibid., p. 405.

18. Laffont, *Dictionnaire historique*, p. 399. This number is according to François Marcot, professor of history at the Sorbonne.

19. Collins Weitz, *Sisters in the Resistance*, p. 10.

20. Paxton, *Vichy France*, p. 294.

21. Again, according to François Marcot.

22. Paxton, *Vichy France*, p. 294.

23. Large, *Contending with Hitler*, p. 72.

24. Hamerow, p. 95. From the essay "Resistance and Opposition: The Example of the German Jews" by Konrad Kwiet. In the country at large, the number was apparently closer to 2 percent.

25. Obviously none of this is unique to the Third Reich. People commonly accept those in power even when they are directly attacked or punished by those in power. Orlando Figes examined this at work under Stalin in his book *The Whisperers*. One kulak sent to Siberia, Dmitrii Streletskii, still adored Stalin, explaining: "I believed in everything. I believed the trials. My family believed, too. We weren't the only ones. Everyone believed in them. If that's what they tell you in school, if that's what you hear on the radio, if that's what you read in the newspapers, you couldn't help believing in them. We believed in Stalin. . . . Perhaps it was a form of self-deception, believing in the justice of Stalin. It made it easier for us to accept our punishments, and took away a bit of our fear."

26. It is important to identify the reasons why some people refuse to give in to conformity, obedience, learned helplessness, and inaction. But understanding those reasons still doesn't allow us to convert a majority of people into active resisters. These psychological phenomenon (and others) are so interlocking, so complex, so constantly reinforced, and so deeply entrenched, that most people will never shake them off.

27. Of course, some of the shift in public opinion in the decades following the war was demographic—those who were retroactively opposed to resistance died off, and young people were brought up and educated about the atrocities of the Nazi regime.

Chapter 8: Organizational Structure

1. Jackson, *France*, pp. 387–388. Emphasis added.
2. Definitions from Dictionary.com and the *American Heritage Dictionary* 4th ed., respectively.
3. Think, for example, of the outside union organizer Reuben Warshowsky in the film *Norma Rae.*
4. Jackson, *France*, p. 406.
5. Ibid., pp. 408–409.
6. Ibid., pp. 408–409. Emphasis added.
7. Ibid., p. 409.
8. Ibid.
9. Ibid., p. 476.
10. Ibid., pp. 411–412.
11. Cocteau, *Journal*, 113–114. Quoted in Jackson, *France*, p. 387.
12. Jackson, *France*, pp. 478–479.
13. Quoted in Jackson, *France*, p. 479.
14. Ibid., p. 477.

Chapter 9: Decision Making

1. Of course, strict consensus systems can break down if a group is *too* politically divergent, because there needs to be a basic political common ground.

Chapter 10: Recruitment

1. See, for example Soon, Brass, Heinze, and Haynes, "Unconscious Determinants of Free Decisions in the Human Brain."
2. McCurley, *101 Tips*, p. 29.
3. Oath of the Greek Democratic Army: "I, a child of the Greek people and a DSE fighter, swear to battle with gun in hand, to shed my blood, and give even my life to banish from the soil of my motherland every last foreign occupier. To banish every trace of fascism. To secure and defend the national independence and territorial integrity of my motherland. To secure and defend democracy, honour, work, fortune, and progress of my people.

    I swear to be a good, brave and disciplined soldier, to carry out all the orders of my superiors, to observe all regulations, and not betray any secrets of the DSE.

    I swear to be a good example to the people, to encourage popular unity and reconciliation, and to avoid any action that reduces and dishonours me, as a person and as a fighter.

    My ideal is a free and strong democratic Greece and the progress and prosperity of the people. And in the service of my ideal I offer my gun and my life.

    If I ever prove to be a liar, and with bad intent violate my oath, let the vengeful hand of the nation, and the hate and scorn of the people, fall upon me implacably."
4. A surviving example of a Luddite Oath reads: "I, of my own voluntary will, do declare and swear that I will never reveal to any person or persons under the canopy of Heaven the names of any of the persons composing the secret committee, either by word, deed, or sign, or by address, marks, or complexion, or by any other thing that may lead to the discovery of the same, under penalty of being put out of the world by the first brother whom I would meet, and of having my name and character blotted out of existence. And I do further swear that I will use my utmost endeavours to punish with death any traitor or traitors who may rise up against us, though he should fly to the verge of existence. So help

me God to keep this oath inviolable." From Hill, *Bygone Stalybridge: Traditional, Historical, Biographical*, pp. 52–53.

5. Graff, *Beyond Police Checks*, p. 9.
6. Davis, *Apartheid's Rebels*, pp. 80–81.

Chapter 11: Security

1. Security Culture, *A Handbook*, p. 6.
2. Ibid., p. 6.
3. Ibid., p. 7.
4. Specifically, he answered in general terms a question about how he had started a fire in an Animal Liberation Front–related arson.
5. Foot, *SOE*, p. 128.

Chapter 12: Introduction to Strategy

1. Albert, p. 397.
2. US Army, *Special Forces Operations* (FM 31-21), pp. 108–109.
3. These limitations are described in the US Army's field manual on guerrilla warfare, *Special Forces Operations* (FM 31-21), p. 13, among other places.
4. Liberals often have a knee-jerk opposition to the idea of destroying anything. It makes them uncomfortable. Of the four choices, they'd prefer to create, change, or monitor just about anything before demolishing it. If you ask them "Aren't you glad segregation ended?" they'll say yes. But if you ask "Aren't you glad segregation was destroyed?" they'll shuffle their feet and look at you sidelong. (It's mostly a semantic difference, but we'd probably be better off if segregation and inequality had been deliberately and systematically destroyed—socially, economically, politically—rather than simply stricken from the lawbooks.)

   Liberals are fond of throwing Buckminster Fuller quotations around: "You never change things by fighting the existing reality. To change something, build a new model that makes the existing model obsolete." Which is fine if that's a working option, but it's naïve to think it's applicable in every situation. Were the indigenous of America wiped out because the invading Europeans made their reality "obsolete"? Even the soldiers, muskets loaded and marching for civilization, didn't believe that. Were the Nazis defeated because the Allies made the Nazi paradigm "obsolete"?
5. Ackerman, *Strategic Nonviolent Conflict*, p. 21.
6. Ibid., p. 24.
7. Ibid., p. 25.
8. Ibid., p. 26.
9. Ibid., p. 30.
10. Ibid., p. 31.
11. Ibid., p. 34.
12. Ibid., p. 38.
13. Ibid., p. 42.
14. Ibid., p. 48.
15. Chaliand, *Guerilla Strategies*, pp. 9–19. Emphasis in original.
16. Foot, *Resistance*, p. 57.
17. Austin, *Up Against the Wall*, p. 157.
18. Jacobs, *The Way the Wind Blew*, p. 70.
19. Gillies, "The Last Radical."
20. Barber, "Leading the Vanguard."

21. Smith, "Sudden Impact."
22. Varon, *Bringing the War Home*, p. 80.
23. Foot, *Resistance*, p. 42.
24. Varon, *Bringing the War Home*, p. 308.
25. Ibid., p. 309.
26. Collins, *Good to Great*, p. 86.
27. Dulles, *Germany's Underground*, p. 32.
28. Kershaw, *Hitler*, p. 665.
29. Jacobs, *The Way the Wind Blew*, p. 123.
30. This has been related many times, but you can read Michael Albert's take on the story in *Remembering Tomorrow*, pp. 166–167.
31. Ibid., p. 165.
32. For further documentation and discussion of this mythology, see Gara, *The Liberty Line*, p. 3.
33. Franklin, *Runaway Slaves*, p. 229.
34. National Parks Service, "Aboard the Underground Railroad: Operating the Underground Railroad."
35. See James Brewer Stewart's essay "From Moral Suasion to Political Confrontation: American Abolitionists and the Problem of Resistance, 1831–1861," in Blight, *Passages to Freedom*.
36. Blight, pp. 70–71.
37. Ibid., p. 75.
38. See "Kidnapping and Resistance: Antislavery Direct Action in the 1850s," in Blight, *Passages to Freedom*.
39. William Lloyd Garrison, *The Liberator*, December 8, 1837.
40. Blight, *Passages to Freedom*, p. 89.
41. Stewart, *Abolitionist Politics and the Coming of the Civil War*, p. 28.
42. DeCaro, "Some White People . . ."
43. Vallandigham, quoted in Henry David Thoreau's "A Plea for Captain John Brown."
44. DeCaro, "Harpers Ferry Raid vs Defense of the Alamo."
45. Ibid.
46. Thoreau, "A Plea for John Brown."
47. DeCaro, "Some White People . . ."
48. Du Bois, *John Brown*, p. 186.
49. From Blight, *Passages to Freedom*, pp. 224–225.
50. Bruce Levine, "Flight and Fight: The Wartime Destruction of Slavery, 1861–1865." In Blight, *Passages to Freedom*, p. 225.
51. Hochschild, *Bury the Chains*, p. 2.

Chapter 13: Tactics and Targets

1. There's a whole related concept in political science called the Overton Window, which is worth reading about.
2. Strain, *Pure Fire*, pp. 91–92.
3. Ibid.
4. Biswas, "India's 'Pink' Vigilante Women." Also, Prasad, "Banda Sisters."
5. The July 20 plot is actually a good example of how lack of coordination by resistance elements can be counterproductive. Failed assassination plots like July 20 caused Hitler to tighten security and greatly reduce the number of his public appearances, meaning that it became much harder for other groups to attempt to assassinate him.

6. Deborah Gray White, "Simple Truths: Antebellum Slavery in Black and White." In Blight, *Passages to Freedom*, p. 65.

7. Blight, *Passages to Freedom*, p. 100.

8. This is discussed in the book *Dangerous Patriots: Canada's Unknown Prisoners of War* by Kathleen M. Repka and William Repka.

9. See the photo online at http://en.wikipedia.org/wiki/File:Mexico.Chis.EZLN.01.jpg.

10. US Army, *Guerrilla Warfare*, pp. 111–114.

11. Berger, *Outlaws of America*, p. 163.

12. The BLA Coordinating Committee, *Message to the Black Movement: A Political Statement from the Black Underground*, online at http://archive.lib.msu.edu/AFS/dmc/radicalism/public/all/messageblackmovement/AAL.pdf?CFID=327481&CFTOKEN=390641 16.

Chapter 14: Decisive Biological Warfare

1. Even the US military now recognizes this. See Macalister, "US Military Warns Oil Output May Dip Causing Massive Shortages by 2015."

2. Aric and Derrick explored the relationships between collapse, carrying capacity, racism, and the Nazis in the closing chapters of *What We Leave Behind*.

3. Shortly after this was written, the government of Spain cancelled $24 billion worth of solar energy investments to avoid spiraling into a national debt crisis that they worried would collapse their economy.

4. See Kevin Bales's important book *Disposable People: New Slavery in the Global Economy*.

5. See International Union of Forest Research Organizations, "Adaptation of Forests and People to Climate Change." Also, the conversation of forests into carbon emitters because of warming, disease, logging, and fires is already happening (Kurz et al., "Mountain Pine Beetle").

6. *Science Daily*, "Regional Nuclear War Could Devastate Global Climate."

7. *Science Daily*, "Regional Nuclear Conflict Would Create Near-Global Ozone Hole, Says Study."

8. Cobalt bombs are nuclear bombs with a cobalt jacket. They were the "doomsday device" in the film *Dr. Strangelove*. Regular fallout has a half-life of days, but cobalt bomb fallout would have a half-life in excess of five years. Some experts believe that cobalt bombs could literally destroy all life on Earth.

9. Novacek et al., "The Current Biodiversity Extinction Event."

10. See Lovelock, *The Ages of Gaia: A Biography of Our Living Earth*.

11. Core samples from the floor of the Arctic Ocean show that about fifty-five million years ago the region was tropical because of a spike in atmospheric CO_2. The biota ringing the ocean was swampy with dense sequoia and cypress trees, and "mosquitoes the size of your head." The year-round average temperature was about 23°C (74°F). Since the Arctic Circle has twenty-four-hour sunlight for most of the summer and twenty-four-hour dark for most of the winter, this average must have been associated with remarkable temperature extremes. Most of the planet was virtually uninhabitable by our standards. The growth of heat-tolerant ferns eventually sequestered carbon and returned the planet to a cooler state, but that took almost a million years to occur. See Associated Press, "Arctic Circle—Ancient Vacation Hotspot?"

12. Congressional Research Service, "Energy Use in Agriculture: Background and Issues."

13. Energy Information Administration, "EIA Annual Energy Review 2008," p. 3.

14. Remember that even now, with plenty of surplus food and housing available, there are tens of millions of unsettled refugees in various parts of the world (not counting those who have been uprooted from traditional landbases and resettled in urban slums).

15. That is net population growth, the number of daily births minus the number of daily deaths.

16. For example, Joseph Tainter writes that "[a] society has collapsed when it displays a rapid, significant loss of an established level of sociopolitical complexity."

17. Again, criteria here based on Tainter.

18. Quotation from a speech by Dimitry Orlov, "Social Collapse Best Practices," given in San Francisco on February 13, 2009, online at http://cluborlov.blogspot.com/2009/02/social-collapse-best-practices.html.

19. Markusen, *The Holocaust and Strategic Bombing*, p. 152.

20. Transcripts of the trial are a matter of public record. See "The Proceedings of the Trial of the Major War Criminals before the International Military Tribunal at Nuremburg," vol. 15, p. 350, at http://www.loc.gov/rr/frd/Military_Law/NT_major-war-criminals.html.

21. Collins, *Grand Strategy*, p. 214.

22. Ibid., p. 230.

23. Pape, *Bombing to Win*, pp. 77–78.

24. Ibid., p. 317.

25. Pape discusses how his preferred strategy of transportation disruption might play out in different settings. "Against an exceptionally import-dependent economy," he writes, "such as Japan in World War II, disruption of transportation can best be accomplished by blockading sea routes, using air power less for bombing than for shipping attack and mining. If imports can be totally cut off, the target economy will collapse when domestic stockpiles are exhausted; the Japanese merchant marine was essentially destroyed by the end of 1944, leading to the collapse of war production by the middle of 1945." Even increasing the cost of imports would have a beneficial effect. The pirates of Somalia are currently doing an excellent job of increasing the cost of international shipping, through delays, ransoms, increased insurance costs, and military expenses for defending the ships. So far, piracy off the coast of Somalia doesn't even require fund raising—it's a self-sufficient business enterprise.

     Conversely, Pape writes: "Against a relatively resource-rich economy, such as Nazi-controlled Europe, strategic interdiction requires stopping the flow of commerce along domestic railroad, highway, and canal systems by destroying key nodes (bridges, canal locks, and railroad marshalling yards), moving traffic, and rolling stock and cargo vessels. This mission is hard because commercial transportation systems are large and redundant and are rarely used to full capacity. Thus, the United States could not bring the German economy to quick collapse even though U.S. air forces were vastly superior."

26. Laffont, *Dictionnaire historique*, p. 399. This number is according to François Marcot, professor of history at the Sorbonne.

27. Collins Weitz, *Sisters in the Resistance*, p. 10.

28. Paxton, *Vichy France*, p. 294.

29. Again, according to François Marcot.

30. Paxton, *Vichy France*, p 294.

31. Jefferies, "The UK Population."

32. BBC News, "Guide: Armed Groups in Iraq."

33. Barrell, "Conscripts to Their Age," p. 495. Interview with Mac Maharaj, *IV/Maharaj*.

34. Britannica, http://www.britannica.com/EBchecked/topic/68134/Black-Panther-Party.

35. Demma, "The U.S. Army," chapter 28.

36. These being very approximate numbers based on Mackenzie, *Shoulder to Shoulder*.

Chapter 15: Our Best Hope

1. Obomsawin, *Kanehsatake*.

2. Bancroft, *Why Does He Do That*, p. 75.
3. Robinson, "Nigeria's Deadly Days," p. 1.
4. Hanson, "MEND."
5. Ibid.
6. Eboh, "Nigeria Shuts Refineries."
7. Ibid.
8. Ibid.
9. Howden, "Shell May Pull Out."
10. Bookchin, *The Spanish Anarchists*, p. 117.
11. We do not mean this as a threat to anyone, but as a reference to the "cold, dead hands" speech of Charlton Heston for the NRA and similar slogans by other gun rights groups.
12. Camp for Climate Action, "About Us."
13. Earth First Appreciation Society, "Mainshill Coal Site."
14. Etnier, "Trnasition Times," p. 9.
15. Online at http://www.johnjeavons.info/.
16. Online at http://www.oildrum.com/node/3541.
17. See Manning, "The Amazing Benefits."
18. Ibid.
19. Markegard and Osofsky, "Living the Local Food Life," p. 14.
20. Bane, "Storing Carbon," p. 57.
21. Markegard and Osofsky, "Living the Local Food Life," p. 14.
22. Hopkins, p. 75.
23. Chamberlin, *The Transition Timeline*, p. 55.
24. Kraussman, "Global Human Appropriation."
25. Dworkin, *Right-Wing Women*, p. 219.
26. Mann, *Iroquoian Women*.
27. *Rain and Thunder*, "Number of U.S. Extremist Groups," p. 3.
28. Elshtain, "Women Under Siege."
29. Walt, "In France, Pension Protests."
30. Lakey.
31. Varon, *Bringing the War Home*, p. 308.
32. Oxfam International, "Suffering the Science."
33. Taylor, *The Age of Terror*.
34. Ibid.
35. Raines, "Terrorism."
36. Kunstler, "Making Other Arrangements."
37. Shah, "Poverty Facts."
38. The backbones of the Internet all run on fiber optics now, which (unlike phone lines) require power all along the transmission route.
39. Lance, "Football Field."
40. Greenbelt Movement, "Climate Change."
41. Shah and Arshad, "Land Degradation."
42. Kunstler, "Making Other Arrangements."

Bibliography

Abramson, Michael, et al. *Palante: Young Lords Party*. New York: McGraw-Hill Book Company, 1971.

Ackerman, Peter, and Christopher Kruegler. *Strategic Nonviolent Conflict: The Dynamics of People Power in the Twentieth Century*. Westport, CT: Praegar Publishers, 1994.

Aitkenhead, Decca. "Enjoy Life While You Can." *The Guardian*, March 1, 2008. http://www.guardian.co.uk/theguardian/2008/mar/01/scienceofclimatechange.climatechange/print (acccessed on June 8, 2008).

Albert, Michael. *Remembering Tomorrow: From SDS to Life After Capitalism*. New York: Seven Stories Press, 2006.

Allen, Norm R., Jr. "Reactionary Black Nationalism: Authoritarianism in the Name of Freedom." *Free Inquiry*, Fall 1995, 10.

Allport, Susan. *The Primal Feast: Food, Sex, Foraging and Love*. New York: Harmony Books, 2000.

American Academy of Pediatrics. "Media Violence." *Pediatrics* 108, no. 5 (November 2001). http://aappolicy.aappublications.org/cgi/content/full/pediatrics;108/5/1222 (accessed July 14, 2009).

American Psychological Association. "Report of the APA Task Force on the Sexualization of Girls," 2007. http://www.apa.org/pi/women/programs/girls/report.aspx (accessed July 27, 2009).

Amin, Massoud. "Security Challenges for the Electricity Infrastructure." *Computer Magazine* 35, no. 4 (April 2002). http://www.computer.org/portal/web/csdl/doi/10.1109/MC.2002.10042 (accessed April 14, 2010).

Angervall, Thomas, Britta Florén, and Friederike Ziegler. "Vilken bukett broccoli väljer du?" *Konsument Foreningen Stockholm*, November 2006. www.konsumentforeningenstockholm.se/upload/Konsumentfrågor/Broccolirapporten.pdf (accessed January 15, 2010).

Anonymous. "Health Consequences of Cuba's Special Period." *Canadian Medical Association Journal* 29, no. 3: 257. http://www.ncbi.nlm.nih.gov/pmc/articles/PMC2474886/#r14-19.

Asch, S. E. "Effects of Group Pressure Upon the Modification and Distortion of Judgment." In H. Guetzkow, ed., *Groups, Leadership and Men*. Pittsburgh, PA: Carnegie Press, 1951.

Associated Press. "Arctic Circle—Ancient Vacation Hotspot?" May 21, 2006. http://www.komonews.com/news/archive/4187966.html (accessed July 29, 2009).

Austin, Curtis J. *Up Against the Wall: Violence in the Making and Unmaking of the Black Panther Party*. Fayetteville: University of Arkansas Press, 2006.

Australian Marine Conservation Society. "Orange Roughy: Down and Out." http://www.amcs.org.au/default2.asp?active_page_id=358 (accessed March 2, 2010).

Bakunin, Mikhail. "Letter to a Frenchman on the Present Crisis." In Sam Dolgoff, trans. and ed., *Bakunin on Anarchy*. New York: Vintage Books, 1971.

Baldwin, Ian. "Be There: The Date is January 15," *Vermont Commons* 33 (Winter 2010).

Bancroft, Lundy. *Why Does He Do That? Inside the Minds of Angry and Controlling Men*. New York: G.P. Putnam's Sons, 2002.

Bane, Peter. "Storing Carbon in the Soil: The Possibilities of a New American Agriculture," *Permaculture Activist* 65 (Autumn 2007).

Barber, David. *A Hard Rain Fell: SDS and Why It Failed*. Jackson: University Press of Mississippi, 2008.

——. "Leading the Vanguard: White New Leftists School the Panthers on Black Revolution." In Jama Lazerow and Yohuru Williams, eds., *In Search of the Black Panther Party: New Perspectives on a Revolutionary Movement*. Raleigh, NC: Duke University Press, 2006.

Barrell, Howard. "Conscripts to Their Age: ANC Operational Strategy, 1976–1986." PhD dissertation, Oxford University, online at http://www.sahistory .org.za/pages/library-resources/thesis/barrel_thesis/index.htm.

Bass, Jonathan S. *Blessed Are the Peacemakers: Martin Luther King, Jr., Eight White Religious Leaders, and the 'Letter from Birmingham Jail.'* Baton Rouge, LA: Louisiana State University Press, 2001.

BBC. "The Story of Africa: Tax Wars." http://www.bbc.co.uk/worldservice/africa/ features/storyofafrica/11chapter10.shtml (accessed July 12, 2009).

BBC News. "Guide: Armed Groups in Iraq." August 15, 2006. http://news.bbc .co.uk/2/hi/middle_east/4268904.stm (accessed August 11, 2008).

——. "Scientists Highlight Fish 'Intelligence.'" August 31, 2003. http://news.bbc.co.uk/2/hi/uk_news/england/west_yorkshire/3189941.stm (accessed March 12, 2010).

Begley, Sharon. "Climate-Change Calculus. Why It's Even Worse Than We Feared." *Newsweek*. July 24, 2009. http://www.newsweek.com/id/208164 (accessed July 30, 2009).

Berger, Dan. *Outlaws of America: The Weather Underground and the Politics of Solidarity*. Oakland, CA: AK Press, 2006.

Biehl, Janet, and Peter Staudenmaier. *Ecofascism: Lessons from the German Experience*. San Francisco: AK Press, 1995.

Biswas, Soutik. "India's 'Pink' Vigilante Women." BBC News, November 26, 2007. http://news.bbc.co.uk/2/hi/south_asia/7068875.stm (accessed July 12, 2009).

Black, Richard. "'Only 50 Years Left' for Sea Fish." BBC News, November 2, 2006. http://news.bbc.co.uk/2/hi/science/nature/6108414.stm (accessed March 9, 2009.)

Blackborn, David. *History of Germany 1780–1918: The Long Century*. Malden, MA: Blackwell Publishing, 2003.

Blackfriars Communications. "U.S. Marketing Spending to Exceed $1 Trillion in 2005." June 1, 2005. http://blackfriarsinc.com/sizing-release.html (accessed April 12, 2008.)

Blakemore, Bill. "NASA: Danger Point Closer Than Though From Warming." ABC News, May 29, 2007. http://abcnews.go.com/Technology/ GlobalWarming/story?id=3223473&page=1 (accessed April 1, 2009.)

Blatt, Thomas. "Dragnet." Sobibór—The Forgotten Revolt.
http://www.sobibor.info/dragnet.html (accessed June 2, 2009).
——. "The Hero." Sobibór—The Forgotten Revolt.
http://www.sobibor.info/hero.html (accessed June 2, 2009).
Blight, David W., ed. *Passages to Freedom: The Underground Railroad in History and Memory.* Washington: Smithsonian Books, 2006.
Bookchin, Murray. "Introductory Essay." In Sam Dolgoff, ed., *The Anarchist Collectives: Workers' Self-management in the Spanish Revolution (1936–1939).* New York: Free Life Editions, 1974.
——. *The Spanish Anarchists: The Heroic Years.* New York: Free Life Editions, 1977.
Bridges, Ana, Robert Wosniter, and Michelle Chang. "Mapping the Pornographic Text: Content Analysis Research of Popular Pornography." http://www.stoppornculture.org/online-resources/
Brown, Culum. "Animal Minds: Clever Fish." *New Scientist,* June 12, 2004.
Brown, Lester. "Chapter 4: Emerging Water Shortages." In *Plan B 3.0: Mobilizing to Save Civilization.* New York: W. W. Norton & Company, 2008.
http://www.earth-policy.org/Books/Seg/PB3ch04_ss2.htm
——. *Plan B 4.0: Mobilizing to Save Civilization.* New York: W. W. Norton & Company, 2009.
Bryan, Frank. "Vermont's Genetic Code: Toward a Decentralist Manifesto." *Vermont Commons* 33 (Winter 2010).
Bullard, Robert D., Paul Mohai, Robin Saha, and Beverly Wright. *Toxic Wastes and Race at Twenty: 1987–2007.* A Report Prepared for the United Church of Christ & Witness Ministries. Online at http://www.ejrc.cau.edu/TWARTlight.pdf (accessed March 28, 2008).
Bureau of Industry and Security. "Antiboycott Compliance." http://www.bis.doc.gov/complianceandenforcement/antiboycottcompliance.htm (accessed April 2, 2008).
Bureau of Labor Statistics. "Consumer Expenditures 2008." October 6, 2009.
http://www.bls.gov/news.release/cesan.nro.htm (accessed November 12, 2009).
Bushman, Richard L. "Massachusetts Farmers and the Revolution." In Richard M. Jellison, ed., *Society, Freedom and Conscience: The American Revolution in Virginia, Massachusetts and New York.* New York: W. W. Norton, 1976.
Butler, Rhett A. "Climate Change Claims a Snail." Mongabay.com. August 13, 2007. http://news.mongabay.com/2007/0813-snail.html (accessed June 4, 2008).
——. "Deforestation in the Amazon." Mongabay.com. http://www.mongabay.com/brazil.html (accessed August 9, 2008)
——. "Soybeans May Worsen Drought in the Amazon Rainforest." Mongabay.com April 18, 2007. http://news.mongabay.com/2007/0418-amazon.html (accessed March 21, 2009).
C40 Cities Climate Leadership Group website. "Renewables: Växjö, Sweden." http://www.c40cities.org/bestpractices/renewables/vaxjo_fossilfuel.jsp (accessed on December 13, 2009).
Caldicott, Helen. "Nuclear Power is the Problem, Not a Solution." *The Australian,* April 15, 2005.
Camp for Climate Action. "About us." http://www.climatecamp.org.uk/about (accessed May 12, 2010).

Camus, Albert. *The Rebel.* Trans. Anthony Bower. New York: Random House, 1954.

Capra, Fritz. *The Turning Point: Science, Society and the Rising Culture.* New York: Simon & Schuster, 1982.

Carpenter, C. C. J., et al. "A Call for Unity," April 12, 1963. http://www.library .spscc.ctc.edu/electronicreserve/eng9697/instructors/PublicStatementbyEight AlabamaClergymen.pdf

Carter, Miguel. *The MST and Democracy in Brazil.* Working Paper CBS-60-05, Centre for Brazilian Studies, University of Oxford, 2005. http://www.brazil.ox.ac.uk/workingpapers/Miguel%20Carter%2060.pdf, and http://www.mstbrazil.org.

Catholic Encyclopedia. "Adamites." Online at http://www.newadvent.org/cathen/ 01135b.htm (accessed January 30, 2009).

Catton, Bruce. *The Civil War.* New York: Mariner Books, 2005.

CBC News. "Pine beetle Outbreak Adds to Greenhouse Gas Woes." April 23, 2008. http://www.cbc.ca/technology/story/2008/04/23/tech-beetle-carbon.html (accessed May 20, 2008.)

Censky, Annalyn. "GE: 7,000 Tax Returns, $0 U.S. Tax Bill." CNN, April 16, 2010. http://money.cnn.com/2010/04/16/news/companies/ ge_7000_tax_returns/ (accessed April 20, 2010).

Center for Responsive Politics. "Money Wins Presidency and 9 of 10 Congressional Races in Priciest U.S. Election Ever." November 5, 2008. https://opensecrets.org/news/2008/11/money-wins-white-house-and.html (accessed January 22, 2010).

Chaliand, Gérard. *Guerrilla Strategies: An Historical Anthology from the Long March to Afghanistan.* Berkeley, CA: University of California Press, 1982.

Chamberlin, Shaun. *The Transition Timeline for a Local Resilient Future.* White River Junction, VT: Chelsea Green Publishing, 2009.

Civil Rights Movement Veterans. "Birmingham Segregation Laws." http://www.crmvet.org/info/seglaws.htm (accessed April 14, 2008.)

Cleaver, Eldridge. *Soul On Ice.* New York: Dell Publishing, 1968.

Collins, Jim. *Good to Great.* New York: HarperCollins, 2001.

Collins, Patricia Hill. *Black Feminist Thought: Knowledge, Consciousness, and the Politics of Empowerment.* New York: Routledge, 2000.

Collins Weitz, Margaret. *Sisters in the Resistance—How Women Fought to Free France 1940–1945.* New York: John Wiley & Sons, Inc., 1995.

Commonwealth of Vermont Working Group. "A 21st Century Statement of Principles (Draft—Winter 2010)." *Vermont Commons* 33 (Winter 2010).

Congressional Public Health Summit. "Joint Statement on the Impact of Entertainment Violence on Children." July 26, 2000. www.aap.org/advocacy/releases/jstmtevc.htm (accessed March 3, 2009).

Congressional Research Service. "Energy Use in Agriculture: Background and Issues." The Library of Congress. November 19, 2004. http://www.nation-alaglawcenter.org/assets/crs/RL32677.pdf (accessed October 23, 2009).

Connor, Steve. "Chernobyl: Lost World." *The Independent,* Dec. 12, 2007. http://www.independent.co.uk/news/science/chernobyl-lost-world-764528.html (accessed January 24, 2010).

Conser, Walter H., Jr., Ronald M. McCarthy, David J. Toscano. "The American Independence Movement, 1765–1775: A Decade of Nonviolent Struggles," In

Walter H. Conser Jr., Ronald M. McCarthy, David J. Toscano, and Gene Sharp, eds., *Resistance, Politics, and the American Struggle for Independence, 1765–1775.* Boulder, CO: Lynne Rienner Publishers, 1986.

Cotman, John. *Birmingham, JFK, and the Civil Rights Act of 1963: Implications for Elite Theory.* New York: Peter Lang Publishing, 1989.

Council on Foreign Relations. "Hamas: Background Q&A." March 16, 2006. http://www.cfr.org/publication/8968/ (accessed January 12, 2009).

Crossman, R. H. S. "Introduction." In Walter Z. Laqueur, *Young Germany: A History of the German Youth Movement.* New York: Basic Books, 1962.

Cuban Health Statistics Bureau. "Annual Health Statistics Reports 1973–2006." Havana City: Ministry of Public Health, 1974–2007. http://www.sld.cu/servicios/estadisticas/ (accessed 2008 June 12).

Danson, Ted. "Industrial Fishing Is Killing Our Oceans." CNN, April 2, 2010. http://www.cnn.com/2010/TECH/03/22/eco.ted.danson.oped/index.html (accessed May 2, 2010).

Darley, J. M., and B. Latane. "Bystander Intervention in Emergencies: Diffusion of Responsibility," *Journal of Personality and Social Psychology,* 8: 377-83,1968. http://www.wadsworth.com/psychology_d/templates/student_resources/0155060678_rathus/ps/ps19.html.

Darley, John."This Week's Classic Citation." February 2, 1981. http://garfield.library.upenn.edu/classics1981/A1981KY95400001.pdf (accessed November 12, 2008).

Daumas, François. *Ägyptische Kultur im Zeitalter der Pharaonen.* Munich: Knaur Verlag, 1969.

Davis, Burke. *The Civil War, Strange and Fascinating Facts.* New York: The Fairfax Press, 1988.

Davis, Stephen. *Apartheid's Rebels: Inside South Africa's Hidden War.* New Haven, CT: Yale University Press, 1987.

DeCaro, Louis A. "Harpers Ferry Raid vs Defense of the Alamo." John Brown the Abolitionist. October 18, 2006. http://abolitionist-john-brown.blogspot.com/2006/10/harpers-ferry-raid-vs-defense-of-alamo.html (accessed November 21, 2009).

———. "Some White People Just Cannot Get Past . . ." John Brown the Abolitionist. March 31, 2008. http://abolitionist-john-brown.blogspot.com/2008/03/some-white-people-just-cannot-get-past.html (accessed November 21, 2009).

de Longbhuel, Máirtín Pilib. "The Gaelic Revival's Influence on the Making of the Nationalist Movement." http://irelandsown.net/revival.html.

Demma, Vincent H. "The U.S. Army in Vietnam." *American Military History.* http://www.ibiblio.org/pub/academic/history/marshall/military/vietnam/short.history/chap_28.txt.

Deutsch, M., and H. B. Gerard. "A Study of Normative and Informational Social Influences upon Individual Judgment." *Journal of Abnormal and Social Psychology* 51: 629–636.

Diamond, Jared. "The Worst Mistake in the History Of The Human Race." *Discover,* May 1987, 64–66.

Diamond, Stanley. *In Search of the Primitive.* New Brunswick, NJ: Transaction Publishers, 1974.

Díaz-Briquets, Sergio, and Jorge F. Pérez-López. "The Special Period and the Environment." http://lanic.utexas.edu/la/cb/cuba/asce/cuba5/FILE23.PDF.

Dines, Gail. *Pornland.* Boston: Beacon Press, 2010.

Dines, Gail, and Robert Jensen. "Pornography Is a Left Issue." *ZNet,* Dec 6, 2005. http://www.zmag.org/znet/viewArticle/4868.

Du Bois, W. E. B. *John Brown: A Biography.* Armonk, NY: M. E. Sharpe, 1997.

Dulles, Allen Welsh. *Germany's Underground.* New York: Macmillan Company, 1957.

Duncan, Richard C. "The Peak of World Oil Production and the Road to the Olduvai Gorge," November 13, 2000. http://www.dieoff.org/page224.htm (accessed May 12, 2008).

Dworkin, Andrea. *Letters from a War Zone.* New York: E. P. Dutton, 1988.

——. *Pornography: Men Possessing Women.* New York: G. P. Putnam's Sons, 1979.

——. *Right-Wing Women.* New York: Perigee Books, 1983.

——. "The Making of a Radical Feminist." *On the Issues* 9, 1988. http://www.ontheissuesmagazine.com/1988vol9/vol9_1988_interview.php (accessed February 6, 2009).

——. "Woman-Hating Right and Left." In Dorchen Leidholdt and Janice G. Raymond, eds., *The Sexual Liberals and the Attack on Feminism.* Elmsford, NY: Pergamon Press, 1990.

Earth First Appreciation Society. "Mainshill Coal Site Sabotaged," *Northern Indymedia.* April 15, 2010. http://northern-indymedia.org/articles/638.

Eberstadt, Nicholas. "Russia, The Sick Man of Europe." *Public Interest,* Winter 2005.

Eboh, Camillus. "Nigeria Shuts Refineries After Pipeline Attacks." Reuters, December 22, 2010. http://www.reuters.com/article/2010/12/22/nigeria-oil-idUSLDE6BL187201012222 (accessed January 28, 2011).

Eicher, David J. *The Longest Night: A Military History of the Civil War.* New York: Simon & Schuster, 2001.

Ekberg, Gunilla. "The Swedish Law that Prohibits the Purchase of Sexual Services: Best Practices for Prevention of Prostitution and Trafficking in Human Beings." *Violence Against Women* 10, no. 10, 1187–1218. http://www.prostitutionresearch.com/laws/000164.html.

Eley, Geoff. *From Unification to Nazism.* Winchester, MA: Allen & Unwin, 1986.

Elshtain, Jean Bethke. "Women Under Siege, Let's Finally Right the Wrongs: Rape is a War Crime." http://www.ontheissuesmagazine.com/1993summer/Summer1993_2.php (accessed April 8, 2010).

Energy Information Administration, "EIA Annual Energy Review 2008." http://www.eia.doe.gov/emeu/aer/pdf/aer.pdf

Ericsson, Staffan, Lars Östlund, and Anna-Lena Axelsson. "A Forest of Grazing and Logging: Deforestation and Reforestation History of a Boreal Landscape in Central Sweden." *New Forests* 19, no. 3.

Eskew, Glenn. *But for Birmingham: The Local and National Movements in the Civil Rights Struggle.* Chapel Hill: University of North Carolina Press, 1997.

Etnier, Carl. "Transition Times: Indiana Transition Plan Sets Bar for Vermont Efforts." *Vermont Commons* 33 (Winter 2010).

EurekAlert. "Bottom Trawling Impacts, Clearly Visible From Space." February 15, 2008. http://www.eurekalert.org/pub_releases/2008-02/s-bti021508.php (accessed March 14, 2009).

Facilitation Committee of the Global Mechanism of the United Nations Convention to Combat Desertification. "Issues Paper for CSD-12 Side Event, Interlinkages between Drought, Desertification and Water." http://www.cyen.org/innovaeditor/assets/Interlinkages_between_drought,_desertification_and_water.pdf.

FAIR. "TV: The More You Watch, the Less You Know." http://www.fair.org/index.php?page=1517 (accessed July 13, 2009).

Fairclough, Adam. To Redeem The Soul of America: The Southern Christian Leadership Conference And Martin Luther King, Jr. Athens: University of Georgia Press, 1987.

Fanon, Franz. The Wretched of the Earth. New York: Grove Press, 1968.

Farnsworth, Alexander. "Sweden's Biggest Hydroelectric Plant Going Strong After 56 Years." http://www.reliableplant.com/Read/18560/sweden%27s-biggest-hydroelectric-plant-going-strong-after-56-years.

Fasulo, Jennifer. "The Revolution Within the Revolution: Feminist Organizing in the Young Lord's Party: An Interview with Iris Morales." Rain and Thunder 46 (Spring Equinox 2010).

Fendrich, James M. "Radicals Revisited: Long Range Effects of Student Protest." Nonprofit and Voluntary Sector Quarterly 2, no. 3: 161–168.

Festinger, Leon, Henry Riecken, and Stanly Schachter. When Prophecy Fails: A Social and Psychological Study of A Modern Group that Predicted the Destruction of the World. New York: Harper Torchbook, 1956.

Fitz, Don. "Energy, Environment and Exhortationism, " Synthesis/Regeneration: A Magazine of Green Social Thought, 49 (Spring 2009).

Flomenhoft, Gary. "Vermont's Common Assets: From Banana Republic to Sovereign Commonwealth." Vermont Commons, 33 (Winter 2010).

Foot, M.R.D. Resistance: An analysis of European Resistance to Nazism 1940–45. London: Methuen, 1976.

——. SOE: An Outline History of the Special Operations Executive 1940-46. London: British Broadcasting Corporation, 1984.

Foster, R. F. Modern Ireland. New York: Penguin Books, 1990.

Franklin, John Hope, and Loren Schweninger, Runaway Slaves: Rebels on the Plantation. New York: Oxford University Press, 1999.

Fromm, Erich. The Anatomy of Human Destructiveness. New York: Holt, Rinehart and Winston, 1973.

Frye, Marilyn. The Politics of Reality. Trumansberg, NY: Crossing Press, 1984.

Gado, Mark. "Bombingham." CrimeLibrary.com/Court TV Online. 2007. http://www.trutv.com/library/crime/terrorists_spies/terrorists/birmingham_church/3.html (accessed September 12, 2008).

Gara, Larry. The Liberty Line: The Legend of the Underground Railroad. Lexington: University of Kentucky Press, 1967.

Garrison, William Lloyd. The Liberator. December 8, 1837.

Garrow, David. Bearing the Cross: Martin Luther King, Jr., and the Southern Christian Leadership Conference. New York: William Morrow, 1986.

Gibbon, Edward. Decline and Fall of the Roman Empire. London: J.B. Bury, 1909.

Gillies, Kevin. "The Last Radical." Vancouver Magazine, November 1998.

Glendinning, Chellis. My Name is Chellis and I'm in Recovery from Western Civilization. Boston: Sambhala, 1994.

Gordon, R. B., M. Bertram, and T. E. Graedel. "Metal Stocks and Sustainability." *Proceedings of the National Academy of Sciences* 13, no. 5: 1209–1214.

Gordon, Raymond G., Jr., ed. *Ethnologue: Languages of the World,* 15th ed. Dallas: SIL International, 2005. http://www.ethnologue.com/.

Grady, Denise. "Chernobyl's Voles Live But Mutations Surge." *New York Times,* May 7, 1996.

Graff, Linda L. *Beyond Police Checks: The Definitive Volunteer & Employee Screen Guidebook.* Dundas, ON: Graff and Associates, 2005.

Graham, Dee. *Loving to Survive: Sexual Terror, Men's Violence, and Women's Lives.* New York: New York University Press, 1994.

"Green Guru." *Audubon,* May–June 2008.

Greenbelt Movement, The. "Climate Change." http://greenbeltmovement.org/w.php?id=98 (accessed on May 6, 2010).

Greenpeace Canada. "Industrial Fishing: Emptying Our Seas." June 17, 2008. http://www.greenpeace.org/canada/en/campaigns/More/safeguard-our-oceans/stop-bottom-trawling/industrial-fishing-emptying-o/ (accessed May 4, 2010).

Greer, John Michael, *The Long Descent: A User's Guide to the End of the Industrial Age.* Gabriola Island, BC: New Society Publishers, 2008.

Guilbert, George-Claude. *Madonna as Postmodern Myth: How One Star's Self-Construction Rewrites Sex.* Jefferson, NC: McFarland & Company, 2002.

Haddow, Douglas. "Pornocalypse Now." *Adbusters,* April 13, 2009. https://www.adbusters.org/magazine/83/pornocalypse_now.html.

Hamerow, Theodore S. *On the Road to the Wolf's Lair: German Resistance to Hitler.* Cambridge, MA: Harvard University Press, 1997.

Hanley, Charles J. "Climate Trouble May Be Bubbling Up in Far North." Yahoo News. August 31, 2009. http://news.yahoo.com/s/ap/20090831/ap_on_re_ca/cn_climate_09_troubling_bubbles (accessed September 3, 2009).

Hansen, James, et al. "Target Atmospheric CO2: Where Should Humanity Aim?" *The Open Atmospheric Science Journal* 2: 217–231.

Hanson, Stephanie. "MEND: The Niger Delta's Umbrella Militant Group." Council on Foreign Relations. March 22, 2007. http://www.cfr.org/publication/12920/.

Harrison, Barbara Grizzuti. *Visions of Glory.* New York: Simon and Schuster, 1978.

Hart, B. H. Liddell. *Strategy,* 2nd edition. New York: Praeger, 1967.

Heinberg, Richard. *Peak Everything: Waking Up to the Century of Declines.* Gabriola Island, Canada: New Society Publishers, 2007.

———. *Powerdown: Options and Actions for a Post-Carbon World.* Gabriola Island, Canada: New Society Publishers, 2004.

Held, Virginia. *Feminist Morality: Transforming Culture, Society, and Politics.* Chicago, IL: University of Chicago Press, 1993.

Hellman, Christopher. "The Runaway Military Budget: Analysis." Center for Arms Control and Nonproliferation, March 2006. http://www.fcnl.org/now/pdf/2006/mar06.pdf (accessed February 8, 2009).

Hill, Samuel. *Bygone Stalybridge: Traditional, Historical, Biographical.* Stalybridge, England: 1907.

Hirschfeld, Katherine. *Health, Politics and Revolution in Cuba since 1898.* New Brunswick, NJ: Transaction Press, 2007.

——. "Re-examining the Cuban Health Care System: Towards a Qualitative Critique." *Cuban Affairs* 2, no. 3.

Hobsbawm, E. J. *Revolutionaries: Contemporary Essays*. New York: Pantheon Books, 1973.

Hochschild, Adam. *Bury the Chains: Prophets and Rebels in the Fight to Free an Empire's Slaves*. New York: Houghton Mifflin, 2005.

Hodge, Scott. "Number of Americans Paying Zero Federal Income Tax Grows to 43.4 Million." The Tax Foundation, March 30, 2006. http://www.taxfoundation.org/research/show/1410.html (accessed October 9, 2008).

Hoffman, Abbie. *Revolution for the Hell of It*. New York: The Dial Press, 1968.

Holocaust Education and Archive Research Team. "The Sobibór Death Camp." Holocaust Research Project. http://www.holocaustresearchproject.net/ar/sobibor.html (accessed June 3, 2009).

Hopwood, Nick, and Jordan Cohen. "Greenhouse Gases and Society." http://www.umich.edu/~gs265/society/greenhouse.htm (accessed February 10, 2010).

Howden, Daniel. "Shell May Pull Out of Niger Delta After 17 Die in Boat Raid." *The Independent*, January 17, 2006. http://www.independent.co.uk/news/world/africa/shell-may-pull-out-of-niger-delta-after-17-die-in-boat-raid-523341.html.

Hubbert, Marion King. "Nuclear Energy and the Fossil Fuels." Paper presented before the Spring Meeting of the Southern District, American Petroleum Institute, Plaza Hotel, San Antonio, Texas, March 1956.

Intergovernmental Panel on Climate Change. "Summary for Policymakers," *Climate Change 2007: The Physical Science Basis. Contribution of Working Group I to the Fourth Assessment Report of the Intergovernmental Panel on Climate Change*. February 2007. http://ipcc-wg1.ucar.edu/wg1/Report/AR4WG1_Print_SPM.pdf.

International Union of Forest Research Organizations. "Adaptation of Forests and People to Climate Change—a Global Assessment Report," 2009. www.iufro.org/download/file/3580/3985/Full_Report.pdf.

Ismael.org. "Is There Hope for the Future?" October 15, 2005. http://www.ishmael.com/Education/Writings/bioneers.cfm (accessed on June 14, 2009).

Jackson, Julian. *France: The Dark Years, 1940–1944*. Oxford: Oxford University Press, 2003.

Jackson, Maggie. *Distracted: The Erosion of Attention and the Coming Dark Age*. Amherst, NY: Prometheus Books, 2008.

Jacobs, Ron. *The Way the Wind Blew: A History of the Weather Underground*. London: Verso, 1997.

James, Oliver. *Affluenza*. London: Vermillion, 2007.

James, William. *Varieties of Religious Experience*. New York: MacMillan Publishing, 1985.

Jefferies, Julie. *The UK Population: Past, Present And Future. Focus on People and Migration*. 2005. http://www.statistics.gov.uk/downloads/theme_compendia/fom2005/01_FOPM_Population.pdf (accessed 23, 2009).

Jeffreys, Sheila. *The Lesbian Heresy*. North Melbourne, Australia: Spinifex Press, 1993.

——. *The Industrial Vagina: The Political Economy of the Global Sex Trade*. London: Routledge, 2009.

Jensen, Derrick. *Endgame*. New York: Seven Stories, 2006.

———. *Resistance Against Empire*. Crescent City, CA: Flashpoint Press, 2010.

Jensen, Derrick, and Aric McBay. *What We Leave Behind*. New York: Seven Stories, 2008.

Jensen, Derrick, and Stephanie McMillan. *As the World Burns: 50 Simple Things You Can Do To Stay in Denial*. New York: Seven Stories, 2007.

Jensen, Robert."Pornography Is What the End of the World Looks Like." In Karen Boyle, ed., *Everyday Pornography*. New York: Routledge, 2010.

Johnson, Tim. "Crisis Cubana Cobra Alto Precioa Ancianos." *El Nuevo Herald*, July 12, 1993.

Josephy, Alvin M., Jr. *500 Nations: An Illustrated History of North American Indians*. New York: Alfred A. Knopf, 1994.

Jubilee Debt Campaign. "Debt and Public Services." October 2007. http://www .jubileedebtcampaign.org.uk/Debt20and20Public20Services+3704.twl.

Just, Adolph, and Benedict Lust. *Return to Nature: The True Natural Method of Healing and Living And the True Salvation of the Soul*. Whitefish, MT: Kessinger Publishing, 2007.

Keith, Lierre. *The Vegetarian Myth: Food, Justice, and Sustainability*. Oakland, CA: PM Press, 2009.

Kennedy, Florence. "Institutionalized Oppression vs. The Female." In Robin Morgan, ed., *Sisterhood Is Powerful: An Anthology of Writings from the Women's Liberation Movement*. New York: Random House, 1970.

Kennedy, Gordon, and Kody Ryan. "Hippie Roots and the Perennial Subculture." Hippyland. http://www.hippy.com/php/article-243.html (accessed January 27, 2009).

Kent, Stephen A. *From Slogans to Mantras: Social Protest and Religious Conversion in the Late Vietman War Era*. Syracuse, NY: Syracuse University Press, 2001.

Kerouac, Jack. *On the Road*. New York: Penguin Classics, 1991.

Kershaw, Ian. *Hitler: 1936 Nemesis*. New York: Norton, 2000.

Kinzer, Stephen. *All the Shah's Men: An American Coup and the Roots of Middle East Terror*. New York: John Wiley and Sons, 2003.

Kraussman, Fridolin. "Global Human Appropriation of Net Primary Production (HANPP)." *The Encyclopedia of Earth*, December 10, 2008. http://www.eoearth.org/article/Global_human_appropriation_of_net_primary _production_%28HANPP%29.

Kuhn, Thomas S. *The Structure of Scientific Revolutions*. Chicago, IL: University of Chicago Press, 1962.

Kunstler, James Howard. "Making Other Arrangements: A Wake-Up Call to a Citizenry in the Shadow of Oil Scarcity." *Orion Magazine*, January/February 2007. http://www.orionmagazine.org/index.php/articles/article/7/ (accessed on May 5, 2010).

Kurz, W. A., C. C. Dymond, G. Stinson, G. J. Rampley, E.T. Neilson, A. L. Carroll, T. Ebata, and L. Safranyik. "Mountain Pine Beetle and Forest Carbon Feedback to Climate Change." *Nature* 452: 987–990. doi: 10.1038/71190xa

Laffont, Robert. *Dictionnaire historique de la Résistance*. Paris: Bouquins, 2006.

Lakey, George. "Nonviolent Action as the Sword that Heals: Challenging Ward Churchill's *Pacifism As Pathology*." *Training for Change*, March 2001. http://trainingforchange.org/nonviolent_action_sword_that_heals (accessed on April 23, 2010).

Lance, Jennifer. "Football Field Sized Trucks Head to Canadian Tar Sands with Superloads." Gas 2.0. February 1, 2009. http://gas2.org/2009/02/01/football-field-sized-trucks-headed-to-canadian-tar-sands-with-superloads/ (accessed on May 25, 2010).

Laqueur, Walter Z., *Young Germany: A History of the German Youth Movement.* New York: Basic Books, 1962.

Large, David Clay, ed. *Contending with Hitler: Varieties of German Resistance in the Third Reich.* Cambridge, MA: Cambridge University Press, 1991.

Latané, B., and J. M. Darley "Group Inhibition of Bystander Intervention In Emergencies." *Journal of Personality and Social Psychology* 10: 215–221.

Laurance, W. F. "Reflections on the Tropical Deforestation Crisis." *Biological Conservation* 91: 109–117.

Leahy, Stephen. "Biodiversity: Earth's Life Support Systems Failing." Inter Press Service, October 13, 2009. http://ipsnorthamerica.net/news .php?idnews=2604 (accessed on January 13, 2010).

Lean, Geoffrey. "Dying Forest: One Year to Save the Amazon." *The Independent,* July 23, 2006. http://www.independent.co.uk/environment/dying-forest-one-year-to-save-the-amazon-408926.html (accessed May 14, 2009).

Lean, Geoffrey, and Fred Pearce. "Amazon Rainforest Could Become a Desert." *The Independent,* July 23, 2006. http://www.independent.co.uk/environment/ amazon-rainforest-could-become-a-desert-408977.html (accessed May 14, 2009).

Leber, Jessica. "Trash Course." *Audubon,* November–December 2008.

Lehmann, Ernst. *Biologischer Wille. Wege und Ziele biologischer Arbeit im neuen Reich.* München, 1934.

Lovelock, James. The Ages of Gaia; A Biography of Our Living Earth. New York: W. W. Norton, 1995.

Lunn, Eugene. *Prophet of Community: The Romantic Socialism of Gustav Landauer.* Berkeley, CA: University of California Press, 1973.

Macalister, Terry. "US Military Warns Oil Output May Dip Causing Massive Shortages by 2015." *The Guardian,* April 11, 2010. http://www.guardian.co.uk/ business/2010/apr/11/peak-oil-production-supply (accessed April 12, 2010).

Mackenzie, Midge. *Shoulder to Shoulder.* New York: Alfred A. Knopf, Inc., 1975.

MacKinnon, Catharine. *Feminism Unmodified: Discourses on Life and Law.* Cambridge, MA: Harvard University Press, 1987.

Maier, Pauline. *From Resistance to Revolution: Colonial Radicals and the Development of American Opposition to Britain, 1765–1776.* New York: Alfred A. Knopf, 1972.

Malcolm X. "Message to the Grass Roots." Speech, November 1963, Detroit. Published in George Breitman, ed., *Malcolm X Speaks: Selected Speeches and Statements.* New York: Grove Press, 1965.

Maltz, Wendy, and Larry Maltz. *The Porn Trap: The Essential Guide to Overcoming Problems Caused by Pornography.* New York: HarperCollins, 2008.

Mann, Barbara Alice. *Iroquoian Women: The Gantowisas.* New York: Peter Lang, 2006.

Manning, Richard. "The Amazing Benefits of Grass-Fed Meat," *Mother Earth News,* April–May 2009. http://www.motherearthnews.com/Sustainable-Farming/Grass-Fed-Meat-Benefits.aspx?page=7.

Margolies, Dan. "North Kansas City Company Settles Charge Related To Boycott Of Israel." *Kansas City Star,* June 25, 2003.

Markegard, Doniga, and Susan Osofsky. "Living the Local Food Life," *Permaculture Activist*, no. 75 (Spring 2010).

Markusen, Eric, and David Kopf. *The Holocaust and Strategic Bombing: Genocide and Total War in the Twentieth Century*. Boulder, CO: Westview Press, 1995.

Marufu, et al. "The 2003 North American Electrical Blackout: An Accidental Experiment in Atmospheric Chemistry." *Geophysical Research Letters* 31: L13106.

Marx, Karl. "No Tax Payments!" *Neue Rheinische Zeitung*, no. 145 (November 1848).

Mather, Anne. "A History of Feminist Periodicals, Part II." *Journalism History* 1, no. 4: 109.

McCurley, Stephen, and Sue Vineyard. *101 Tips for Volunteer Recruitment*. Downers Grove, IL: Heritage Arts Publishing, 1988.

McDermott, Patrice. *Politics and Scholarship: Feminist Academic Journals and the Production of Knowledge*. Urbana: University of Illinois Press, 1994.

McGinn, Anne Platt, Christopher Flavia, and Hilary French. "Promoting Sustainable Fisheries." In L. R. Brown, ed., *State of the World*. New York: W. W. Norton, 1998.

McGourty, Christine. "Global Crisis 'to Strike by 2030.'" March 19, 2009. http://news.bbc.co.uk/2/hi/uk_news/7951838.stm.

Melville, Keith. *Communes in the Counter Culture: Origins, Theories, Styles of Life*. New York: William Morrow and Company, 1972.

Milanovic, Branko. "True World Income Distribution, 1988 and 1993." *The Economic Journal* 112, no. 476: 51.

Milgram, Stanley. "Behavioral Study of Obedience." *Journal of Abnormal and Social Psychology* 67: 371–378.

Millennium Ecosystem Assessment. *Millennium Ecosystem Assessment: Ecosystems and Human Well-Being*. Covelo, CA: Island Press, 2005. http://www.millenniumassessment.org/.

Mollison, Bill. *Permaculture: A Designer's Manual*. New South Wales: Tagari Publications, 1988.

Mongabay.com. "Africa May Be Able to Feed Only 25% of Its Population by 2025." December 14, 2006. http://news.mongabay.com/2006/1214-unu.html (accessed June 12, 2009).

Mongabay.com. "Swedish Deforestation Rates and Related Forestry Figures." http://rainforests.mongabay.com/deforestation/2000/Sweden.htm (accessed January 14, 2009).

Mongabay.com. "Two-Thirds of Polar Bears At Risk of Extinction by 2050." September 7, 2007. http://news.mongabay.com/2007/0907-polar_bears.html (accessed June 9, 2008).

Morgan, Robin. *The Demon Lover: On the Sexuality of Terrorism*. New York, NY: W. W. Norton & Company, 1989.

Mrasek, Volker. "Melting Methane: A Storehouse of Greenhouse Gases Is Opening in Siberia." *Spiegel* Online, April 17, 2008. http://www.spiegel.de/international/world/0,1518,547976,00.html (accessed April 20, 2008).

Mumford, Lewis. *The City in History: Its Origins, Its Transformations, and Its Prospects*. New York: Harcourt, Brace & World, Inc., 1961.

Murphy, Pat. *Plan C: Community Survival Strategies for Peak Oil and Climate Change*. Gabriola Island, BC: New Society Publishers, 2008.

Myers, Ransom A., and Boris Worm. "Rapid Worldwide Depletion of Predatory Fish Communities," *Nature* 423 (2003): 280–283.

Nassil, Alberta J., and Stephen I. Abramowitz. "Transition or Transformation? Personal and Political Development of Former Berkeley Free Speech Movement Activists." *Journal of Youth and Adolescence* 8, no. 1 (March, 1979).

National Parks Service. "Aboard the Underground Railroad: Operating the Underground Railroad." http://www.nps.gov/history/Nr/travel/underground/opugrr.htm (accessed March 20, 2009).

Naylor, Thomas. "Rebel Against the Empire." http:www.*vermontrepublic.org/rebel-against-the-empire.*

Neal, Mark Anthony. *New Black Man: Rethinking Black Masculinity.* New York: Routledge, 2005.

Neely, Mark E., Jr. "Was the Civil War a Total War?" *Civil War History* 50 (2004): 434.

Newman, Richard J., "Hybrids Aren't So Green After All." http://www.usnews.com/usnews/biztech/articles/060331/31hybrids.htm.

Newsweek. "Birmingham USA: Look at Them Run." May 13, 1963, 27.

Novacek, Michael J., and Elsa E. Cleland. "The Current Biodiversity Extinction Event: Scenarios for Mitigation And Recovery." *Proceedings of the National Academy of Sciences,* May 8, 2001. http://www.pnas.org/content/98/10/5466.full

Nunnelley, William. *Bull Connor.* Tuscaloosa: University of Alabama Press, 1991.

Obomsawin, Alanis. *Kanehsatake: 270 Years of Resistance.* http://www.archive.org/details/kanehsatake.

Office of Global Analysis. "Cuba's Food & Agriculture Situation Report." March 2008. http://www.fas.usda.gov/itp/cuba/CubaSituation0308.pdf.

Office of Strategic Services. *Simple Sabotage Field Manual.* Washington, DC: OSS, 1944.

Ogbar, Jeffrey O. G., "Rainbow Radicalism." In Peniel E. Joseph, ed., *The Black Power Movement: Rethinking the Civil Rights-Black Power Era.* New York: Routledge, 2006.

Olson, Dan. "Species Extinction Rate Speeding Up." MPR, February 1, 2005. http://news.minnesota.publicradio.org/features/2005/01/31_olsond_biodiversity/ (accessed June 5, 2008).

Omestad, Thomas. "Cuba Plans New Offshore Drilling in Search for Big Oil Finds in the Gulf of Mexico." *US News and World Report,* February 3, 2009. http://www.usnews.com/news/energy/articles/2009/02/03/cuba-plans-new-offshore-drilling-in-search-for-big-oil-finds-in-the-gulf-of-mexico.html.

Orlov, Dmitry. *Reinventing Collapse: The Soviet Example and American Prospects.* Gabriola Island, BC: New Society Publishers, 2008.

Oxfam International. "Suffering the Science." July 2009. http://www.oxfam.org/en/policy/bp130-suffering-the-science (accessed on May 3, 2010).

Pakenham, Thomas. *The Scramble for Africa: White Man's Conquest of the Dark Continent from 1876-1912.* London: Abacus, 1992.

Pape, Robert Anthony. *Bombing to Win: Air Power and Coercion in War.* Ithaca, NY: Cornell, 1996.

Paulson, Tom. "The Lowdown on Topsoil: It's Disappearing: Disappearing Dirt Rivals Global Warming As an Environmental Threat." *Seattle Post-*

Intelligencer, January 22, 2008. http://seattlepi.nwsource.com/local/348200_dirt22.html.

Paxton, Robert. *Vichy France: Old Guard and New Order, 1940–1944*. New York: Columbia University Press, 1972.

Pearce, Fred. "Asian Farmers Sucking the Continent Dry." *New Scientist*, August 28, 2004. http://www.newscientist.com/article/dn6321-asian-farmers-sucking-the-continent-dry.html.

Pearson, Hugh. *The Shadow of the Panther: Huey Newton and the Price of Black Power in America*. Cambridge, MA: Da Capo Press, 1995.

Peterson, C., and C. Park. "Learned Helplessness and Explanatory Style." In D. F. Barone, M. Hersen, and V. B. Van Hasselt, eds., *Advanced Personality*, 287–308. New York: Plenum Press, 1998.

Pew Research Center. "A Deeper Partisan Divide Over Global Warming." May 8, 2008. http://people-press.org/report/417/a-deeper-partisan-divide-over-global-warming (accessed May 9, 2008).

Philosophy Dictionary. http://www.answers.com/topic/millenarianism (accessed on August 12, 2009).

Population Institute, The. "Iran's Family Program Is Succeeding." June, 1998. http://www.overpopulation.org/Iran%20Popline%20Jun98.html (accessed on October 23, 2009).

"Post WWI German Communities." Fellowship for Intentional Communities. http://wiki.ic.org/wiki/Post_WWI_german_communities (accessed February 28, 2009).

Prasad, Raekha. "Banda Sisters." *The Guardian*, February 15, 2008. http://www.guardian.co.uk/lifeandstyle/2008/feb/15/women.india (accessed April 9, 2009).

Putnam, Robert D. *Bowling Alone: The Collapse and Revival of American Community*. New York: Touchstone Books, 2000.

Queenan, Joe. *Malcontents*. Philadelphia: Running Press, 2004.

Quinn, Daniel. *Beyond Civilization: Humanity's Next Great Adventure*. New York: Three Rivers Press, 2000.

Rain and Thunder. "Number of US Extremist Groups 'Exploded' in 2009." No. 46 (Spring Equinox 2010).

Radzinsky, Edvard. *Stalin: The First In-depth Biography Based on Explosive New Documents from Russia's Secret Archives*. New York: Anchor, 1997.

Raines, Howell. "Terrorism; With Latest Bomb, I.R.A. Injures Its Own Cause." *New York Times*, November 15, 1987. http://www.nytimes.com/1987/11/15/weekinreview/the-world-terrorism-with-latest-bomb-ira-injures-its-own-cause.html?sec=&spon=&pagewanted=all (accessed April 23, 2010).

Rainforest Action Network. "About the Campaign." http://ran.org/what_we_do/rainforest_agribusiness/about_the_campaign/ (accessed June 9, 2009).

Raphael, Ray. *The First American Revolution: Before Lexington and Concord*. New York: The New Press, 2002.

Rasenberger, Jim. "Nightmare on Austin Street." *American Heritage Magazine* 57, no 5. http://www.americanheritage.com/articles/magazine/ah/2006/5/2006_5_65.shtml.

Ravilious, Kate. "Only Zero Emissions Can Prevent a Warmer Planet." *New Scientist*, February 29, 2008. http://www.newscientist.com/article/

dn13395-only-zero-emissions-can-prevent-a-warmer-planet.html?feedId=climate-change_rss20 (accessed on June 6, 2008).

Repka, Kathleen M., and William Repka. *Dangerous Patriots: Canada's Unknown Prisoners of War*. Vancouver: New Star Books, 1982.

Reuters. "Jane Goodall Says Biofuel Crops Hurt Rain Forests." September 27, 2007. http://www.enn.com/top_stories/article/23414 (accessed January 12, 2009).

Rich, Adrienne. "Twenty-One Love Poems." *The Dream of a Common Language*. New York: W. W. Norton, 1978.

Robbins, Richard H. *Global Problems and the Culture of Capitalism*. Boston: Allyn and Bacon, 1999.

Robinson, Simon. "Nigeria's Deadly Days." *Time*, May 14, 2006. http://www.time.com/time/magazine/article/0,9171,901060522-1193987-1,00.html.

Rosenberg, Matt. "Negative Population Growth: 20 Countries Have Negative or Zero Natural Increase." About.com. http://geography.about.com/od/populationgeography/a/zero.htm.

Rosenthal, Elizabeth. "An Amazon Culture Withers as Food Dries Up." *New York Times*, July 24, 2009. http://www.nytimes.com/2009/07/25/science/earth/25tribe.html (accessed March 2, 2009).

Ross, Julia. *The Mood Cure: The 4-Step Program to Rebalance Your Emotional Chemistry and Rediscover Your Natural Sense of Well Being*. New York: Viking Adult, 2002.

Roszak, Theodore. *The Making of a Counter Culture*. Garden City, NY: Doubleday & Co., Inc, 1968.

Rousseau, Jean-Jacques. *Émile: or, On Education*. New York: Dutton Book, 1975.

Sale, Kirkpatrick. "Dispersions: How to Get an Independent Vermont, Starting NOW." *Vermont Commons*, 32 (Stick Season/Holiday 2009).

Salonius, Peter. "A 10,000 Year Misunderstanding." *ScienceAlert*, April 30, 2008. http://www.sciencealert.com.au/opinions/20083004-17256.html (accessed March 21, 2009).

———. "Intensive Crop Culture for High Population Is Unsustainable." *Culture Change*. http://www.populationpress.org/publication/2008-2-salonius.html (accessed March 21, 2009).

Sample, Ian. "Warming Hits 'Tipping Point.'" *The Guardian*, August 11, 2005. http://www.guardian.co.uk/environment/2005/aug/11/science.climatechangı (accessed March 12, 2009).

Sanderson, et al. "The Human Footprint and the Last of the Wild." *BioScience* 52: 891–904.

Science Daily. "Empathy: College Students Don't Have as Much as They Used To, Study Finds." May 29, 2010. http://www.sciencedaily.com/releases/2010/05/100528081434.htm (accessed May 30, 2010).

Science Daily. "Pollution Causes 40 Percent of Deaths Worldwide, Study Finds." August 14, 2007. http://www.sciencedaily.com/releases/2007/08/070813162438.htm (accessed August 4, 2008).

Science Daily. "Regional Nuclear Conflict Would Create Near-Global Ozone Hole, Says Study." April 8, 2009. http://www.sciencedaily.com/releases/2008/04/080407172710.htm (accessed July 28, 2009).

Science Daily. "Regional Nuclear War Could Devastate Global Climate." December 11, 2006. http://www.sciencedaily.com/releases/2006/ 12/061211090729.htm (accessed July 28, 2009).

Security Culture. *A Handbook for Activists.* November 2001. http://security.resist .ca/personal/securebooklet.pdf.

Seligman, M. E. P., and S. F. Maier. "Failure to Escape Traumatic Shock." *Journal of Experimental Psychology* 74: 1–9.

Shah, Anup. "Poverty Facts and Stats." March 28, 2010. http://www.globalissues .org/article/26/poverty-facts-and-stats (accessed on May 12, 2010).

———. "Structural Adjustment—A Major Cause of Poverty." July 2, 2007. http://www.globalissues.org/TradeRelated/SAP.asp (accessed August 4, 2008).

———. "World Military Spending." September 13, 2009. http://www.globalissues .org/Geopolitics/ArmsTrade/Spending.asp (accessed August 4, 2008).

Shah, Zia-ul-Hassan, and Dr. Muhammad Arshad. "Land Degradation in Pakistan: A Serious Threat to Environments and Economic Sustainability." *Green Pages,* July 2006. http://www.eco-web.com/edi/index.htm (accessed on May 4, 2010).

Sharp, Gene. *Social Power and Political Freedom.* Boston: Extending Horizons Books, 1980.

———. *The Politics of Nonviolent Action.* Boston: Porter Sargent Press, 1973.

Sheridan, C. L., and K. G. King. "Obedience to Authority with an Authentic Victim." *Proceedings of the 80th Annual Convention of the American Psychological Association* 7: 165–166.

Sherman, William T. *"Military Division of the Mississippi Special Field Order 120."* November 9, 1864. America's Civil War Documents. http://www.sewanee .edu/faculty/Willis/Civil_War/documents/Sherman120.html.

Shuttlesworth, Fred. "Interview with Fred Shuttlesworth." Birmingham Civil Rights Institute Online. December 10, 1996. http://www.bcri.org/resource _gallery/interview_segments/index.htm (accessed March 12, 2008).

Smedley, Audrey. *Race in North America: Origin and Evolution of a Worldview.* Boulder, CO: Westview Press, 2007.

Smith, Anthony. *The Ethnic Origins of Nations.* Oxford: Blackwell Publishers, 1988.

Smith, Bryan. "Sudden Impact." *Chicago Magazine.* December 2006. http://www.chicagomag.com/Chicago-Magazine/December-2006/ Sudden-Impact/index.php?cp=2&si=1#artanc

Smith, J. W. In Derrick Jensen, ed., *Resistance Against Empire.* Crescent City, CA: Flashpoint Press, 2010.

Soon, Chun Siong, Marcel Brass, Hans-Jochen Heinze, and John-Dylan Haynes. "Unconscious Determinants of Free Decisions in the Human Brain." *Nature Neuroscience* 11, no. 5: 543–545.

Spender, Stephen. *World Within World.* London: Hamish Hamilton, 1951.

Stearns, Richard. *The Hole in Our Gospel.* Nashville, TN: Thomas Nelson, 2009.

Stewart, James Brewer. *Abolitionist Politics and the Coming of the Civil War.* Boston: University of Massachusetts, 2008.

Strain, Christopher B. *Pure Fire: Self-Defense as Activism in the Civil Rights Era.* Athens: University of Georgia Press, 2005.

Swedish Institute. "The Sami People in Sweden (Factsheet)." February, 1999. http://www.samenland.nl/lap_sami_si.html.

Taylor, Peter. *The Age of Terror, Part Two.* BBC Radio World Service.
http://www.bbc.co.uk/worldservice/documentaries/2008/05/080610_age_of
_terror_two.shtml (accessed April 11, 2009).

Thompson, Denise. *A Discussion of the Problem of Horizontal Hostility.*
http://users.spin.net.au/~deniset/alesfem/mhhostility.pdf.

Thoreau, Henry David. "A Plea for Captain John Brown." The Thoreau Society.
http://thoreau.eserver.org/plea1.html.

Time. "Dogs, Kids, and Clubs." May 10, 1963.

Toy, Mary-Anne. "China Covers up Pollution Deaths." *The Age,* July 5, 2007.
http://www.theage.com.au/news/world/china-covers-up-pollution-
deaths/2007/07/04/1183351291152.html.

Trainer, Ted. *Renewable Energy Cannot Sustain a Consumer Society—Toward a Sustainable Economy: The Need for Fundamental Change.* Sydney, Australia:
Envirobook, 1996.

Transition Network. "12 Ingredients." http://www.transitionnetwork.org/com-
munity/support/12-ingredients.

Tye, Larry. *Rising from the Rails: Pullman Porters and the Making of the Black
Middle Class.* New York: Henry Holt and Company, 2004.

United States Army. *Operations* (FM 3-0). 2001.

United States Army. *Special Forces Operations* (FM 31-21). 1961.

United States Holocaust Memorial Museum. "Sobibór" *Holocaust Encyclopedia.*
http://www.ushmm.org/wlc/article.php?lang=en&ModuleId=10005192
(accessed August 12, 2009).

Varon, Jeremy. *Bringing the War Home: The Weather Underground, the Red Army
Faction, and Revolutionary Violence in the Sixties And Seventies.* Berkeley, CA:
University of California Press, 2004.

Växjö kommun. "Sustainable Development." http://www.vaxjo.se/ VaxjoTem-
plates/Public/Pages/Page.aspx?id=1661 (accessed February 10, 2010).

Walsh, David. *Why Do They Act That Way? A Survival Guide to the Adolescent
Brain for You and Your Teen.* New York: Free Press, 1984.

Walt, Vivienne. "In France, Pension Protests Spark Oil Shortages." *Time,* October
20, 2010. http://www.time.com/time/world/article/0,8599,2026553,00
.html?iid=sphere-inline-bottom (accessed Nov 1, 2010).

Ward, Peter D. *Under a Green Sky: Global Warming, the Mass Extinctions of the
Past, and What They Can Tell Us about Our Future.* New York: Collins, 2007.

Watters, Peter. "When Prophecies Fail." Freeminds.org. http://www.freeminds
.org/doctrine/prophecy/when-prophecies-fail.html (accessed August 24,
2009).

Weindling, Paul. *Health, Race and German Politics Between National Unification
and Nazism, 1870–1945.* Cambridge, MA: Cambridge University Press, 1989.

Wielgolaski, F. E. P., Staffan Karlsson, Seppo Neuvonen, and Dietbert
Thannheiser. *Plant Ecology, Herbivory, and Human Impact in Nordic Mountain
Birch Forests.* New York: Springer, 2005.

Wikipedia contributors. "Diggers." *Wikipedia, The Free Encyclopedia.*
http://en.wikipedia.org/wiki/Diggers (accessed February 2, 2009).

———. "Ghost Dance," *Wikipedia, The Free Encyclopedia.*
http://en.wikipedia.org/w/index.php?title=Ghost_Dance&oldid=363928036
(accessed August 5, 2009).

——. "Johann Gottfried Herder." *Wikipedia, The Free Encyclopedia.* http://en.wikipedia.org/wiki/Johann_Gottfried_Herder (accessed February 12, 2009).

——. "Narcissism." *Wikipedia, The Free Encyclopedia.* http://en.wikipedia.org/w/index.php?title=Narcissism&oldid=364421808 (accessed June 29, 2008).

Williams, Laurie, and Allan Zabel. "An Open Letter to Congress." *Carbon Fees,* May 4, 2008. http://www.carbonfees.org/home.

Wilson, Edward O. *The Future of Life.* New York: Vintage, 1992.

Wilson, Elizabeth. *Bohemians: The Glamorous Outcasts.* New Brunswick, NJ: Rutgers University Press, 2000.

Wittig, Monique. *Les Guérillères.* New York: Avon Bard Books, 1973.

Wolfenstein, Martha. *Disaster: A Psychological Essay.* Glencoe, IL: The Free Press, 1957.

Worldwatch Institute. *Vital Signs 2007–2008.* New York: W. W. Norton, 2007.

The Xhosa Cattle-Killing Movement. "Overview." http://www.xhosacattlekilling .net/overview.asp (accessed August 24, 2009).

York, Geoffrey, and Loreen Pindera. *People of the Pines; The Warriors and the Legacy of Oka.* Toronto, ON: Little, Brown, & Co., 1992.

Zablocki, Benjamin. "Foreword." In Stephen A. Kent, *From Slogans to Mantras: Social Protest and Religious Conversion in the Late Vietman War Era.* Syracuse, NY: Syracuse University Press, 2001.

ABOUT THE AUTHORS

Derrick Jensen is the best-known voice of the growing deep ecology movement. Winner of numerous awards and honors including the Eric Hoffer Book Award, *USA Today*'s Critic's Choice, and Press Action's Person of the Year, Jensen is the author of over fifteen books, including *Endgame, A Language Older Than Words, What We Leave Behind* (with Aric McBay), and *Deep Green Resistance* (with McBay and Lierre Keith). Philosopher, teacher, and radical activist, he regularly stirs packed auditoriums across the country with revolutionary spirit. Jensen holds degrees in creative writing and mineral engineering physics. He lives in Crescent City, California.

Lierre Keith is a writer, small-scale farmer, and radical feminist activist. She is the author of two novels, as well as *The Vegetarian Myth: Food, Justice, and Sustainability*. She's been arrested six times. She lives in Humboldt County, California.

Writer, activist, and small-scale organic farmer **Aric McBay** is the coauthor of *What We Leave Behind*, with Derrick Jensen, author of *Peak Oil Survival: Preparation for Life after Gridcrash*, and creator of "In the Wake: A Collective Manual-in-progress for Outliving Civilization" (www.aricmcbay.org).